T0139782

Studies in Systems, Decision and Control

Volume 259

Series Editor

Janusz Kacprzyk, Systems Research Institute, Polish Academy of Sciences, Warsaw, Poland

The series "Studies in Systems, Decision and Control" (SSDC) covers both new developments and advances, as well as the state of the art, in the various areas of broadly perceived systems, decision making and control–quickly, up to date and with a high quality. The intent is to cover the theory, applications, and perspectives on the state of the art and future developments relevant to systems, decision making, control, complex processes and related areas, as embedded in the fields of engineering, computer science, physics, economics, social and life sciences, as well as the paradigms and methodologies behind them. The series contains monographs, textbooks, lecture notes and edited volumes in systems, decision making and control spanning the areas of Cyber-Physical Systems, Autonomous Systems, Sensor Networks, Control Systems, Energy Systems, Automotive Systems, Biological Systems, Vehicular Networking and Connected Vehicles, Aerospace Systems, Automation, Manufacturing, Smart Grids, Nonlinear Systems, Power Systems, Robotics, Social Systems, Economic Systems and other. Of particular value to both the contributors and the readership are the short publication timeframe and the world-wide distribution and exposure which enable both a wide and rapid dissemination of research output.

** Indexing: The books of this series are submitted to ISI, SCOPUS, DBLP, Ulrichs, MathSciNet, Current Mathematical Publications, Mathematical Reviews, Zentralblatt Math: MetaPress and Springerlink.

More information about this series at http://www.springer.com/series/13304

Alla G. Kravets · Alexander A. Bolshakov ·
Maxim V. Shcherbakov
Editors

Cyber-Physical Systems: Advances in Design & Modelling

Editors
Alla G. Kravets
Volgograd State Technical University
Volgograd, Russia

Alexander A. Bolshakov
Peter the Great St. Petersburg
Polytechnic University
St. Petersburg, Russia

Maxim V. Shcherbakov
Volgograd State Technical University
Volgograd, Russia

ISSN 2198-4182 ISSN 2198-4190 (electronic)
Studies in Systems, Decision and Control
ISBN 978-3-030-32581-7 ISBN 978-3-030-32579-4 (eBook)
https://doi.org/10.1007/978-3-030-32579-4

This Springer imprint is published by the registered company Springer Nature Switzerland AG
The registered company address is: Gewerbestrasse 11, 6330 Cham, Switzerland

Preface

This book dwells with cyber-physical systems' design and defines digital tools eventually broadening the scope of application domains of such systems. In this book, the authors substantiate the scientific, practical, and methodological approaches to the modeling and simulation of cyber-physical systems, the design of adaptive optimization algorithms, problem analysis, and data preprocessing. The authors both analyzed and improved existing approaches to solving problems and demonstrated new ways by showing their new view on the subject matter. Implemented breakthrough systems, models, programs, and methods that could be used in industrial processes to predict and lead cyber-physical systems' functions, states, and evolution.

The authors determine key industrial challenges and the main features of modeling processes. The implementations of the developed prototypes, including testing in real industries, which have collected and analyzed big data and proved their effectiveness, are presented.

This book is devoted to the simulation of cyber-physical systems and their modules based on the concept of a digital twin. In particular, the digital twin digitally copies physical assets that can be used for various purposes, such as understanding, predicting, and optimizing system performance to achieve better results. The authors describe the advanced approaches to the digital design of production systems, the main features of the design of digital twins, and the use of various models, methods, and technologies, such as artificial intelligence, neural networks, for condition assessment, diagnostics, prognostication, and proactive maintenance of a cyber-physical system.

Edition of the book is dedicated to the 120th Anniversary of Peter the Great St. Petersburg Polytechnic University and technically supported by the Project Laboratory of Cyber-Physical Systems of Volgograd State Technical University.

Volgograd, Russia — Alla G. Kravets
St. Petersburg, Russia — Alexander A. Bolshakov
Volgograd, Russia — Maxim V. Shcherbakov
August 2019

Contents

Cyber-Physical Systems and Digital Twins

Cyber-Physical Systems Design

Flow Analysis and Its Applications for Equipment Design

Vadim Shakhnov, Elena Rezchikova, Lyudmila Zinchenko and Natalia Sergeeva

Abstract An approach based on flow analysis of matter, energy and information is discussed. The cognitive approach matter-energy-information allows to analyze complex systems in different modes, including dynamic situations. It is shown that flow diagrams are universal, common to many complex technical systems including electronic equipment design and technologies, communication systems, and etc. Restrictions of the proposed method is discussed. It is remarkable that visual analytics of dynamic process is available using weighted directed graphs. Based on the available experience, a classification of the concepts of flow analysis of systems is proposed. The chapter discusses flow analysis practical applications. Finally, conclusions are derived.

Keywords Systems · Subsystems · Metasystems · Flows · Transformations of flows · Flow analysis · Cognitive models

1 Introduction

Design methods are crucial in engineering activities. Many design techniques have been proposed to formalize innovation and invention processes [1–5]. However, till now a method that is suitable for any technical problem decision does not exist.

The correct definition for method is discussed [6]. In [6], the following definition of design method has been given as follows: "a specification of how a specified result

V. Shakhnov (✉) · E. Rezchikova · L. Zinchenko · N. Sergeeva
Bauman Moscow State Technical University, 2 Baumanskaya, 5, Moscow 105005, Russia
e-mail: shakhnov@mail.ru

E. Rezchikova
e-mail: rezc-elena@yandex.ru

L. Zinchenko
e-mail: lyudmillaa@mail.ru

N. Sergeeva
e-mail: snataliaa@yandex.ru

A. G. Kravets et al. (eds.), *Cyber-Physical Systems: Advances in Design & Modelling*, Studies in Systems, Decision and Control 259,
https://doi.org/10.1007/978-3-030-32579-4_1

is to be achieved. This may include specifications of how information is to be shown, what information is to be used as inputs to the method, what tools are to be used, what actions are to be performed and how, and how the task should be decomposed and how actions should be sequenced".

Therefore, information flow representation and analysis are important in engineering design. In [7], cognitive information flow analysis has been introduced. The technique exploits both goal-directed task analysis and a modified cognitive work analysis. The main focus is on information flows. However, some researchers (Koller, Hubka et al.) [8, 9] have considered, that three flows (information, energy, matter) have to be estimated in design process.

In [10], the Information-Matter-Energy model has been proposed. It is important that information, matter and energy are linked. It is impossible to estimate an element of the model independently [11].

In the chapter, we discuss the method Cognitive Flow Analysis that can be applied for flow analysis of different types of flows: energy, information, and matter. It is important that different elements of the model have common features and can be considered using the general method.

The rest of the chapter is structured as follows. The next section reviews the related works in the field of engineering design. Section 3 presents our cognitive flow analysis. We discuss its application features for electronic equipment design as well. Finally, conclusions are derived in Sect. 4.

2 Review of Related Works

Engineering design is an important step in engineering activities. Different discussions about understanding of engineering design as art or as science continue till now [12]. Some researchers (e.g. Hubka [9]) treat design as a scientific discipline and seek general methods, while other researchers (e.g. Leyer) argue that designing is not a scientific activity and creative approaches have to be used [12]. In [13], system design and creativity were used simultaneously.

In [14], a general approach to engineering design was proposed. The main idea is that each system contains flows of energy/matter/information and therefore, all systems can be classified either a system of energy transformation or matter transformation or information transmission. Koller proposed a knowledge base related to the corresponding functions. However, current systems are much more complex and combine flows of energy, matter, and information to achieve the required functionality. Microelectromechanical systems (MEMs, microsystems) are examples of the systems [10]. Therefore, a novel approach is required to design the systems.

In [10], the Information-Matter-Energy model for Cognitive Informatics has been proposed. It is important that Matter (M) and Energy (E) belong to the physical world (PW), while Information (I) can be used to model the abstract world (AW). In the Information-Matter-Energy model [10], the natural world represents a dual world. The natural world model is given as follows [10]:

$$NW \equiv PW \| AW = F(I, M, E), \tag{1}$$

where $\|$ denoted a parallel relation between two worlds: the physical and the abstract worlds;

F is a function that determines the natural world.

In [6], the links between I-M-E are given as follows:

$$I = f_1(M), \tag{2}$$

$$M = f_2(I), \tag{3}$$

$$I = f_3(E), \tag{4}$$

$$E = f_4(I), \tag{5}$$

$$E = f_5(M), \tag{6}$$

$$M = f_6(E). \tag{7}$$

Wang [10] have used the famous equation proposed by A. Einstein

$$E = Mc^2, \tag{8}$$

where c is the speed of the light

$$c = 3 \times 10^8 \text{ m/c}$$

for functions f_5 and f_6.

Cognitive information flow analysis [7] integrates goal-directed task analysis and a modified cognitive work analysis. The first method is able to satisfy dynamic situation awareness requirements. Cognitive work analysis allows to analysis the human work. A term *information item* has been proposed as a discrete data element abstraction. It is important that the abstraction includes both physical objects and information reports. The proposed information flow diagrams show the flows of information through the system. Cognitive information flow analysis is a directed graph. Its nodes correspond to functions. The four information consumption types are represented by different edges styles. Edges are shown either solid lines or dashed lines. A direction of edges is shown either a single solid arrowhead or a double solid arrowhead. However, energy and matter transformation are not considered. Therefore, a cognitive approach that can be able to represent flows of energy, matter, and information simultaneously is useful.

3 Cognitive Flow Analysis

Cognitive flow analysis can be applied for different systems. Only time is independent variable. Cognitive flow analysis is a weighed directed graph

$$G = (V, E), \tag{9}$$

where

V is a set of vertices;
E is a set of edges.

Its vertex represents a subsystem or a transformation point. Its edge must have the source and the drain and represent a flow of energy/matter/information. The edge weight represents a transformation coefficient. Figure 1 illustrates some particular cases. Figure 1a shows a linear cognitive flow analysis model for the case of similar flows. Figure 1b shows a parallel cognitive flow analysis model for the case of similar flows. Figure 1c demonstrates a model, containing a transformation node. It should be noted that a color of an edge can be used to distinguish flows of energy, matter, and information. All models represent a directed graph. The linear model represents a 2-regular graph. A parallel model represents a union of 2-regular graphs.

In despite of the generality of the proposed method, some restrictions exist. The method can be applied if:

- the total energy of the system is constant;
- the mass of the system must remain constant;
- flows should be continuous;
- chaotic changes are not allowed;
- transformations between flows should follow the law of conservation of mass and the law of conservation of energy. The corresponding transformation coefficients are used.

The general operations that can be used are given as follows:

- discretization of continuous flows;
- division of a single flow to parallel flows;

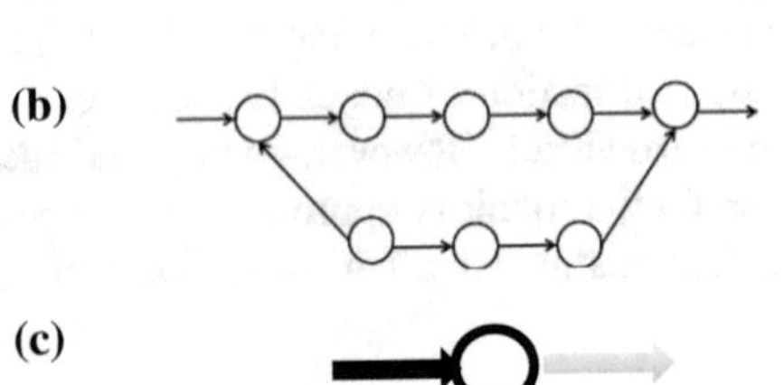

Fig. 1 Examples of cognitive flow analysis models: **a** a linear flow analysis model, **b** a parallel flow analysis model, **c** a transformation node

– a speed of flow can be changed;
– a direction of flow can be changed.

It should be noted, that in order to avoid energy losses a number of transformation should be minimal.

An algorithm of cognitive flow analysis application for systematic design is given as follows.

Step 1. Define all subsystems with flows through the subsystems.
Step 2. Classify types of flows (energy, matter, information).
Step 3. Define subsystems that should be controlled.
Step 4. Define flows that hinder the desired functionality.
Step 5. Transform flows and achieve the functionality.

The electronic equipment design was analyzed using the cognitive flow analysis method. Table 1 shows energy flows between subsystems.

The corresponding equations are given as follows:

$$\begin{aligned}(P_1 \oplus P_2)/\omega &= E_1(\eta_1 \oplus \eta_{12} \oplus \eta_{13} \oplus \eta_{14} \oplus \eta_{15}) - E_2\eta_{21} \\ &\quad - E_3\eta_{31} - E_4\eta_{41} - E_5\eta_{51}; \\ 0 &= E_2(\eta_2 \oplus \eta_{21} \oplus \eta_{23}) - E_1\eta_{12} - E_3\eta_{32}; \\ 0 &= -E_1\eta_{13} - E_2\eta_{23} \oplus E_3(\eta_3 \oplus \eta_{31} \oplus \eta_{32} \oplus \eta_{34}) - E_4\eta_{43}; \\ 0 &= -E_1\eta_{14} - E_3\eta_{34} \oplus E_4(\eta_4 \oplus \eta_{41} \oplus \eta_{43} \oplus \eta_{45}) - E_5\eta_{54}; \\ 0 &= -E_1\eta_{15} - E_4\eta_{45} \oplus E_5(\eta_5 \oplus \eta_{51} \oplus \eta_{54}). \end{aligned} \tag{10}$$

Table 1 Energy flow analysis in electronic equipment

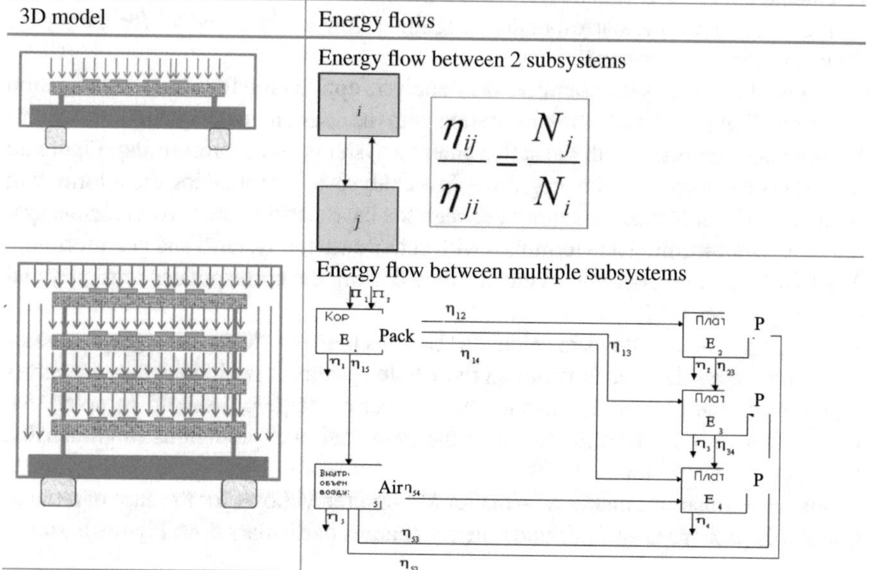

3D model	Energy flows
	Energy flow between 2 subsystems
	Energy flow between multiple subsystems

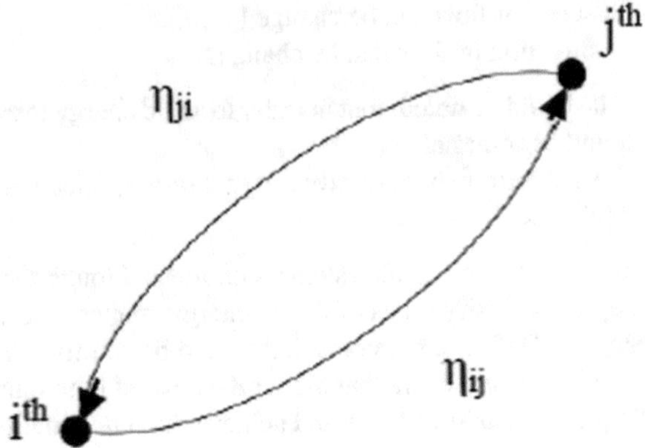

Fig. 2 Examples of cognitive flow analysis models

In the general case, the corresponding equations are given as follows:

$$\mathrm{E}_i = f\left[(P_1 \oplus P_2);\, \eta_i;\, \eta_{ij}\right];\, \eta_{ij} = f(N_i);$$

where

$\mathrm{E_i}$ is the energy of a subsystem;
P_i is a power of input signal;
η_i is a coefficient of energy losses in the flow;
η_{ij} is a transformation coefficient of flows from ith subsystem to jth subsystem;
N_i is a modal density of ith subsystem.

It is important that Eigen frequencies are important in the analysis. They can be calculated using CAD tools, e.g. ANSYS [15].

Figure 2 shows the correspondent cognitive flow analysis model for the matter flow between 2 subsystems.

Figure 3 illustrates the cognitive flow analysis application for 5G communication networks. Figure 3a shows the base station with the active antenna systems and beamforming and terminals with the active antenna systems and beamforming. Figure 3b represents the corresponding cognitive flow analysis model showing the information I—matter M—information I flows between the base station with active antenna systems and beamforming and terminals with active antenna systems and beamforming. The transformation coefficients are calculated using the correspondent circuitry and antenna patterns [16].

The graph model allows to calculate relations between flows using graph metrics and compare a role of each vertex in the whole system. It is obvious that the vertex representing the base station has the highest vertex degree. It should be noted that the number depend from the mode of the base station exploitations (high density, low density of rural region [16]).

Figure 4 illustrates matter M—matter M—matter M flows for the case of a radiation shield properties analysis under heavy charges particulars flow. Figure 4a shows

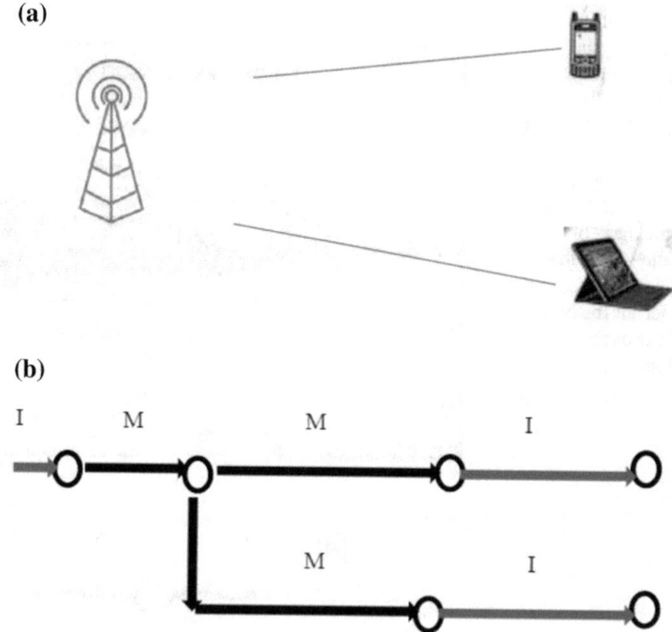

Fig. 3 Examples of cognitive flow analysis application for the 5G base station with active antenna systems and beamforming and terminals with active antenna systems and beamforming. **a** The 5G base station with active antenna systems and beamforming and terminals with active antenna systems and beamforming, **b** the cognitive flow analysis model

the simulation results of radiation shield properties. Figure 4b represents the corresponding cognitive flow analysis model for the case of the thickness of the radiation shield is less than 4.5 mm. Figure 4c shows the corresponding cognitive flow analysis model for the case of the thickness of the radiation shield is more than 4.5 mm.

However, flows of different physics should be calculated separately. Therefore, a hierarchical and multiphysics approach has to be used. A union of 2-regular graphs are used for multiphysics applications. A hierarchical approach allows to expand cognitive flow analysis methods for nanosystems design [17] using multi-level representation [18].

The proposed approach can be used for radiation-hardened equipment design [19]. It should be noted that in the case flow of matter and flow of energy should be evaluated in the correspondent models.

Finally, our approach is in line with physical effects and the unified theory of intelligence [20].

The approach simplifies system design and visualize relations between different subsystems and can be easily applied for a digital twin design [21].

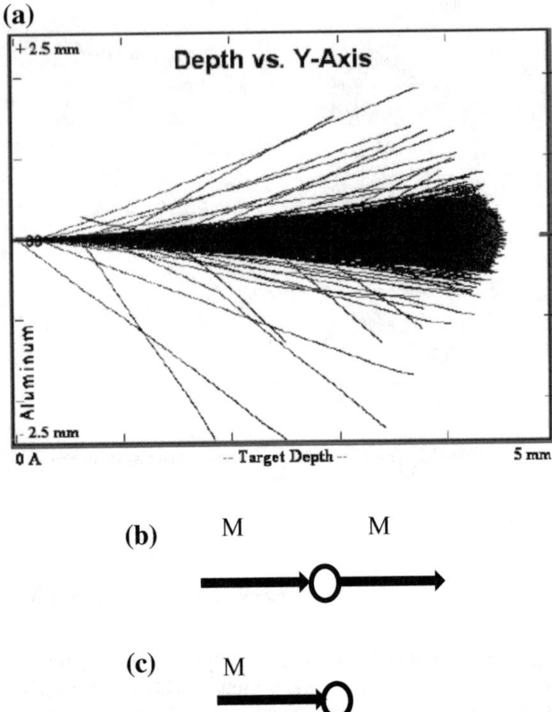

Fig. 4 Examples of the cognitive flow analysis application for radiation shield properties analysis. **a** Aluminum radiation shield properties under 30 MeV protons flow, **b** the cognitive flow analysis model for the case of the thickness of the radiation shield is less than 4.5 mm, **c** the cognitive flow analysis model for the case of the thickness of the radiation shield is more than 4.5 mm

4 Conclusion

In the chapter, we have discussed our approach to formalization of engineering design. Several approaches to engineering design have been discussed in detail. It was shown that flows of energy, matter, and information have to be considered simultaneously.

The cognitive flow analysis model based on graph theory has been reviewed. An application of the method for electronic equipment design is discussed. It is obvious that visualization of relations between subsystems allow to review the system as the whole object. The bottlenecks can be defined by a visual control and some design efforts to modify the bottlenecks can be applied in a scientific manner.

It is remarkable that in despite of differences in physic effects flows of energy, matter, and information can be managed as an abstract item. Therefore, the *information item* [7] can be expanded as a *flow item* that can be used in multiphysics applications.

A visual representation of a graph with many vertices is not trivial task. Therefore, for many subsystems a hierarchical approach is preferable.

These areas will be investigated in our future research.

This work was supported by RFBR 18-29-18043.

References

1. Wilson, J., Sharples, S.: Evaluation of Human Work. CRC Press, Boca Raton (2015)
2. Eggleston, R.G.: Work-centered design: a cognitive engineering approach to system design. In: Proceedings of the Human Factors and Ergonomics Society Annual Meeting (2003)
3. Militello, L.G., Klein, G.: Decision-centered design. In: The Oxford Handbook of Cognitive Engineering (2013)
4. Weber, C., Husung, S., Cantamessa, M., et al.: Effectiveness of the Systematic Engineering Design Methodology. Lund University, Lund (2015)
5. Ovalle Fresa, R., Rothen, N.: Training enhances fidelity of color representations in visual long-term memory. J. Cogn. Enhancement **3**, 315–327 (2019)
6. Gericke, K., Eckert, C., Stacey, M.: What do we need to say about a design method? In: Proceedings of ICED17, pp. 101–110 (2017)
7. Humphrey, C.M., Adams, J.A.: Cognitive information flow analysis. Cogn. Tech. Work **15**(2), 133–152 (2013)
8. Koller, R.: Design Method for Machine, Device and Apparatus Construction. Springer, Berlin/Heidelberg (1979)
9. Hubka, V., Eder, W.E.: Engineering Design. Heurista, Zürich (1992)
10. Wang, Y.: On the informatics laws and deductive semantics of software. IEEE Trans. Syst. Man Cybern. (Part C) **36**(2), 161–171 (2006)
11. Shivhare, R., Cherukuri, A.K., Li, J.: Establishment of cognitive relations based on cognitive informatics. J. Cogn. Comput. **9** (2017)
12. Heymann, M.: “Art” or science? Competing claims in the history of engineering design. In: Engineering in Context. Academica, Aarhus (2009)
13. Pahl, G., Beitz, W.: Engineering Design. Springer, London (1977)
14. Koller, R.: Konstruktionslehre für den Maschinenbau. Grundlagen zur Neuund Weiterentwicklung technischer. Springer, Berlin (1998)
15. Shakhnov, V., Makarchuk, V., Zinchenko, L., Verstov, V.: Heterogeneous knowledge representation for VLSI systems and MEMS design. In: Proceedings of IEEE CogInfoCom’13, pp. 189–194 (2013)
16. Dahlman, E., Parkvall, S., Sköld, J.: 5G NR: The Next Generation Wireless Access Technology. Academic Press, Cambridge (2018)
17. Shakhnov, V.A., Zinchenko, L.A., Rezchikova, E.V.: Simulation and visualization in cognitive nanoinformatics. Int. J. Math. Comput. Simul. **8**, 141–147 (2014)
18. Shakhnov, V., Zinchenko, L., Rezchikova, E., Kosolapov, I.: Information representation and processing in cognitive nanoinformatics. In: Proceedings of 2014 5th IEEE Conference on Cognitive Infocommunications (CogInfoCom), pp. 43–47 (2014)
19. Khanna, V.: Extreme-Temperature and Harsh-Environment Electronics. IOP Publishing, UK (2017)
20. Wang, Y.: On abstract intelligence: toward a unified theory of natural, artificial, machinable, and computational intelligence. Int. J. Softw. Sci. Comput. Intell. **1**(1), 1–17 (2009)
21. Detzner, A., Eigner, M.: A digital twin for root cause analysis and product quality monitoring. In: Proceedings of the DESIGN 2018 15th International Design Conference (2018)

Cyber-Physical Control System of Hardware-Software Complex of Anthropomorphous Robot: Architecture and Models

Mikhail Stepanov, Vyacheslav Musatov, Igor Egorov, Svetlana Pchelintzeva and Andrey Stepanov

Abstract Autonomous anthropomorphous robots represent complicated hardware-software complexes designed to functioning in a changing external environment. Additional features of any particular robot are defined by its scope. Educational activity imposes tight restrictions to ensure safe work and study environment at an institution. It is required to solve many different tasks in a real-time mode. The efficiency of their solutions is defined by availability of computing resources, as well as by thorough organization of the hardware-software complex oriented toward a specialized class of the autonomous robotics tasks. With this goal in mind, we analyzed the complex of the tasks for the teaching assistant robot. Among those, one of the most important was the task of obtaining information about the environment. We analyzed the task of a trainee status examination and possible ways of its solution, and offered the architecture of a hardware-software complex of the anthropomorphous robot assistant. The set-theoretic model of a hardware-software complex was constructed. Its use would further allow defining an optimum configuration of the offered hardware-software complex architecture for anthropomorphous robots. The distributed computing system of a hardware-software complex for anthropomorphous robot assistant facilitated parallel solving of the tasks related to situation analysis, as well as planning and control of the robot operations.

Keywords Anthropomorphous robotics · Hardware-software complex · Set-theoretic model · Brain activity · Digital signals · Pedagogy · Wavelet analysis · EEG test · Neuroscience · Brain-computer interface

1 Introduction

Currently, educational robotics gains an increasing importance both at a high-school stage, and in higher education. As for secondary schools, much attention is paid to inclusive education, specifically to disabled students' adaptation to studying in

M. Stepanov (✉) · V. Musatov · I. Egorov · S. Pchelintzeva · A. Stepanov
Yury Gagarin State Technical University of Saratov, Saratov, Russia
e-mail: mfstepanov@mail.ru

A. G. Kravets et al. (eds.), *Cyber-Physical Systems: Advances in Design & Modelling*, Studies in Systems, Decision and Control 259,
https://doi.org/10.1007/978-3-030-32579-4_2

a mainstream class. Despite some available positive experiences, there are many children with difficulties to learn in a conventional group owing to specific features of their physical health or emotional status.

Studies on the autism conducted in 2014 by the group of Christina Kostesku and Zachary Vorren, published in a Journal of Autism and Developmental Disorders, showed that children demonstrated a positive feedback to robotic systems, paid more attention to the robot rather than a person, better performed repetitive tasks with participation of the robot, while their cognitive abilities were identical no matter trained by the robot or a person.

On the basis of those results, the following requirements were imposed to teaching assistant robots:

- Small dimensions: height of the automated robotic complex (ARC) needs to be about 155 cm, which subconsciously associates ARC with a teenager;
- Compliance of behavioral modification, which represents training challenging skills (such as speech, game playing, or ability to look in the face) but can be conducted by ARC;
- Readiness for communication, creation of friendship bonds, and satisfaction of displayed curiosity;
- Psychologically appropriate design to exclude any possible harm to a child even in the case of his or her aggressive behavior;
- Ability to react to aggression and other suspicious and potentially harmful actions from the child, and to inform the operator about those.

We suggest two possible directions of using a hardware-software complex for anthropomorphous robot assistant.

- First, it is appropriate for work with disabled children. The anthropomorphous robot can be useful to disabled school students as a teaching assistant. It can be beneficial both for teaching children with mental retardation and for working with autistic students, helping to improve their social communication skills. In these situations, the methods of a biological feedback can be especially helpful. These methods include analyzing the electroencephalogram (EEG) signals, which helps specifying an emotional and psychological state of a child, concentration of the child's attention via the brain activity analysis in alpha, beta and delta ranges, and therefore, defining the strategy of operating the robot assistant.
- Another direction includes an android assistance to different age groups. The android robot can be useful at different stages of school training. For elementary school students, and for those in the 5th–6th grades, the android robot acts as an assistant (the so-called "supplementary child sitter") who would approve and adjust their behavior on the basis of a biological feedback therapy. Students of the 7th–8th grades belong to the awkward age category when similar system can be used for identification of their psychosocial and emotional problems, which can be further adjusted by using the android assistant acting in this case as a "senior companion". In high school years, especially at the specialized study schools, the connection of trainees with a robotic complex is quite possible. For example,

within the framework of their scientific activity, they can develop original control algorithms for android robots using brain-computer interface technology.

Thus, a variety of android robot applicability fields necessitates conducting in-depth studies of the tasks and architecture of anthropomorphous robot assistant hardware-software complex (ARA HSC) for a teaching professional.

2 Defining Study Task

A distinctive feature of prospective application of the robotics in educational process is the variety of tasks needed to be solved in a real-time mode.

The goal of our study includes an analysis of tasks, and determination of architecture and the methods of solving the major problems providing an efficiency of functioning in a real-time mode. We should specify that our efficiency estimates are based on conventional criteria of the real-time mode: time of a task solving and requirements to hardware resources in terms of used memory volume.

3 Tasks and Structure of ARA HSC

The ARA HSC for teaching professionals represents the cyber-physical system intended for active support of educational process on the basis of assessing an extent of mastering a training material by means of a solving the following tasks:

- Investigating brain activity patterns of a trainee on the basis of EEG examination;
- Analyzing an emotional status on the basis of recognition of a trainee face video image;
- Evaluating physical activity of a trainee (movements of a head, hands, fingers, eyelids and eyes);
- Analyzing written answers of a trainee to educational tasks;
- Assessing speech activity of a trainee (oral answers to educational tasks);
- Measuring an extent of a training material mastering by a trainee on the basis of solution results' generalization for the tasks 1–5;
- Evaluating the environment on the basis of video images of a classroom obtained by ARA video cameras;
- Choosing the standard procedure of ARA actions adequate to a current status for activating a trainee for a deeper mastering of a training material;
- Planning actions (dialog options, ARA movement) on activation of a trainee in the direction of deepening the mastering of a training material taking into account the current status;
- Planning actions (dialog options, ARA movement) on encouragement of a trainee taking into account the current status;

- Planning actions for data treatment of a trainee's current status taking into account an availability of unwanted artifacts as a part of the analyzed data set;
- Planning managing impact of the drives of a hardware component of ARA for implementation of the planned physical activity taking into account safety of interaction with subjects and objects of a surrounding situation;
- Controlling the course of implementing ARA planned physical activity and its adjustment to the status change of a surrounding situation;
- Self-diagnosing ARA subsystems for identifying a possibility of executing the objectives, or, otherwise, making a transition to a "recovery" status;
- Other tasks arising in the process of expansion of the purposes and opportunities of ARA HSC.

The provided task list conducted by ARA HSC is far from being full. It is also necessary to consider a research scope of ARA HSC under development, and, therefore, among ARA users, besides those directly participating in educational process, the following should be taken into account: a teaching professional, computing system administrator, psychologist (optionally), developers of application software, developers of the system software, investigators, and methodologists.

Considering rather limited energy resources of autonomous ARA, the tasks demanding large computing resources for their solution are subject to a transfer onto the stationary server placed in a close proximity to the classroom. Such placement would allow avoiding big tasks on the traffic. Alternatively, there are well-known problems of organizing a network interaction between the computing system of autonomous ARA and a server component of ARA HSC. Consequently, a wide variety of the problems to be solved requires the distributed organization of the computing environment in ARA HSC. At the present stage of information technology development, the so-called cloud computing approach is frequently employed in similar situations [1]. It allows using additional resources located on other servers.

Cloud computing services presume control of the cloud-computing software via conventional web browsers [1]. Cloud computing is a dynamically developing technology of information infrastructure use.

The concept of cloud computing includes a set of the following notions [1]:

- Infrastructure as a Service (IaaS) is a computer infrastructure provided, as a rule, in the form of virtualization. IaaS is the service within the concept of a cloud data treatment;
- Platform as a Service (PaaS) is an integrated platform for development, expansion, testing and support of web applications. It is presented in the service form based on the cloud computing concept;
- Software as a service (SaaS) represents a business model of software licensed presuming software development and support by the supplier. Customers are given an opportunity of its paid use, as a rule, by means of the Internet;
- Desktop as a Service (DaaS) is another business model of software licensed use, which represents slightly improved SaaS model, generally assuming simultaneous use of several services required for comprehensive work.

Besides above-mentioned cloud computing notions concept, the ideas of Big Data as a Service (BDaaS) and Everything as a Service (EaaS) are rather widespread. Both notions show that, by means of a World Wide Web with cloud computing, it is possible to satisfy any needs for information processing. The latter is a primary benefit of cloud computing within the IT solutions for the business.

Taking into account our previous experiences, we selected the option of implementing GAMMA-3 system of the automated task solution with cloud computing use as a basis for developing ARA HSC [2].

The enlarged structure of ARA HSC system of the automated task solution with use of the cloud computing concept is presented in Fig. 1. In the simplified look, the offered structure consists of the interconnected system of specialized servers complemented with the server redirector (proxy server). The purpose of the server redirector is simplification of the interactions among client applications and cloud service resources. It is achieved by encapsulation, i.e. by means of concealing information from the client on the physical location of the data or the server providing required services.

For ARA HSC system, it is possible to define the following categories of the users:

- ARA per se making requests to specialized servers for data treatment, action planning, etc.;
- Teaching professionals, setting the tasks for ARA HSC at their lessons on execution and control of implementing the stages of a lesson, assessment of its results, making the amendments for the course of a lesson, etc.;
- Methodologists developing the techniques of conducting the lessons, creating standard procedures for conducting lessons, feedback procedures to possible deviations from the conventional course of the educational process, etc.;

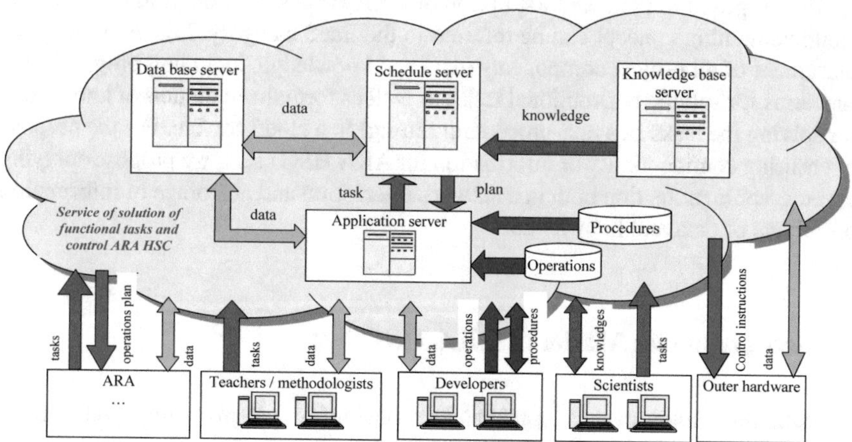

Fig. 1 The integrated skeleton diagram of the automated solution system of ARA HSC tasks with a cloud computing use

- Developers of the application software implementing computing processes on solving functional tasks;
- Developers of the system software of the bottom level providing interactions with external equipment (encephalographs, video cameras, microphones, gyroscopes, accelerometers, moment sensors, encoders, etc.);
- Investigators creating and studying the components of ARA HSC providing possibilities for solving intellectual tasks;
- Head of development exercising control of the course of executing the project;
- Accidental users (students of higher education institutions enrolled in ARA HSC project development).

For each user category, we developed customized applications giving the opportunities corresponding to the category. As the advantages in comparison with local or intranet options of implementing the automation equipment for solving design tasks, it is possible to identify:

- Lack of participation need of ultimate users in filling available local automation equipment with components implementing new methods of the management theory for solving new tasks;
- Lack of constant capacity expansion requirement of the memory devices for placement of increasing design data volumes;
- Reduction of financial expenses due to use of software product rent instead of their full-cost acquisition;
- Reduction in the total time of solving the tasks of design due to decrease in unproductive costs of designers' work hours for execution of unusual work.

The specified effect is achieved due to centralization of support and development for the automation equipment of a design task solving by the ARA HSC developer.

The proposed implementation option of ARA HSC within the framework of the cloud computing concept can be referred to the EaaS category. It is defined by the placement of all project components (data and knowledge, including programs and functions for solving conventional tasks, as well as formalized models of knowledge for solving the tasks in a non-procedural setting) in a cloud service. For the purpose of ensuring confidentiality of information for ARA HSC users, we propose applying cryptographic protection both in a network interaction and at storage of information on servers of data and knowledge.

4 Set-Theoretic Model of ARA HSC

Information technology development allows raising the system organization of information support for developing management systems to a new level. Among priority directions [3], top two places are taken by virtualization and cloud computing. The generalized ARA HSC system structure can be presented in the form:

$$S = \langle F, A, G, Rl, T, D, O \rangle, \quad (1)$$

where $F = \{f_i\}$ is a set of functional components, $A = \{a_i\}$ is a set of algorithms of component functioning $\dim(A) = n_a$; $G = \{\langle f_i, f_j, r_{ij}(f_i, f_j)\rangle | f_i \in F,\ f_j \in F, r_{ij}(f_i, f_j) = \{0|r_k\}, r_k \in Rl, n_R = \dim(Rl),\ k = \overline{1, n_R}, i = \overline{1, n_f}, j = \overline{1, n_f}\}$ is a scheme (graph) of component interrelations; $Rl = \{r_1, r_2, \ldots, r_{n_R}\}$ is a set of component interrelations for the system; $D = \{d|d = \langle p, c, m\rangle\}, d \cdot m \in C^{n_{m_1}} \times C^{n_{m_2}} \times C^{n_{m_3}} \times C^{n_{m_4}}, d \cdot c \in C^{n_{c_1}} \times C^{n_{c_2}} \times C^{n_{c_3}}, d \cdot p \in \{true, false\}$ is a set of types of data information structures used for solving the tasks, where C^x is a x-dimensional space of complex numbers; $P = \{p_i | p_i \in \{true, false\},\ p_i = f_{p_i}(d_x, d_y) d_x, d_y \in D\}$ is a set of the relations used in settings of the tasks being solved; $O = \{o|o = \langle o_c, o_s, o_r, o_d\rangle,\ o_s = \{d|d \in D\},\ o_r = \{d|d \in D\},\ o_c \subset \{\{\{d \cdot p\}^*\} \times Rl\}, o_d \subset Rl\}$ is a set of actions (operations) used for solving the tasks; $T = \{t|t = \langle t_S, t_R, t_D\rangle\}, t_S \subset D^*, t_R \subset D^*, t_D \subset \{\{\{d \cdot p\}^*\} \cup Rl\}, d \in D\}$ is a set of tasks, where $X^* = X^0 \cup X^1 \cup X^2 \cup \ldots \cup X^N \cup \ldots$, $X^0 = \varnothing$, $t_S = Src(t)$ are input data, $t_R = Rqr(t)$ are required results, $t_D = Dmnd(t)$ are requirements to the results of a $t \in T$ task solution.

For the purpose of optimizing the structure of considered information support service for the automated task solution, it is necessary to construct mathematical models and to conduct modeling to identify the system characteristics. It is reasonable to solve such problems using the methods of the queuing theory [4]. Its tools include analytical and simulation modelling of queuing systems and networks.

Among the tasks to be solved, there are declarative (non-procedural) tasks, solution automation of which requires artificial intelligence methods [5]. The cloud service of the automated solution of intellectual tasks (SITaaS, or "Solving of Intellectual Tasks as a Service") has the construction tool (planning) [6] for the sequence of actions (design operations). Execution (for example, with use of the software "Instrument-3m-I" [7] as environments of actions execution for a tasks solution) of the latter would lead to obtaining required task results. Not only tasks of data processing, but also tasks of actions planning [6] included into tasks area for solving by ARA HSC. The big share of the tasks solved by PAK RAAT is related with processing and data analysis about a trainee's status. Often analysis of EEG [8–11], of physical activity data [10, 12], of video images streams for assessment of a psychophysical and emotional status of trainees carry out by artificial neural networks [13–15] as the fast means of parallel information processing. The corresponding methods are implemented as procedures and operations (modules) of different hierarchy level of ARA HSC' knowledge model. ARA HSC is using cluster of NVIDIA Jetson TX2 for big data processing. High processing rate of information by ARA HSC is the key to quickly decision making on the basis of objective data.

Set-theoretic SITaaS model is described by the expression:

$$M_{SITaaS} = \langle D, A, Rl, T, Pl, S, I_P, I_{Pl}, I_S\rangle,$$
$$D = \{d|d = \langle p, c, m\rangle\},$$
$$d.p \in \{true, false\},$$

$$
\begin{aligned}
&d.c \in C^{n_{c1}} \times C^{n_{c2}} \times C^{n_{c3}},\\
&d.m \in C^{n_{m1}} \times C^{n_{m2}} \times C^{n_{m3}} \times C^{n_{m4}},\\
&A = \{a_i | a_i = \langle \mathrm{Cnd}(a_i), \mathrm{Src}(a_i), \mathrm{Rst}(a_i), \mathrm{Dmnd}(a_i)\rangle\},\\
&\forall a_i \in A\{(\mathrm{Cnd}(a_i) \cup \mathrm{Src}(a_i) \cup \mathrm{Rst}(a_i)) \subset D*\},\\
&\mathrm{Cnd}(x) \in \{\{d.c\} \times \{r\}\},\\
&\mathrm{Src}(x) \in \{\{d.c\} \times \{d.m\}\},\\
&\mathrm{Rst}(x) \in \{\{d.p\} \times \{d.c\} \times \{d.m\} \times \{r\}\},\\
&\mathrm{Dmnd}(x) \in Rl, d \in D, r \in R,
\end{aligned}
$$

$$
T = \left\{ t_i | t_i = \langle \mathrm{Src}(t_i), \mathrm{Rst}(t_i)\rangle, t_i \subseteq (\bigcup_{s_j \in S} \mathrm{tasksOf}(s_j)) \cap (\bigcup_{pl_i \in Pl} \mathrm{tasksOf}(pl_i)) \right\},
$$

$$
Pl = \left\{ pl_{\mathrm{i}} | pl_{\mathrm{i}}(t_{\mathrm{i}}) = \{a_j^{pl_i} | a_j^{Pl_i} \in A\}, \dim(pl_i) = n_{pl_i}, \bigcup_{a_{ij} \in pl_{\mathrm{i}}} \mathrm{Rst}(a_{ij}) \supseteq \mathrm{Rst}(t_i), \bigcup_{a_{ij} \in pl_{\mathrm{i}}} \mathrm{Src}(a_{ij}) \backslash \bigcup_{a_{ij} \in pl_{\mathrm{i}}} \mathrm{Rst}(a_{ij}) \subseteq \mathrm{Src}(t_i) \right\},
$$

$$
S = \{s_j | \mathrm{tasksOf}(s_j) \in T\}, \forall pl_{\mathrm{i}} \in Pl\left(\mathrm{Src}(t_i) \xrightarrow[\mathrm{I}_{pl}(pl_i(t_i))]{} \mathrm{Rst}(t_i) \right),
$$

$$
\forall t_i \in \bigcup_{s_j \in S} \mathrm{tasksOf}(s_j) \left(\mathrm{Src}(t_i) \xrightarrow[\mathrm{I_S(s_j)}]{} \mathrm{Rst}(t_i) \right),
$$

$$
\forall t_i (\mathrm{Rst}(t_i) \subset \bigcup_{a_j \in A} \mathrm{Rst}(a_j), t_i \xrightarrow[\mathrm{I_{Pl}}(D, A, Rl)]{} pl_i),
$$

where $\mathrm{I_P}(t_i)$ is planning of the actions on a $t_i \in T$ tasks solution, for which, out of the elements of a $a_j \in A$ elementary operations set, building the procedure $pl_{\mathrm{i}}(t_{\mathrm{i}}) = \{a_{ij} | a_{ij} \in A\}$, $\dim(pl_i) = n_i$, $j = \overline{(1, n_i)}$ is possible, such that the following conditions are satisfied: $\forall a_j \in pl_{\mathrm{i}}((\cup \mathrm{Src}(a_j) \backslash \cup \mathrm{Rst}(a_j)) \subseteq \mathrm{Src}(t_i))$, $\forall a_j \in pl_{\mathrm{i}}((\cup \mathrm{Rst}, (a_j) \supseteq \mathrm{Rst}(t_i))$; $\mathrm{I_S}(s_j)$ are the means for task solving of the software $s_j \in S$ providing required result $\mathrm{Rst}(t_i)$ on the set input data $\mathrm{Src}(t_i)$ for all tasks t_i belonging to the class of solvable tasks $t_i \in \bigcup_{pl_{\mathrm{i}} \in Pl} \mathrm{taskOf}(s_j)$ of the software $s_j \in S$; $I_{Pl}(pl_i)$ are the means for executing procedures $pl_i \in Pl$ of a task solution $t_i \in \bigcup_{pl_i \in Pl} \mathrm{tasksOf}(pl_i)$ providing required result $\mathrm{Rst}(t_i)$ on the set input data $\mathrm{Src}(t_i)$.

For implementation specified in the offered $\mathrm{I_S}(s_j)$ model of means (a cure of tasks of the $s_j \in S$ software), $\mathrm{I}_{Pl}(pl_i)$ (means of execution of $pl_i \in Pl$ procedures), $\mathrm{I_P}(t_i)$ (means of planning of actions on a $t_i \in T$ tasks solution) it is supposed to use modifications of the corresponding subsystems of the INSTRUMENT-3m-I systems [16] and the GAMMA-3 systems [17, 18]. Modification is caused by need of their placement for server components of a cloud service.

The constructed set-theoretic model will be used for carrying out researches by methods of the queuing theory for the purpose of optimization of components of system.

5 Design of a Classroom for Correctional Training

Correctional training differs in specific requirements to the conditions of conducting educational process. In this regard, the scheme of an educational group of two trainees for correctional training is offered in the Fig. 2.

In Fig. 2, the following components of a robotic system supporting the educational process of correctional training for two trainees are presented:

- *R* is a robot assistant of anthropomorphous type;
- *the classroom* (4 video cameras in the corners, 2 TV sets for reproduction of educational information, 2 workplaces for trainees, each of which includes a desk, a chair/armchair, a computer, a frontal video camera with a microphone, an audio system);
- *workplace of the teacher* (a desk, a chair/armchair, a computer, a frontal video camera with a microphone, an audio system);
- *workplace of the methodologist* (a desk, a chair/armchair, a computer, an audio system);
- *workplace of the administrator* of the information system server (a desk, a chair/armchair, a server computer).

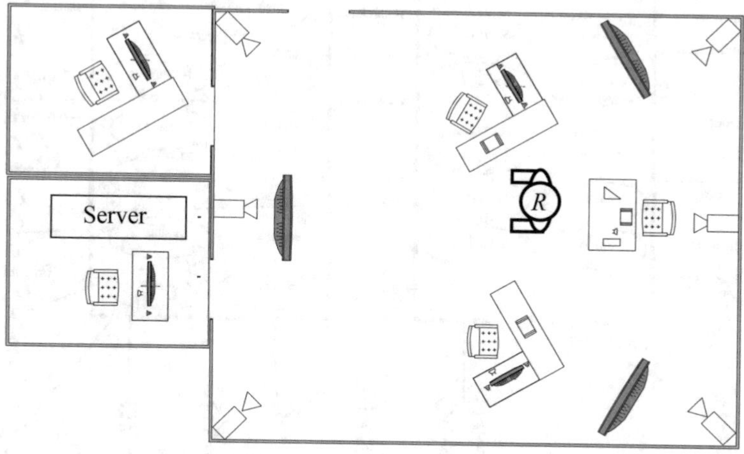

Fig. 2 The scheme of a classroom for two trainees for correctional training

6 Scheme of Component Interactions for Robotic System Supporting the Correctional Training Educational Process

The scheme of component interactions for the robotic system supporting the educational process of correctional training for two trainees is presented in Fig. 3.

Some of the depicted in Fig. 3 includes the following:

- *subjects of educational process* (teaching professional, methodologist, administrator, instructor);
- *active equipment* of a classroom: means of receiving and displaying information (the encephalograph, the microphone, the video camera, the TV set);
- *robot assistant* components, managing which is among the functions of the control system (head, motors of manipulator, pedipulator, range finder, video controller, gripper).

The control of classroom equipment and robot components, depending on the current situation, is carried out:

- according to conventional scenarios;
- according to the program (action plan) constructed by the planning server subsystem of ARA HSC (Fig. 1).

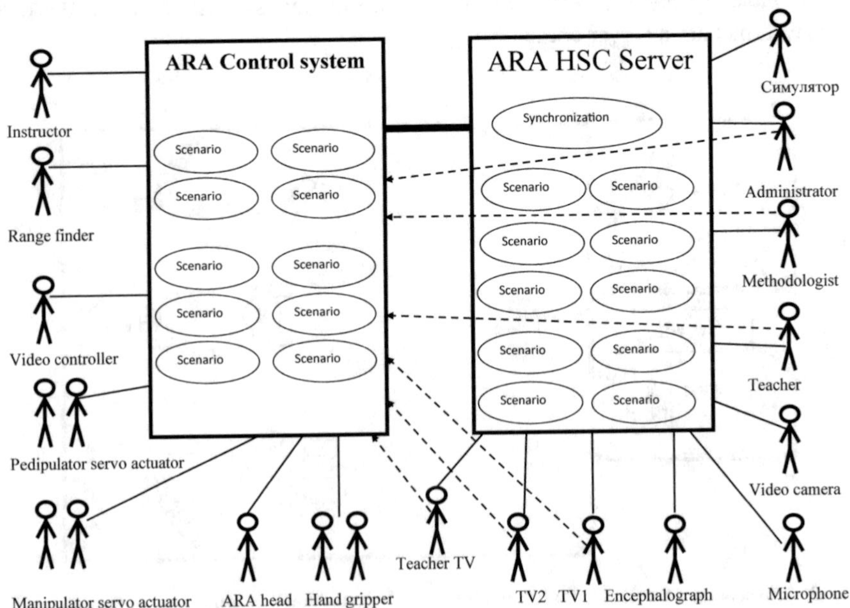

Fig. 3 Scheme of component interactions for robotic system supporting the correctional training educational process for two trainees

7 Conclusion

The study is devoted to the issue of using anthropomorphous robots in education. The distinctive feature of this field of robotics use is a necessity of a large number of solutions for computationally difficult tasks in a real-time mode. The hardware-software complex of anthropomorphous robot assistant (ARA HSC) for a teaching professional is proposed. The architecture is developed and the set-theoretic model of ARA HSC is constructed. The scheme of components' interaction for ARA HSC is provided.

The developed set-theoretic model ARA HSC will be further used for conducting the studies using the queuing theory methods for optimizing the architecture and structure of ARA HSC components.

Acknowledgements The study is performed with assistance of the RF Ministry of Science and Education (the agreement on providing a subsidy No. 14.577.21.0282 from 1 October, 2017, the unique project identifier is RFMEFI57717X0282, the Federal Target Program "Research and Advances on the Priority Directions of Development of a Scientific and Technological Complex of Russia for 2014–2020".

References

1. http://www.inoventica.ru/informacionnyj_centr/tehnologii
2. Aleksandrov, A.G., Mikhailova, L.S., Bragin, T.M., Stepanov, A.M., Stepanov, M.F.: Aspects of application of cloud computing for computer-aided control system design by GAMMA-3 system. Design systems, production technology preparation and control of stages of life cycle of industrial product (CAD/CAM/PDM—2012). In: Artamonov, E.I. (ed.) Proceedings of 12th International Conference, pp. 111–115. LLC "Analitik", Moscow (2012)
3. Demidov, M.: "Clouds" turn into consumer goods. http://www.cnews.ru/reviews/free/infrastructure2009/articles/smb.shtml
4. Aliev, T.I.: Bases of Discrete Systems Modelling, 363 p. SPbSU ITMO, St. Petersburg (2009)
5. Stepanov, M.F.: Automated Solution of Formalized Tasks of Automatic Control Theory, 376 p. Saratov State Technical University, Saratov (2000)
6. Stepanov, M.F., Stepanov, A.M.: Mathematical modelling of intellectual self-organizing automatic control system: action planning research. Procedia Eng. **201**, 617–622 (2017). https://doi.org/10.1016/j.proeng.2017.09.657
7. Aleksandrov, A.G., Mikhailova, L.S., Bragin, T.M., Stepanov, M.F., Stepanov, A.M.: About development of concept of automated solution of control theory tasks by GAMMA-3 system. Mechatron. Autom. Control **9**, 14–19 (2011)
8. Birbaumer, M., et al.: A spelling device for the paralyzed. Nature **398**, 297 (1999)
9. Ma, T., et al.: The extraction of motion-onset VEP BCI features based on deep learning and compressed sensing. J. Neurosci. Methods **275**, 80 (2017)
10. Quitadamo, L.R., et al.: Support vector machines to detect physiological patterns for EEG and EMG-based human-computer interaction: a review. J. Neural Eng. **14**, 011001 (2017)
11. Blankertz, B., Dornhege, G., Krauledat, M., Muller, K.R., Curio, G.: The non-invasive Berlin brain-computer interface: fast acquisition of effective performance in untrained subjects. NeuroImage **37**, 539–550 (2007)

12. Ferrante, A., Gavriel, C., Faisal, A.: Data-efficient hand motor imagery decoding in EEGBCI by using Morlet wavelets & common spatial pattern algorithms. In: 7th International IEEE/EMBS Conference on Neural Engineering (NER), p. 948 (2015)
13. Bishop, C.: Neural Networks for Pattern Recognition. University Press, Oxford (1995)
14. Carling, A.: Introducing Neural Networks. Sigma Press, Wilmslow, UK (1992)
15. Fausett, L.: Fundamentals of Neural Networks. Prentice Hall, New York (1994)
16. Stepanov, M.F.: Analysis and synthesis of automatic control systems by software "INSTRUMENT-3mI". J. Instrum. Eng. **47**(6), 27–30 (2004)
17. Aleksandrov, A.G., Mikhailova, L.S., Stepanov, M.F.: GAMMA-3 system and its applications. Autom. Remote Control **72**(10), 2023–2030 (2011). © Pleiades Publishing, Ltd., 2011. Original Russian Text © Aleksandrov, A.G., Mikhailova, L.S., Stepanov, M.F.: Avtomatika i Telemekhanika (10), 19–27 (2011)
18. Stepanov, M.F., Stepanov, A.M., Pakhomov, M.A., Salikhova, A.P., Mikhailova, L.S.: Development tools of the intellectual self-organized systems of automatic control. In: CEUR Workshop Proceedings, vol. 1638, pp. 674–680 (2016). https://doi.org/10.18287/1613-0073-2016-1638-674-680
19. Aleksandrov, A.G., Mikhailova, L.S., Stepanov, M.F.: GAMMA system—integrated environment for control systems design. Design systems, production technology preparation and control of stages of life cycle of industrial product (CAD/CAM/PDM—2009). In: Artamonov, E.I. (ed.) Proceedings of 9th International Conference. Institute of Control Sciences Academician VA Trapeznikov, Moscow (2009)
20. Aleksandrov, A.G., Mikhailova, L.S., Stepanov, M.F.: Structure hardware and software of distributed gamma system for control systems design. Design systems, production technology preparation and control of stages of life cycle of industrial product (CAD/CAM/PDM—2011). In: Artamonov, E.I. (ed.) Proceedings of 11th International Conference, Institute of Control Sciences Academician VA Trapeznikov, Moscow (2011)

Method of the Exoskeleton Assembly Synthesis on the Base of Anthropometric Characteristics Analysis

Anna Matokhina, Stanislav Dragunov, Svetlana Popova and Alla G. Kravets

Abstract The article describes the synthesis module of design solutions for passive and active exoskeletons, taking into account the anthropometric parameters of the operator, the requirements and limitations imposed on the exoskeleton. Exoskeletons, presented on the market of their functional-parametric structure and technical characteristics, are investigated. The method of generating design solutions is considered by the example of an exoskeleton. An algorithm has been developed for the operator upper and lower extremities exoskeleton assembly synthesis, taking into account the anthropometric characteristics, requirements, and limitations of the exoskeleton, and also developed a system prototype for generating the assembly of the exoskeleton with regard to the anthropometric characteristics. The system for generating assemblies of different types of exoskeletons makes it easy to adapt to emerging markets and select the most suitable model taking into account various parameters.

Keywords Exoskeleton · Construing · Modeling · CAD system · Prototyping · Adaptive modeling

---

A. Matokhina (✉) · S. Dragunov · S. Popova · A. G. Kravets
Volgograd State Technical University, 28 Lenin av., Volgograd 400005, Russia
e-mail: matokhina.a.v@gmail.com

S. Dragunov
e-mail: dragunov.stanislav.e@gmail.com

S. Popova
e-mail: lorbrinil@gmail.com

A. G. Kravets
e-mail: agk@gde.ru

A. G. Kravets et al. (eds.), *Cyber-Physical Systems: Advances in Design & Modelling*, Studies in Systems, Decision and Control 259,
https://doi.org/10.1007/978-3-030-32579-4_3

1 Introduction

In biology, the concept of "exoskeleton" is used to denote the external type of skeleton in some invertebrates. It supports, protects the body of the animal from damage, and is a mechanical barrier that serves as the first stage of protection against infection [1–3].

In robotics, the exoskeleton has other functions depending on the field of application: medical, military, industrial, and consumer. Despite a sufficient number of the ongoing development of exoskeletons, information about them is scattered.

The field of exoskeleton systems is constantly evolving. They are also called "wearable robots" (exoskeleton from the Greek "external skeleton") because he repeats the human biomechanics. First, we consider in general what exoskeletons are [4].

Exoskeletons are wearable devices, i.e. are placed on the user's body and act as supporting devices that increase, enhance or restore human performance. In the human exoskeleton system, part of the functions, for example, maintaining balance, is left to the person, while the weight of the load or large efforts falls on the exoskeleton mechanism [5].

Wearable robots are designed to help people with various military, medical, and industrial purposes [6]. But there are still many problems associated with exoskeletons and their orthopedic design, which is constantly being improved. Exoskeleton robots are widespread in areas related to rehabilitation, tactile interaction, and increased human power [7].

The following features and characteristics of the exoskeleton can be distinguished [8]. Improving human productivity: an exoskeleton should increase the user's strength, endurance and/or speed, allowing them to perform tasks that they previously could not perform. Low resistance: the exoskeleton should not interfere with the natural movement. Natural interface: the exoskeleton should provide a natural, intuitive, simple interface, so that the user feels that the exoskeleton is really an extension of his body, and not something that moves the user. Long life: the exoskeleton must have a sufficient duration of use between a full charge of the power system and a quick and easy charge. Convenience: the exoskeleton should be comfortable and safe to wear and easy to turn on and off.

The mechanical design of the exoskeleton should take into account various design criteria. Design criteria set forth by researchers from Selcuk University include:

- ergonomic and comfortable design;
- high maneuverability;
- lightweight and durable design;
- adaptability to different users;
- user security.

The exoskeleton of the lower limbs is a mobile device, worn on a person, which provides at least part of the energy that is necessary for the movement of the limb.

The synthesis of the exoskeleton, taking into account the anthropometric characteristics of the operator was carried out in [9–12]. However, the works provide examples of the synthesis of either individual parts of the exoskeleton or individual mechanisms used in rehabilitation tasks.

2 Description of the Method of Anthropometric Characteristics Analysis

2.1 Exoskeleton Classification

In recent years, studies of lower limb exoskeletons have become a hot topic. Several organizations around the world have developed impressive exoskeletons for power, varying significantly in performance and technology used. The main use of exoskeleton robots in the modern market is focused on rehabilitation services in the field of medicine - training muscle movements and helping to repair injuries in a more precise and effective way than was previously possible. They are also used in the army to combat fatigue and injuries of soldiers in battle. Other exoskeletons may provide ergonomic support to workers in industry and medicine who perform repetitive or strenuous work [13].

Exoskeletons can be classified according to the energy source and the principle of the drive [14–16] or by user interface [17]:

- joystick: for exoskeletons that provide 100% of the energy for the movement required by the owner. The great advantage of the joystick is that no movements or nerve functions are required to use the exoskeleton [18];
- buttons or control panels: the exoskeleton is in different programmed modes;
- mind control: using an electrode cap for the skull;
- sensors: modern exoskeleton designs can have up to 40 different built-in sensors that control rotation, torque, tilt, pressure, and can capture nerve signals in the hands and feet. An integrated 5 sensor network transmits data back to the microcomputer to interpret and correct movement [19];
- no control: some passive exoskeletons do not have control buttons or switches.

All classification signs of exoskeleton and types are shown in Fig. 1.

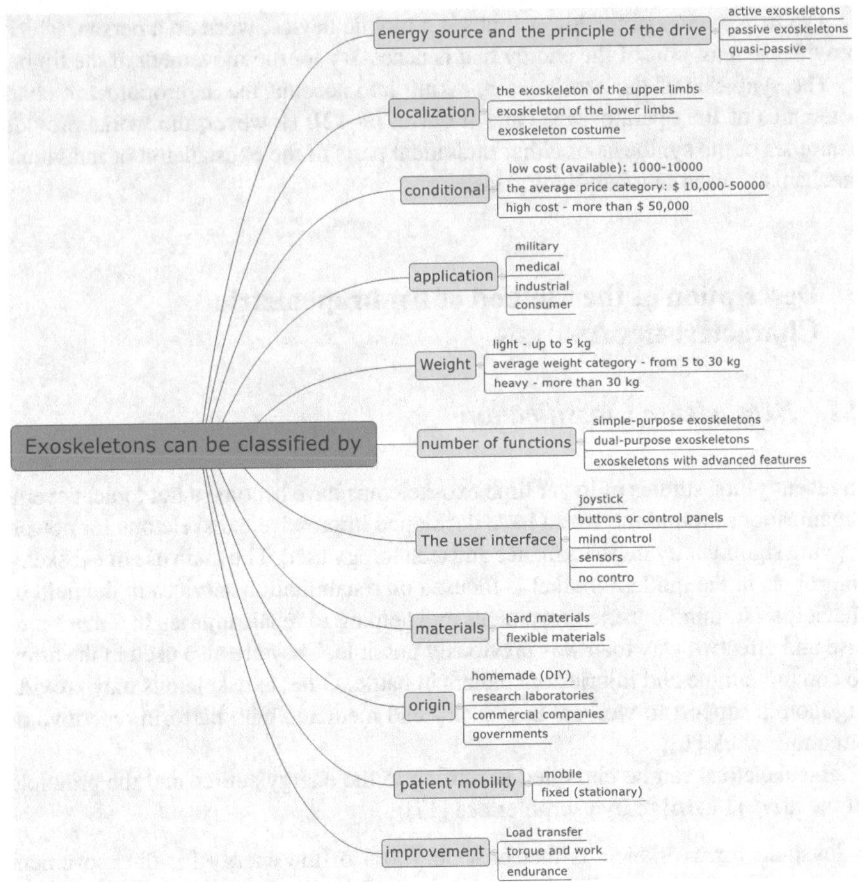

Fig. 1 Classification of exoskeleton

2.2 *Collection of Technical Characteristics and Parameters of the Existing Exoskeleton*

The main parts of the exoskeleton are (Fig. 2):

- the external frame is a rigid, stationary frame of the exoskeleton, made in the form of a three-dimensional frame structure that defines the space for changing the position of the orthopedic modules of the internal structure [20];
- the inner frame is a controlled, mobile frame of the exoskeleton, which is kinematically connected with the outer one [20];
- a servo drive is any type of mechanical drive (device, working body) that includes a sensor (position, speed, effort, etc.) and a drive control unit (electronic circuit or mechanical system of rods) that automatically supports the necessary parameters

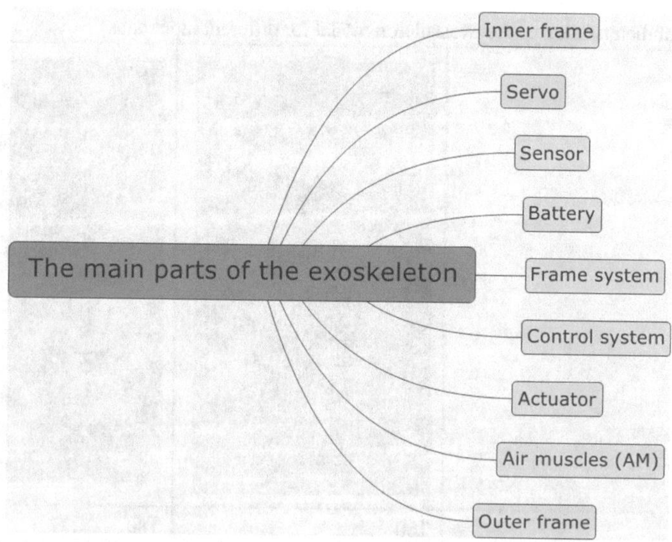

Fig. 2 The main parts of the exoskeleton

on the sensor (and, respectively, on the device) according to a given external value (the position of the control knob or the numerical value of other systems) [21];

- sensors—a measurement tool designed to generate a signal of measurement information in a form convenient for transmission, further transformation, processing and/or storage, but not amenable to direct perception by an observer [22];
- a battery is a device for storing energy for the purpose of its subsequent use;
- frame system—a system of supporting structure and consisting of a combination of linear elements [8];
- control system—a system that controls the movement of the exoskeleton;
- an actuator is an engine that is used to move or control the mechanics of a system [23];
- air muscles (BM) are devices that contract or stretch under the action of air pressure. They represent a hermetic casing in a casing with a braid of inextensible threads [3].

As part of the work, a model for assembling the basic elements of the exoskeleton of the lower extremities was developed. The model is parameterized and adapts to the anthropometric characteristics of the person. Table 1 shows two options for assembling exoskeleton for two people with different anthropometric characteristics [24].

Table 1 Parameterization of the exoskeleton model for different input data

Image	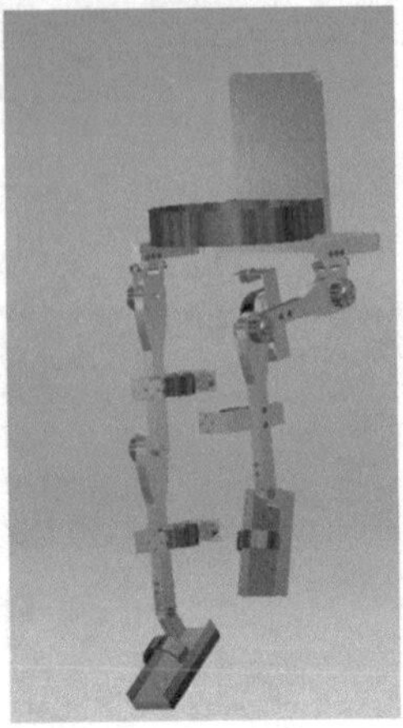	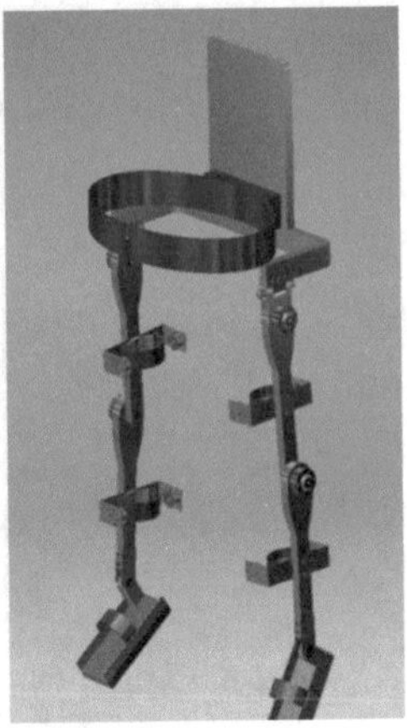
Growth	160	180
Waist girth	65	96
Thigh-length	32	38
Calf length	36	38
Foot length	23	28
Hip girth	43	51
Girth	33	40

2.3 *Description of the Design Subsystem with Parameters*

The exoskeleton design subsystem is designed to create parametric assemblies of the exoskeleton of the lower extremities [5]. It is possible to build two types of exoskeleton: passive and active.

Functions of the subsystem design of the exoskeleton of the lower extremities (Fig. 3):

- the choice of exoskeleton purpose (in what field of activity it will be used);
- selection of components of the exoskeleton (construction materials, types of actuators, joints, batteries, and control);
- the introduction of measurements of a human operator;
- transfer of exoskeleton parameters to a CAD system for automatic change of exoskeleton parameters;
- obtaining a personalized assembly of the exoskeleton.

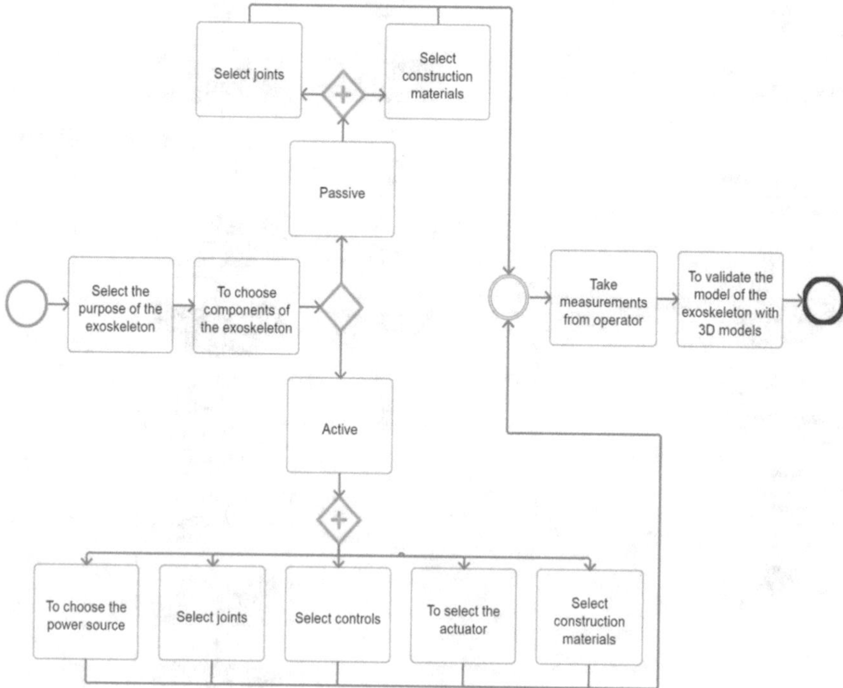

Fig. 3 Exoskeleton design method

2.4 *Designing a Synthesis Module*

Patents [17, 20, 22, 25, 26] were found and analyzed to create parametric components of the assembly of the exoskeleton of the passive and active type. From the drawings of the patents selected parts. Developed models of parts, details are parameterized in the Autodesk Inventor. The general synthesis algorithm is presented in Fig. 4. The algorithm of the processing module is presented in Fig. 5.

Next, the module for generating a parameterized assembly based on its type in the Autodesk Inventor system was developed (Fig. 6). You need to use Inventor's API to design a module for generating a parameterized assembly of an exoskeleton in Autodesk Inventor.

In order to integrate the operator's anthropometric parameters processing module with the parameters of the exoskeleton assembly, a stand-alone EXE program is used, which runs on its own and connects to the Inventor. This type of program is usually used when there is a program that has its own interface and does not require the user to work interactively with Inventor [27, 28].

The parametrization subsystem is part of the exoskeleton design system and allows you to customize the exoskeleton model for a given size of a human operator, taking

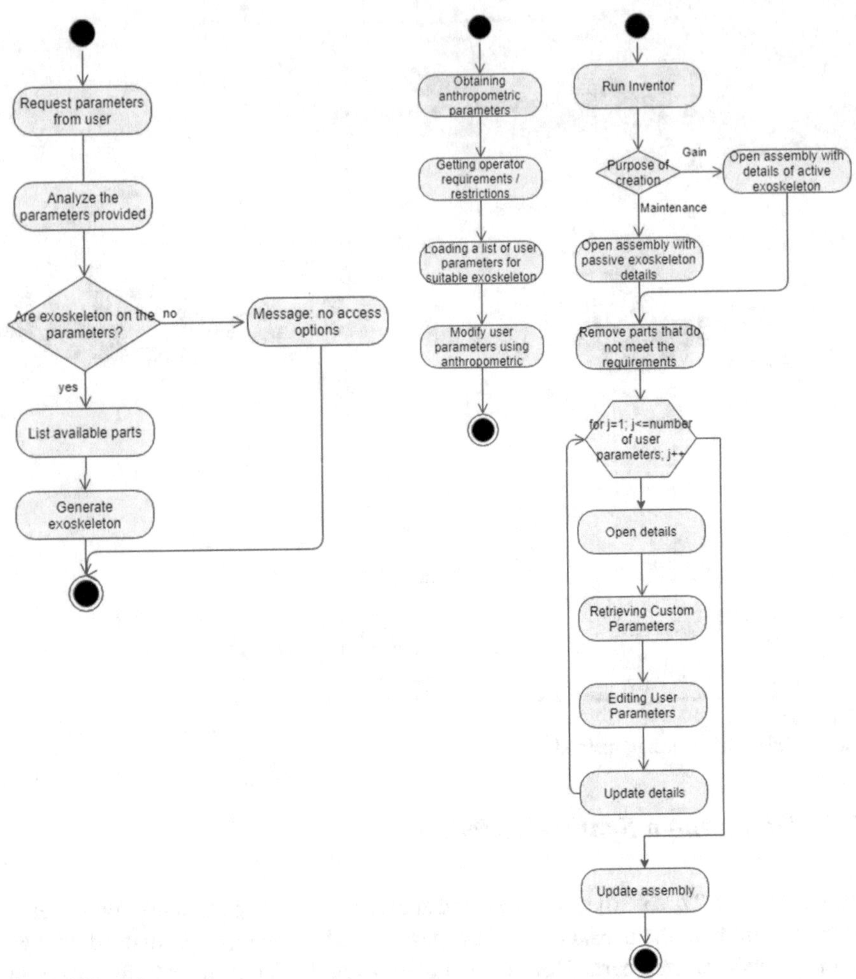

Figs. 4–6 (**4**) A general algorithm for the synthesis of the exoskeleton. (**5**) Diagram of the module processing the anthropometric parameters of the operator. (**6**) Algorithm for generating a parameterized assembly of the exoskeleton according to its type in the autodesk inventor system

into account the anthropometric parameters of a person [29]. The subsystem allows reducing the time spent on inputting the parameters taken into the model.

To link the Inventor with the exoskeleton design subsystem, the Inventor API is used (Fig. 7).

An API, or application programming interface, is a term used to describe the functionality of the Inventor graphical editor provided by the program.

Autodesk Inventor is a common CAD system, which means that it is not intended for any particular industry or is used to model only certain types of products. Having

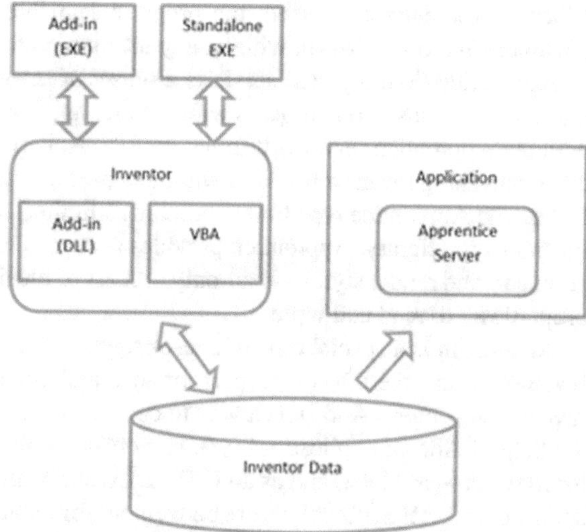

Fig. 7 Inventor API operation diagram

an API allows you to add Inventor-specific functionality that is specific to individual needs.

The Inventor API is implemented based on COM technology and provides various ways to access Inventor data using various types of add-in applications (Plug-in modules).

Components that provide API—Inventor and "Apprentice Server". Inventor Data at the base represents Inventor data accessed by parts, assemblies, drawings.

Add-in (DLL), Standalone EXE, Add-in (DLL), VBA, Application represent programs that are written separately. When one field includes another field, it indicates that the included field is executed in the same process as the field that includes it.

For example, VBA runs in the same process as an Inventor. A program that runs in the same process runs much faster than a program that runs "out of process".

Usually, choosing VBA, guided by the following advantages:

- VBA is a programming environment that is accessible from within Inventor and is not required to purchase an additional programming language;
- you can implement programs within the document for which processing they are designed. You can also save programs in separate files so that they can be used in various documents;
- VBA runs in the same process as Inventor;
- VBA has the same access to all features of the API as any of the other ways to access the API (with the exception of Add-Ins).

Standalone EXE is a program that runs independently and connects to the Inventor. This type of program is usually used when you have a program that uses Inventor but has its own interface and does not require the user to work with Inventor interactively. For example, the program can monitor new records that will be added to the database.

When a new entry is created, the program launches Inventor, opens the desired document and prints it—all without any user interaction.

Apprentice (journeyman) server is a subset of Inventor that does not have a user interface and runs in one process with another application. The only way to interact with the Apprentice server is through its API. Apprentice server is much more efficient than using Inventor because, without a user interface, operations are performed faster. The Apprentice type library contains a limited set of objects supported by the Inventor type library. Apprentice provides access to the assembly structure, B-Rep geometry and render styles—read-only. Access to file links, attributes, and document properties—to read and write.

An Add-In is a special type of add-on application that automatically loads when Inventor is launched, has high performance and appears to the user as part of the Inventor. Inventor's Add-In is a way to connect with Inventor and use its API. With the help of Add-In applications, you can create commands. The Add-In diagram is listed twice—as a DLL and as an EXE. DLL applications work in the same process as Inventor, EXE applications are convenient for debugging.

2.5 *Interfaces of the Exoskeleton Design Subsystem*

The exoskeleton design subsystem opens separately from Inventor, after launching the assembly of the exoskeleton on Inventor and mates with it using the Inventor API. A dialog box is launched in which you can begin work on taking measurements from a human operator. The subsystem interface consists of five tabs with the help of which the collected operator parameters are entered (Fig. 5).

The tabs "Belt", "Hip", "Drumstick", "Stop" contain instructions with the image of individual human limbs with allocated places for taking the size of a person. Also, on the tabs are fields for entering data obtained from measuring a person. The minimum and maximum values are indicated, in the range of which the part of the assembly of the exoskeleton is built. If you enter a value less than the minimum or greater than the maximum, the program will automatically put the minimum or maximum value in the input field. There is a button on the tabs, which saves the entered data in the program and switches to the next tab.

Figure 8 shows the interface tab "Belt" of the exoskeleton design subsystem is responsible for parametrizing the details of the operator's back "frame".

User parameters of the part "Back frame" are changed using the design subsystem program:

Width of the part—S1—the parameter is changed by the user "1. Hip Girth";
The part length—D1—is calculated automatically by the design subsystem D1 = S1/4;
The height of the part—H1—remains unchanged.
The height of the back of the part—HSpin—the parameter is changed by the user "2. Back height" (Fig. 9).

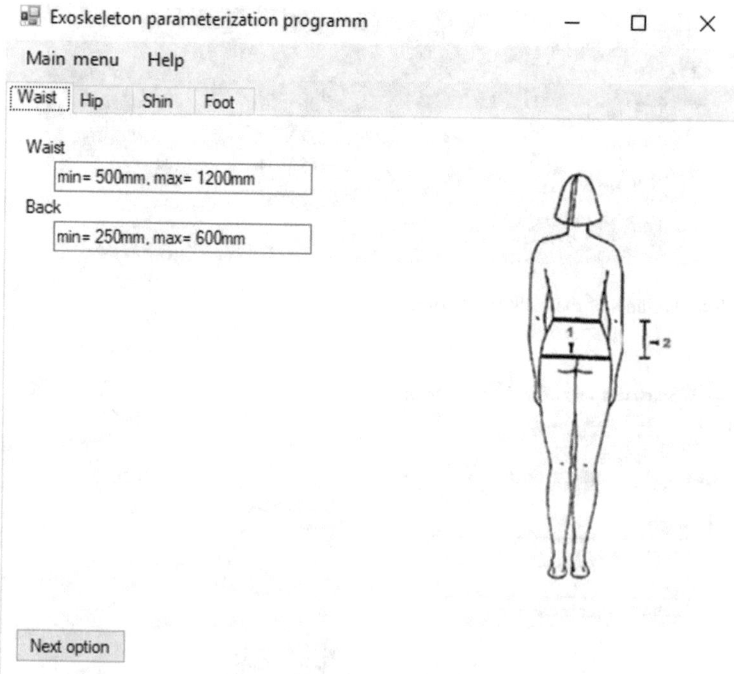

Fig. 8 Design subsystem interface "Belt" tab

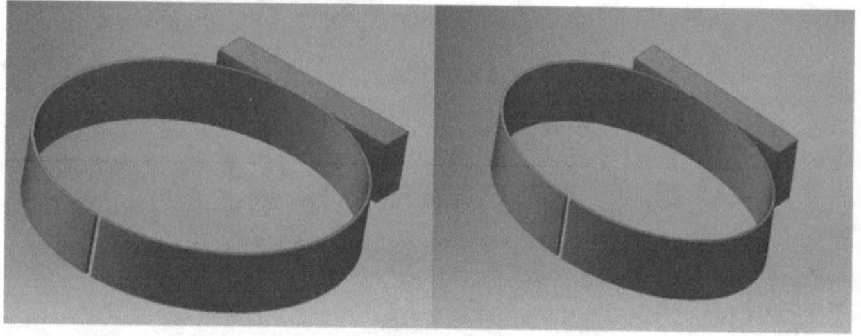

Fig. 9 Two options for the model belt

The remaining parameters of the part are in the parameters of the Inventor, change automatically from the introduction of user parameters, see Fig. 10.

Tab "Thigh" is designed to calculate the parameters of the skeleton of the thigh. Figure 11 shows the interface of the exoskeleton detail design subsystem, the parametrization of which, as well as in the previous tab, takes into account the entered parameters with size restrictions to maximum and minimum values.

The informational image shows in what places the hips should measure the human operator and how it will be done correctly. The tab shows the permissible maximum and minimum values.

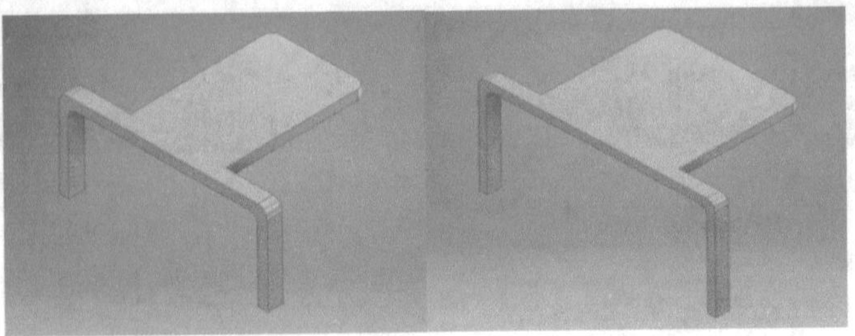

Fig. 10 Two variants of the back frame model

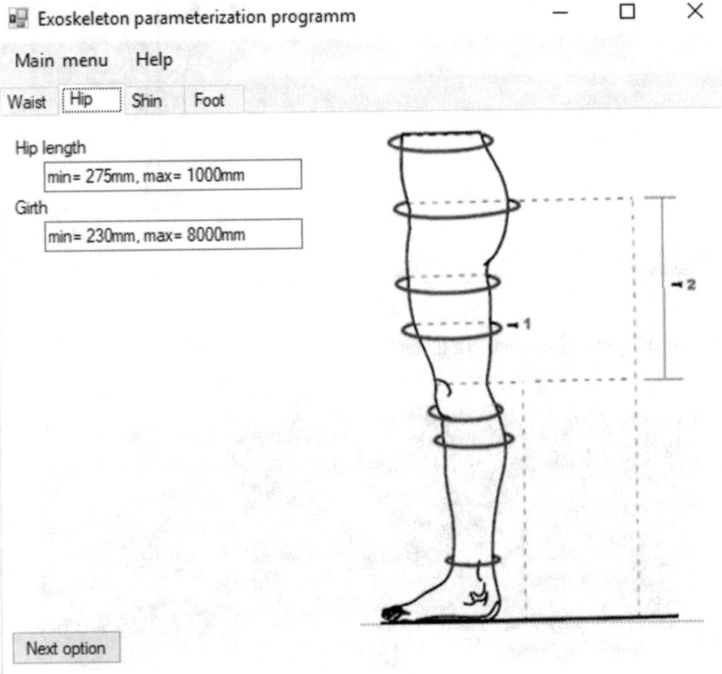

Fig. 11 Design subsystem interface “Belt” tab

The user parameters of the “Thigh carcass” part (Fig. 12) are changed using the design subsystem program:

Thigh-length—divided into two values due to the design features of the part, D1 and D2, therefore, when getting the value from the “1. Thigh Length”, the D2 parameter will receive the cell value divided by 3, and the D1 parameter will receive the D2 * 2 cell value;

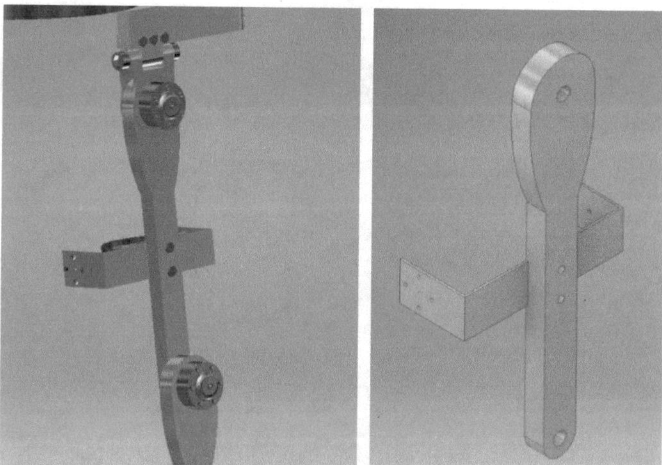

Fig. 12 Detail "The frame of the thigh" in the assembly and a separate part

The length of the location of the sensors is also divided into two values DKrep and DKrep2—and get the values according to the formula: DKrep and DKrep2 = (cell "2. Hip girth"/2 − 50)/2;
The width of the location of the sensors—DKDat—uses the cell "2. Hip Girth"/4;

The remaining parameters of the part are in the parameters of the Inventor, change automatically from the introduction of user parameters.

The "Shin" interface tab informs that the entered parameters will be sent to calculate the "Shin Frame" detail. This tab can be seen in Fig. 13.

The user parameters of the "Shank skeleton" part (Fig. 14) are changed using the design subsystem program:

The length of the tibia is divided into two values due to the design feature of the part, D1 and D2, therefore, when retrieving the value from the "1. Shank length", the D2 parameter will receive the cell value divided by 3, and the D1 parameter will receive the D2 * 2 cell value;
The length of the location of the sensors is also divided into two values DKrep and DKrep2—and get the values according to the formula: DKrep and DKrep2 = (cell "2. Girth of leg"/2 − 50)/2;
The width of the location of the sensors—DKDat—uses the cell "2. Girth of the leg"/4;

The remaining parameters of the part are in the parameters of the Inventor, change automatically from the introduction of user parameters.

The "Foot" tab informs you that the entered parameters will be sent to calculate the "Frame of the foot" part. This tab can be seen in Fig. 15.

The user parameters of the "Foot skeleton" part (Fig. 16) are changed using the design subsystem program:

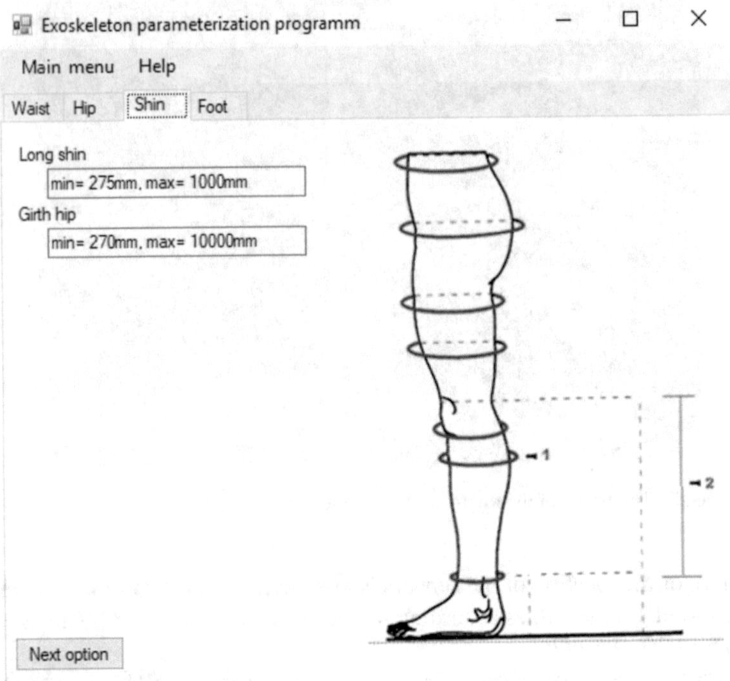

Fig. 13 Design subsystem interface "Shin" tab

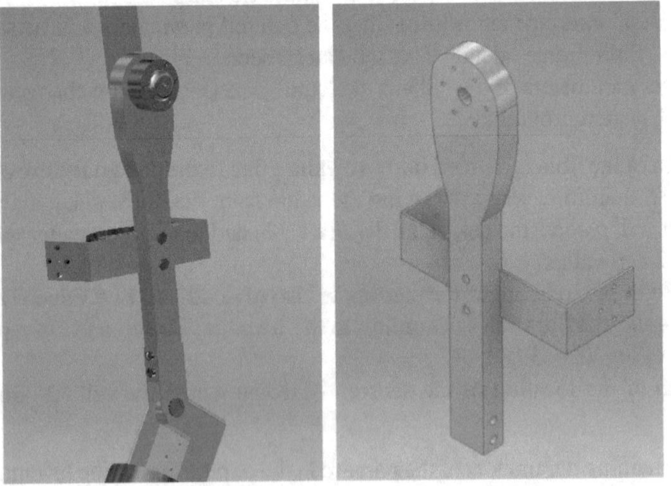

Fig. 14 Detail "The skeleton of the leg" in the assembly and a separate part

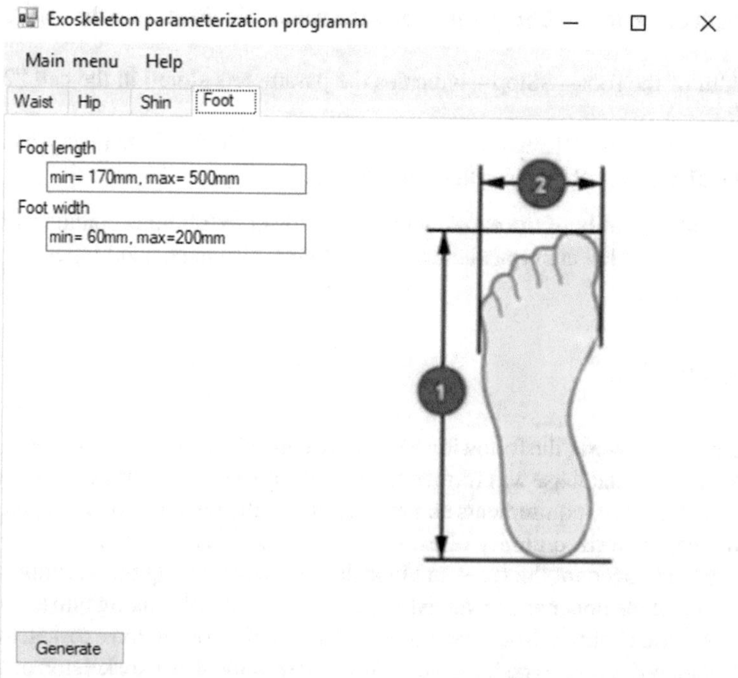

Fig. 15 Design subsystem interface "Foot" tab

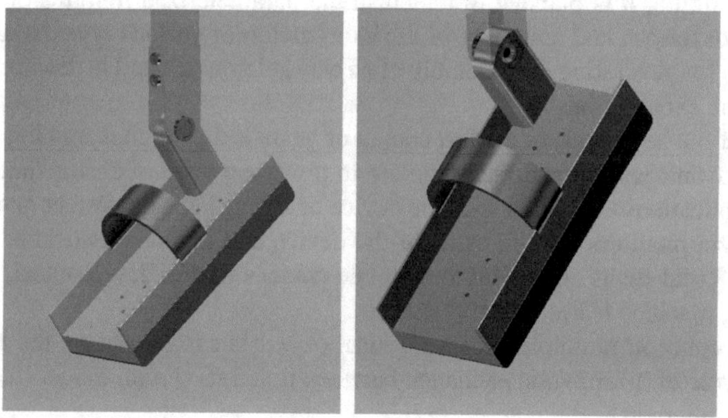

Fig. 16 Detail "The skeleton of the leg" in the assembly and a separate part

The length of the foot—DStop—indicates the values obtained from the cell "1. Foot length";
The width of the foot—SStop—indicates the parameters stored in the cell "2. Foot width";
Attachment height—DKrep—this parameter is calculated automatically using the formula DKrep = cell "1. Length of the foot"/2.7.

The final assembly of the exoskeleton of the lower extremities contains 42 components, 15 of which are typical elements of drives, fasteners, and hinges.

3 Conclusion

As a result of the work, the following tasks were carried out: a review of exoskeletons was conducted, a database was formed to select the type and type of exoskeleton, taking into account the requirements and restrictions of the operator to the exoskeleton; formed a parameterized library of exoskeleton elements and assembly of exoskeleton, taking into account the type; an algorithm for synthesizing the assembly of the exoskeleton of the upper and lower extremities of the operator, taking into account its anthropometric characteristics, requirements, and limitations to the exoskeleton, has been developed; A prototype for generating the assembly of an exoskeleton of human lower limbs was developed taking into account the anthropometric characteristics of the operator.

In the future, it is planned to replenish the parameterized libraries of elements of the exoskeleton and assembly of the exoskeleton of various types, to develop a prototype for generating the assembly of an exoskeleton suit and to develop a digital twin of the exoskeleton.

One of the key trends of the last couple of years is the "digital counterpart".

The advantages of the digital twin are to provide a level of abstraction that will allow applications to interact with the device or devices in a consistent manner. The digital twin monitors the life cycle of the device and the data associated with the device. Digital twins allow you to simulate devices during development, integrate analytics, machine learning, etc.

Basic concept: monitoring of a physical object is carried out on the basis of a closed cycle of information exchange between it and its virtual model (thereby the digital counterpart).

The digital twin can be used as a starting point for a simulation model that may extrapolate how the system will work in the future. The degree and accuracy of these simulations can vary depending on the implementation of the simulation and the desired result.

For example, a CAD drawing package can be used to create a digital model, and then the process control system will use this model as the basis for the digital twin. This software can provide a connection between digital twin sensors and controls with those in the real world.

On the other hand, if the desired results are related to how strong the part will be inside the engine, then the relative level of detail should be greater.

The model used in the simulation may have the characteristics added so that physical modeling is possible. The model may include detailed information about the virtual materials used in the model, which, in turn, will allow the modeling software to replicate the response of the model during the simulation.

Models can have different goals, but they can also have general descriptions, such as information about dimensions, material attributes, etc. Many models can be used by several applications for different purposes, showing the status of the current system to simulate a device that has yet to be built [sixteen].

Since the digital twin will simplify the task of testing, allowing for quick correction and detection of anomalies, it was decided to further design the digital twin and develop recommendations for optimizing the operating mode and maintenance of the exoskeleton.

Acknowledgements The reported study was funded by RFBR according to the research project # 19-07-01200 and technically supported by the Project Laboratory of Cyber-Physical systems of Volgograd State Technical University.

References

1. Kelechava, B.: Powered exoskeletons [Electronic resource]. https://blog.ansi.org/2016/06/powered-exoskeletons/#gref (appeal date: 12.26.2017)
2. Exoskeleton robots are on the verge of exponential market growth [Electronic resource]. Access mode: https://exoskeletonreport.com/what-is-an-exoskeleton/
3. A survey on static modeling of miniaturized pneumatic artificial muscles with new model and experimental results [Electronic resource]. Access mode: https://www.researchgate.net/publication/328209567_A_Survey_on_Static_Modeling_of_Miniaturized_Pneumatic_Artificial_Muscles_With_New_Model_and_Experimental_Results
4. Sobia, A.: What is an exoskeleton? Exoskeleton report. Access mode: http://exoskeletonreport.com/what-is-an-exoskeleton/ (access date: 26.12.2017)
5. Kazerooni, H.: Exoskeletons for human performance augmentation. In: Siciliano, B., Khatib, O. (eds.) Springer Handbook of Robotics. Springer, Berlin (2008)
6. Lee, H., Yu, S., Lee, S., Han, J., Han, C.: Development of human-robot interfacing method for assistive wearable robot of the human upper extremities. In: Proceedings of the SICE Annual Conference, pp. 1755–1760 (2008)
7. Daly, J.J., Hrovat, K., Pundik, S., Sunshine, J., Yue, G.: fMRI methods for proximal upper limb joint motor testing and identification of undesired mirror movement after stroke. J. Neurosci. Methods **175**(1), 133–142 (2008)
8. Frumento, C., Messier, E., Montero, V.: The history and future of rehabilitation robotics. Access mode: https://web.wpi.edu/Pubs/E-project/Available/E-project-031010–112312/unrestricted/HRRIQP_Final.pdf (2010)
9. A novel gait-based synthesis procedure for the design of 4-bar exoskeleton with natural trajectories. Access mode: https://www.researchgate.net/publication/321206172_A_novel_gait-based_synthesis_procedure_for_the_design_of_4-bar_exoskeleton_with_natural_trajectories

10. Meng, Q.: Research on size synthesis optimization design of a bionic exoskeleton for index finger rehabilitation. Adv. Mater. Res. **945–949**, 1447–1450 (2014)
11. Menga, G., Ghirardi, M.: Lower limb exoskeleton for rehabilitation with improved postural equilibrium. Robotics **7**(2), 28 (2018)
12. Song, Z., Tian, C., Dai, J.-S.: Mechanism design and analysis of a proposed wheelchair-exoskeleton hybrid robot for assisting human movement. Mech. Sci. **10**(1), 11–24 (2019). https://doi.org/10.5194/ms-10-11-2019
13. Exoskeleton robots are on the verge of exponential market growth [Electronic resource]. Access mode: https://www.robotics.org/service-robots/exoskeleton-robots
14. Chen, B., Ma, H., Qin, L.-Y, Gao, F., Chan, K.-M, Law, S.-W., Qin, L., Liao, W-H.: Recent developments and challenges of lower extremity exoskeletons. J. Orthop. Transl. **5** (2015). https://doi.org/10.1016/j.jot.2015.09.007
15. Exo robotics. Access mode: http://exo-robotics.ru/ (access date: 26.12.2017)
16. A review of exoskeleton-type systems and their key technologies. Access mode: https://www.academia.edu/6275563/A_review_of_exoskeleton-type_systems_and_their_key_technologies
17. Types and classifications of exoskeletons. Access mode: https://exoskeletonreport.com/2015/08/types-and-classifications-of-exoskeletons/
18. Robotic exoskeleton: for a better quality of life [Electronic resource]. Access mode: https://www.maxonmotor.com/maxon/view/application/Robotic-exoskeleton-For-a-better-quality-of-life
19. Harvard scientists design soft robotic exoskeleton to reduce fatigue and injuries [Electronic resource]. Access mode: https://www.extremetech.com/extreme/190025-harvard-scientists-design-soft-robotic-exoskeleton-to-reduce-fatigue-and-injuries
20. Chernikova, L., Suponeva, N., Klochkov, A., Khizhnikova, A., Lyukmanov, R., Gnedovskaya, E., Yankevich, D., Piradov, M.: Robotic and mechanotherapeutic technology to restore the functions of the upper limbs: prospects for development (review). Sovremennye tehnologii v medicine **8**, 222–230 (2016). https://doi.org/10.17691/stm2016.8.4.27
21. Servo control facts. Access mode: http://www.baldor.com/pdf/manuals/1205–394.pdf (access date: 27.04.2018)
22. Pallas-Areny, R., Webster, J.: Sensors and Signal Conditioning. SERBIULA (Sistema Librum 2.0) (2019)
23. A great combination: pneumatic actuator, pneumatic timer, pneumatic valves, and pneumatic indicators. Access mode: http://www.ekci.com/a-great-combination-pneumatic-actuator-timer-valves-and-indicators-ezp-69.html (access date 14.04.2018)
24. Gopura, R., Kiguchi, K.: Mechanical designs of active upper-limb exoskeleton robots: state-of-the-art and design difficulties, pp. 178–187. https://doi.org/10.1109/icorr.2009.5209630. Access mode: https://www.researchgate.net/publication/224580506_Mechanical_designs_of_active_upper-limb_exoskeleton_robots_State-of-the-art_and_design_difficulties (2009)
25. Rehmat, N., Zuo, J., Meng, W.: Int. J. Intell. Robot Appl. **2**, 283 (2018). https://doi.org/10.1007/s41315-018-0064-8
26. Zang, D.-T.: Adaptive electric drive control exoskeleton [Electronic resource]. Access mode: https://etu.ru/assets/files/nauka/dissertacii/2017/Do-Than-Zang/Dissertaciya-_Do-Than-Zang.pdf
27. Nacy, S., Hussein, N., Abdallh, M.M.: Review of lower limb exoskeletons [Electronic resource]. Access mode: https://www.researchgate.net/publication/313877170_A_Review_of_Lower_Limb_Exoskeletons
28. Design and motion control of a lower limb robotic exoskeleton [Electronic resource]. Access mode: https://www.intechopen.com/books/design-control-and-applications-of-mechatronic-systems-in-engineering/design-and-motion-control-of-a-lower-limb-robotic-exoskeleton
29. Titov, A., Markov, A., Skorikov, A., Tarasov, P., Andreev A.E.: Autonomous locomotion and navigation of anthropomorphic robot. In: Communications in Computer and Information Science, CIT&DS 2017, vol. 754, pp. 242–255 (2017)

Using Special Text Points in the Recognition of Documents

Oleg A. Slavin

Abstract The chapter develops the concept of a textual key point, the detector of which is a certain OCR. The descriptor of a textual key point is determined. Examples of algorithms for analyzing documents, using textual key points, are given. The chapter deals with the tasks of recognized document classification, localization of images of recognized documents and comparison of images of documents for finding differences. The results of the algorithms for the data sets of the documents of the Russian Federation are given. The proposed methods allow achieving high accuracy of complexly structured documents analysis with entering document images in modern cyber-physical systems based on big data technologies.

Keywords Character recognition · Key point · Textual key point · Document classification · Document localization

1 Background

Currently, the use of key points allows solving various image processing and recognition problems, such as document classification, search for document fraud, image normalization and image matching, etc.

A key point is an image point, which differs from points in its locality and satisfies several conditions:

the locality of a key point must contain essential information that determines its properties,
noise resistance,
resistance to some transformations (for example, affine transformations or scaling) [1].

O. A. Slavin (✉)
FRC "Computer Science and Control" RAS, 60-letiya Oktyabrya prosp. 9, Moscow 117312, Russia
e-mail: oslavin@isa.ru

A. G. Kravets et al. (eds.), *Cyber-Physical Systems: Advances in Design & Modelling*, Studies in Systems, Decision and Control 259,
https://doi.org/10.1007/978-3-030-32579-4_4

It is assumed that the key point has an exact mathematical definition that can be used to analyze the information, extracted from the image.

In work [2], the similar properties of key points are listed:

repeatability—a key point must be in the same place as the image object, despite changes in the point of view and illumination,
distinctiveness/informativeness—vicinity of key points should differ,
locality—a key point should occupy a small area of the image,
quantity—the number of detected key points must be large enough so that they are enough to detect objects,
accuracy—the detected key points must be precisely localized both in the original image and on a different scale.
efficiency—the time of key point detection on the image must be valid in time-critical applications.

To match the key points, the similarity must be measured. The measure should take values close to zero in the case of comparing two points, defining one spot on the stage, and large values, when comparing different points of the scene. A descriptor is also defined—the identifier of the key point used when matching the key points. The descriptor's invariance is expected when matching key points with respect to image transformations. The method of extracting key points from an image is called a detector.

The detector ensures the invariance of finding the same key points with respect to image transformations. Corners (a point, the intensity of which varies relative to the center in the vicinity) or blobs (including the center coordinates, scale, and direction) are often used as key points of the image. For the key point of the "corner" type, the following methods of detecting key points are known: Moravek corner detector [3], Harris corner detector [4, 5], Wang and Brady corner detector [6], SUSAN corner detector [7]. For such blob key points, the following methods for detecting key points are known: Laplacian of Gaussian (LoG) and differences of Gaussians (DoG).

Due to the library OpenCV project implementation, SURF descriptors are widely known (uses the Hessian matrix; the image is minimized with sampled second Gaussian derivative filters), SIFT (determined, using the Gaussian difference pyramid), ASIFT [8, 9] (complements SIFT, simulating two parameters, simulating the camera optical axis direction).

The analysis of text documents can be based on global text content descriptors (for example, such as "bag of words" or "vector representation of words" [10]), which is analyzed by classical classifiers. In [11], an architecture was proposed for classifying pages in the banking process. Both graphics and text descriptors are used. Graphics descriptors were calculated, based on the distribution of pixel intensities. Text descriptors were formed, based on latent semantic analysis for thematic classification.

2 Description of the Method of Classification of Recognized Documents Using Text Key Points

For images of text documents, we define a *textual key point*, using a descriptor and a detector. The *descriptor* of a *textual key point* is:

$$(T(W), m_1(W), m_2(W), m_3(W))$$

where

- $T(W)$—the *core* of a textual key point, which is represented by a sequence of characters of a word of a certain alphabet;
- $m_1(W)$—a textual key point *boundary*, consisting of normalized coordinates $m_{4x1}(W), m_{4y1}(W), m_{4x2}(W), m_{4y2}(W)$ in the range [0,1];
- $m_2(W)$—case sensitive feature (case sensitive/insensitive), when comparing characters;
- $m_3(W)$—threshold, when *comparing* two words, $t(W)$ and W^r. To compare two words, we used the Levenshtein distance or a simplified function, which considers only the number of replacements of one character with another. If $d(t(W), W^r) < m_1(W)$, the word W^r and textual key point W are *identical*, otherwise—different. In the simplest case, $m_3(W)$—the maximum number of replacements during the transformation of $t(W)$ to W_r.

A textual key point detector is a process of recognition, using some OCR that extracts the descriptors of special points when it recognizes a document image. The term "textual key point" corresponds to the generally accepted concept of a special (key) point. The above properties of the key points are valid for textual key points in the case of the ability of modern OCR to compensate for different types of image distortions. The uniqueness of textual key point descriptors is determined by the structure of documents (splitting a document into fragments and lines) and the properties of a natural language (a rare match in documents of two adjacent words).

Furthermore, we consider several applied problems, for the solution of which key points can be used.

In [12], document classification methods are divided into document structure analysis methods, text information analysis methods and image analysis methods. The document structure analysis method is focused on such rigidly structured documents as questionnaires or invoices, based on the search for static graphic elements (dividing lines, frames) and static texts.

The authors of the work use image analysis methods to recognize identification documents, since each class of identification documents usually has a characteristic graphic structure in the form of unchanged text areas (field labels: first name, last name, etc.), variable text fields (personal data) and the same background For each class of identification documents, one model was created. When creating a model, field boundaries were first determined. Later, outside the boundaries of the fields, key points are extracted, the presence of which is likely on all document samples. Key points were characterized, using descriptors (SIFT, SURF, ORB). When working with a training set, only key points were saved, corresponding to each training image of one class. The authors of [12] indicate the advantage of the scheme of such training: a new model of the document is added independently of the models already created. The i-th model is denoted by the set M_i of n_i key points, where $M_i = \{D_{i1}, D_{i2},..., D_{i,ni}\}$, and D_{ij} is the set of extracted descriptors for the key point j.

The image classification was made, based on the description of its key points. It consists of finding a winning class that has the maximum number of key points that best match the query key points. Image key points were compared with all studied models, i.e., all models compete in this conformity phase. Sets of m_i direct compliance with the models M_i were constructed. Models were ranked by ratio $r_i = |m_i|/|M_i|$; the model with the highest coefficient $\text{argmax}_i(r_i)$ corresponded to the winning class Further, for all models, a reverse comparison was made between the key points of the models and the key points of the image requested. In this case, the RANSAC algorithm [13] was used to search for a geometric transformation, which displays the largest number of pairs of key points and excludes outliers.

The authors of [12] report that there are no publicly available test data sets of identification documents and describe three of their own data sets for experiments, which show a classification accuracy of about 95%.

The author developed a similar classification algorithm for business documents that do not have a rigid structure and are recognized by OCR Tesseract. In addition to the m_1–m_3 features, several more features were added.

- $m_4(W)$—word length limit
- $m_5(W)$—a sign of *a forbidden word* that cannot be put to the document
- $m_6(W)$—limiting the distance between two words.

When comparing the term W with the word W^r, belonging to the recognized document D_r, the core of the recognized word $t(W^r)$ and the rectangle of the word $F(W^r)$ are used. Comparison of the term W and the word W_r is based on the following condition:

$$d\big(t(W), W^r\big) < m_1(W) \wedge \big(F\big(W^r\big) \cap m_4(W) = F\big(W^r\big)\big). \quad (1)$$

where d—the function of the distance between two words. When comparing, in the case of case insensitive, i.e., with $m_3(W)$ = TRUE both words $t(W)$ and W^r are reduced to the same register. The distance between the term W and the word W^r is defined as $d\,(t(W), W^r)$. For the case $m_5(W) = 0$, we define the predicate $P(W, D_r) = 1$, if the text of the recognized document D_r contains at least one word W^r, identical to the word W, which will be denoted as $W^r \approx W$, and $P(W, D_r) = 0$ otherwise. The search for a match between a text key point and a key point in a query document is carried out by searching the words of the document, included in the $W_r \in T$ dictionary and satisfying the condition (1). Several words {Wr} can be found in document D_r, corresponding to a single text key point of the model.

Let's define the allocation of terms as an ordered set of words $R = W_1, W_2,...$, for which the presence of each of the terms in the recognized document D_r is checked:

$$\mathrm{P}(W_1, D_r) \wedge \mathrm{P}(W_2, D_r) \wedge \cdots \tag{2}$$

and additionally, for each pair of terms W_i and W_{i+1} condition

$$r_{\mathrm{BT}}(W_{i+1}, W_i) < m_6(W),$$

is checked where the function r_{BT} gives is bitermal distance. That is, the parameter $m_6(W)$ determines the distance between adjacent terms in the allocation.

The fulfillment of conditions (1), (2) determines the predicate of the membership of the allocation of R to the recognized document D_r: $\mathrm{P}(R, D_r) = 1$. In the general case, it requires a search of sets identical to W_1, W_2,... .

The evaluation of the correspondence to the recognized document D_r of the allocation $d(R, D_r)$ is defined as

$$\min(d(W_1, D_r), d(W_2, D_r), \ldots).$$

We define the combination as the set of allocations $S = R_1$, R_2,..., for which the presence of each of the allocations in the recognized document Tr is checked:

$$\mathrm{P}(R_1, D_r) \wedge \mathrm{P}(R_2, D_r) \wedge \cdots \tag{3}$$

The order of allocation is not important, as the words in the model "bag of words".

The evaluation of the correspondence to the recognized document D_r of the combination $d(S, D_r)$ is defined as

$$\min(d(R_1, D_r), d(R_2, D_r), \ldots).$$

Finally, we define the model M as the set of combinations S_1, S_2, …, for which the template membership to the recognized document D_r is established by checking the expression

$$\mathrm{P}(M, D_r) = \mathrm{P}(S_1, D_r) \vee \mathrm{P}(S_2, D_r) \vee \cdots \tag{4}$$

The evaluation of the correspondence to the recognized document D_r of the template $d(S, D_r)$ is defined as

$$\max(d(S_1, D_r), d(S_2, D_r), \ldots).$$

In addition to conditions (2), (3), (4), it can be added the check of the hit of words of allocation, combination or template in a certain frame.

To the existing comparisons of the model M with the recognized document D_r, we add tests for the correspondence of some properties of the text (the number of characters on the page, the number of columns of text) with similar properties of the template.

If the set of models $M_1,\ldots, M_n$, is given for n classes, then the task of verifying the correspondence to the class M_i of the recognized document D_r is resulted in calculating the distance $d(M_i, D_r)$ and comparing this distance with the previously known threshold of the distance.

3 Experimental Results of Classification

To classify the flow of documents, consisting of 40 classes, two sets were used:

- T_1—a set, consisting of images of documents of medium and poor digitization quality, selected for the learning stage (1768 pages);
- T_2—a set, consisting of images of documents of average digitization quality, obtained regardless of the training stage (3014 pages).

Two options for the proposed algorithm were considered. In the first variant, forms were used that included terms without geometric characteristics, i.e., without frame m_1, in the second—some terms had frames m_1. This separation was supposed to show the difference between the use of known structures that do not have geometric characteristics and structures that use the geometric characteristics, obtained in the recognition process.

The classification of a selection of q volume was evaluated by the following values:

- n_1—the number of the first pages of documents that were classified correctly,
- n_2—the number of the first pages of documents that were classified incorrectly,
- n_3—the number of the first pages of documents that were not classified,
- k_1—the number of the not first pages of documents that were not classified,
- k_2—the number of the not first pages of documents that were classified incorrectly,

Analysis of classification results was carried out using the following criteria:

- accuracy $a_c = (n_1 + k_1)/q$,
- the fraction of false classification $z_{PF} = (n_2)/q$,
- the fraction of false the classification $z_{NF} = n_3/q$, where $q = (n_1 + n_2 + n_3 + k_1 + k_2)$.

The classifications, obtained by testing algorithm, are shown in Tables 1 and 2.

From the above data, it follows that the described classification method gives an accuracy of 0.86–0.95, while the false classification does not exceed 0.01; the

Table 1 Experimental results of classification for set T_1

	n_1	n_2	n_3	k_1	k_2	a_c (%)	z_{PF} (%)	z_{NF} (%)
Ignore m_1	773	21	229	736	9	85.35	1.19	0.51
Use m_1	768	13	242	743	2	85.46	0.74	0.11

Table 2 Experimental results of classification for set T_2

	n_1	n_2	n_3	k_1	k_2	a_c (%)	z_{PF} (%)	z_{NF} (%)
Ignore m_1	825	13	148	1992	36	93.46	0.43	1.19
Use m_1	837	1	148	2027	1	95.02	0.03	0.03

remaining errors relate to the refusals of classification. The proposed method does not always work, but it rarely offers the wrong class.

The data in Tables 1 and 2 demonstrate an increase in the classification accuracy and a decrease in the false classification fractions and the classification rejection due to the use of geometric characteristics. Note that for the T_2 set, the fraction of the z_{NF} false classification decreased by 36 times due to the use of the term frames.

The accuracy of the obtained method is comparable to the accuracy of the algorithms, mentioned above [11, 12].

4 Description of the Method of Localizing a Document with the Known Structure

Textual key points can be used to localize a document with the known structure. To localize the document in the image, methods, based on the extraction of boundary elements, segments and/or lines, can be used. The resulting quadrilateral is formed, using the detected primitives during traversing the intersection graph [14] or by searching for alternative solutions, using the system of fines and heuristic simplification at different stages. The paper [15] describes the filtering of false correspondences of local features and checking that the obtained solution is well conditioned, which allows improving the classification accuracy. The proposed filtering method requires several samples of each document type. In [12], a generalization of the approach [15] is proposed, which allows for projective document image distortion. The paper also presents a filtering method, based on the preselection of a region, containing variable data and extracting features from these regions. However, the authors of [12] show that this approach allows not only to determine the document type but also to find the document boundaries.

The document model with a flexible structure M is described as follows: there is a set of keywords W and fields for extracting attributes that are grouped into a set ordered according to the height of the strings:

$$S_1 = \{p_{11}, p_{12}, \ldots, p_{1,k(1)}\},$$
$$S_2 = \{p_{21}, p_{22}, \ldots, p_{2,k(2)}\},$$
$$\ldots$$
$$S_i = \{p_{i1}, p_{i2}, \ldots, p_{i,k(i)}\}$$

Each of the elements $p_{ij} = \left\{xp^1_{ij}, yp^1_{ij}, xp^2_{ij}, yp^2_{ij}, tp_{ij}\right\}$ can be a word or a field. The coordinates (rectangle $xp^1_{ij}, yp^1_{ij}, xp^2_{ij}, yp^2_{ij}$) of each element are known in advance. And the string value *tp* of each keyword is known.

For a pair of elements $p = \left\{xp^1, yp^1, xp^2, yp^2, tp\right\}$ and $q = \left\{xq^1, yq^1, xq^2, yq^2, tq\right\}$ define relations:

$$\begin{aligned} &left(p, q) \text{ if } xp^2 > xq^1, \\ &above(p, q) \text{ if } yp^2 < yq^1 \end{aligned}$$

For a pair of strings $S_t = \{p_{t1}, p_{t2}, \ldots\}$ and $S_r = \{p_{r1}, p_{r2}, \ldots\}$ define relation:

$$above\,(S_t, S_r) \text{ if } (\forall p_t \in S_t, \forall p_r \in S_r)\, above\,(p_t, p_r)$$

The following relations should be performed in the model:

$$\begin{aligned} &(\forall t, r\colon t < r)\, above\,(S_t, S_r), \\ &(\forall p_t \in S, \forall p_r \in S_r\colon t < r)\, left(p_t, p_r) \end{aligned} \tag{5}$$

Let us consider an image, part of which is occupied by the document, and set of recognized words W1, extracted from this image. Assuming that the document is a known model *M*, we find a subset of words $W_1 \subset W$ that best matches the model, i.e., we find the maximum number of words $w = \left\{xw^1, yw^1, xw^2, yw^2, tw\right\} \in W_1$, each of which is associated with an element of the model $p = \left\{xp^1, yp^1, xp^2, yp^2, tp\right\} \in M$, so that the words *tw* and *tp* are close in measure *d*, for example, the Levenshtein metric. Moreover, for each pair of words $w_1, w_2 \in W_1$, relations (5) are fulfilled. Placing the recognized words requires a search, the peculiarity of which is the elimination of the conflict outliers of similar words, associated primarily with recognition errors. After establishing a match between W_1 and localization model *M* (binding model), field values may be extracted, using the boundaries of the associated keywords.

The sizes of the elements and the distances between the elements are not constant, for real documents the relations between the corresponding distances can exceed 100%.

The proposed algorithm for extracting attributes from a scanned image of a document with a flexible structure is as follows:

- image normalization (transform to the single-channel image, deskew image),
- finding word boundaries using morphological operations of erosion and dilatation,
- segmentation of the found words into characters, using artificial neural networks,
- recognition of each found character, using the artificial neural networks,
- control the presence of unique keywords from set *W*,
- clustering words into strings by the nearest neighbor's algorithm,
- binding and extracting fields as written above,
- getting results as extracted attributes and document orientation.

Comparison of the recognized word $w = \left\{xw^1_{\text{ij}}, yw^1_{\text{ij}}, xw^2_{\text{ij}}, yw^2_{\text{ij}}, tw\right\}$ and words from the model $p_{ij} = \left\{xp^1_{\text{ij}}, yp^1_{\text{ij}}, xp^2_{\text{ij}}, yp^2_{\text{ij}}, tp\right\}$ must be performed twice, using two metrics. First is the Levenshtein metric $d(tw, tp)$, using which character sequences $tw_1, tw_2, \ldots, tw_{\text{n}}$, and $tp_1, tp_2, \ldots, tp_{\text{m}}$ are compared. The second metric is $d_{\text{rev}}(tw, tp) = d(tp, tw)$, using which sequences $tw_1, tw_2, \ldots, tw_n$ and $tp_m, tp_{m-1}, \ldots, tp_1$ are compared.

5 Experimental Results of Localizing a Document with the Known Structure

For the experiment, a dataset from 2000 marked scans of documents with a resolution greater than 1000 × 1500 pixels was prepared: 1000 scans are in the correct orientation; the rest are their copies, rotated by 180°. The example of the Russian national structured document dataset (PIT—personal income tax) document with found fields is presented in Fig. 1. The localization accuracy for this set was more than 96%; the recognition accuracy was more than 92%.

Textual key points can be used, when comparing two copies of documents to search for possible modifications. Known methods show the high quality of document comparison; for example, the comparison methods, proposed in [16], allow for a test suite, containing bills and medical extracts (40 original documents and 12 fake), to find all the fakes, while 30% of the original documents are suspicious. However, these methods were not tested for documents, printed in Cyrillic, and do not apply to the search for individual modifications, but to test the hypothesis of fake documents.

The proposed comparison method is based on the coordination of pairs of textual key points and the subsequent comparison of the word images of one pair. A similar problem is solved, when coordinating stereo images, using key points [17]. In article [18], a comparison of two digitized images was proposed, based on the combination of several methods for comparing word patterns:

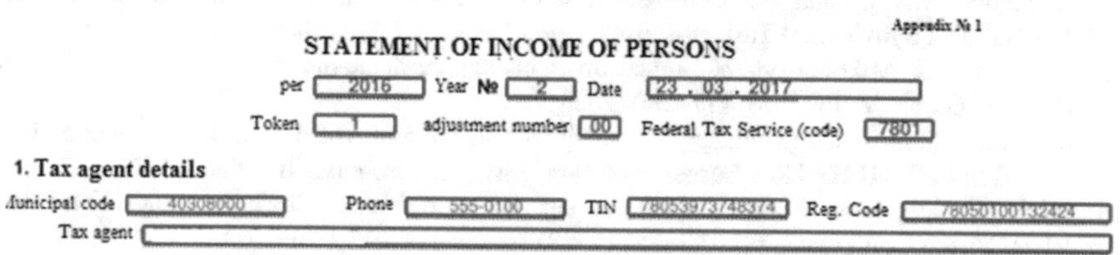
Appendix № 1

STATEMENT OF INCOME OF PERSONS

per 2016 Year № 2 Date 23 . 03 . 2017

Token 1 adjustment number 00 Federal Tax Service (code) 7801

1. Tax agent details

Municipal code 40308000 Phone 555-0100 TIN 78053973748374 Reg. Code 78050100132424

Tax agent

Fig. 1 Example of a PIT document and found fields

– comparison of the cores $T(W_t)$ and $T(W_e)$ of two textual key points W_t and W_e,
– comparison of two sets of RFD points, extracted from textual key point boundaries $m_1(W)$,
– comparison of extended images extracted from textual key point boundaries $m_1(W)$.

The combination of several methods, proposed in the work, allows solving the problem of finding inconsistencies between scanned copies of business documents. Experiments show that the images, scanned with a resolution of 300 dpi, achieved an accuracy of 90.9%; the completeness was 98.2%; for these images, the error of 2nd kind apply ($fn = 0$).

The achieved accuracy and completeness are comparable with the comparison methods, described in [16] and applicable for documents in European languages as well as technical and patent documents [19] for the design of the cyber-physical systems.

6 Conclusion

The chapter describes two methods for analyzing complex-structured documents using special text points obtained using OCR:

– method of classification of recognized documents using text key points;
– method of localizing a document with the known structure.

The proposed methods allow achieving high accuracy of complexly structured documents analysis when entering document images in frames of modern cyber-physical systems design based on big data mining technologies.

References

1. Rodehorst, V., Koschan, A.: Comparison and evaluation of feature point detectors. In: 5th International Symposium Turkish-German Joint Geodetic Days (2006)
2. Tuytelaars, T., Mikolajczyk, K.: Local invariant feature detectors: a survey. Found. Trends Comput. Graph. Vision **3**(3), 177–280 (2008)
3. Moravec, H.: Obstacle avoidance and navigation in the real world by a seeing robot Rover. Tech Report CMU-RI-TR-3 Carnegie-Mellon University, Robotics Institute (1980)
4. Harris, C., Stephens, M.: A combined corner and edge detector. In: Proceedings of the 4th Alvey Vision Conference, pp. 147–151 (1988)
5. Shi, J., Tomasi, C.: Good "features to track". In: 9th IEEE Conference on Computer Vision and Pattern Recognition, pp. 593–600. Springer, Berlin (1994)
6. Wang, H., Brady, M.: Real-time corner detection algorithm for motion estimation. Image Vis. Comput. **13**(9), 695–703 (1995)
7. Smith, S.M., Brady, J.M.: SUSAN—a new approach to low level image processing. Int. J. Comput. Vis. **23**(1), 45–78 (1997)

8. Lowe, D.G.: Object recognition from local scale-invariant features. In: Proceedings of the International Conference on Computer Vision, vol. 2, pp. 1150–1157 (1999). https://doi.org/10.1109/iccv.1999.790410
9. Bay, H., Ess, A., Tuytelaars, T., Van Gool, L.: SURF: speeded up robust features. Comput. Vis. Image Underst. (CVIU) **110**(3), 346–359 (2008)
10. Vorontsov, K.V., Potapenko, A.A.: Tutorial on probabilistic topic modeling: additive regularization for stochastic matrix factorization. In: AIST'2014, Analysis of images, Social networks and Texts, vol. 436. Communications in Computer and Information Science (CCIS), pp. 29–46. Springer International Publishing, Switzerland (2014)
11. Rusiñol, M., Frinken, V., Karatzas, D., Bagdanov, A.D., Lladós, J.: Multimodal page classification in administrative document image streams. IJDAR **17**(4), 331–341 (2014)
12. Awal, A.M., Ghanmi, N., Sicre, R., Furon, T.: Complex document classification and localization application on identity document images. In: Proceedings of 14th IAPR International Conference on Document Analysis and Recognition, pp. 427–432 (2017). https://doi.org/10.1109/icdar.2017.77
13. Chum, O., Matas, J., Kittler, J.: Locally optimized RANSAC. In: DAGM-Symposium, vol. 2781. Lecture Notes in Computer Science, pp. 236–243 (2003)
14. Zhukovsky, A., Nikolaev, D., Arlazarov, V., Postnikov, V., Polevoy, D., Skoryukina, N., Chernov, T., Shemiakina, J., Mukovozov, A., Konovalenko, I., et al.: Segments graph based approach for document capture in a smartphone video stream. In: Proceedings of 14th IAPR International Conference on Document Analysis and Recognition (ICDAR), vol. 1, pp. 337–342, IEEE (2017)
15. Augereau, O., Journet, N., Domenger, J.-P.: Semistructured document image matching and recognition. In: Document Recognition and Retrieval XX, vol. 8658, p. 865804. International Society for Optics and Photonics (2013)
16. Ahmed, A.G.H., Forgery, S.F.: Detection based on intrinsic document contents. In: Proceedings of 11th IAPR International Workshop on Document Analysis Systems (2014). https://doi.org/10.1109/das.2014.26
17. Badino, H., Kanade, T.: A head-wearable "short-baseline stereo system for the simultaneous estimation of structure and motion". In: Proceedings of MVA, pp. 185–189 (2011)
18. Andreeva, E., Arlazarov, V.V., Manzhikov, T., Slavin, O.: Comparison of the scanned pages of the contractual documents. In: Proceedings of SPIE, vol. 10696. Tenth International Conference on Machine Vision (ICMV 2017), Vienna, Austria, 13–15 November 2017. Art. No. 1069605, pp. 106960–106966 (2018). https://doi.org/10.1117/12.2309458
19. Kravets, A.G., Lebedev, N., Legenchenko, M.: Patents images retrieval and convolutional neural network training dataset quality improvement. ACSR-Adv. Comput. Sci. Res. **72**, 287–293 (2017)

Extraction of Cyber-Physical Systems Inventions' Structural Elements of Russian-Language Patents

Sergey S. Vasiliev, Dmitriy M. Korobkin, Alla G. Kravets, Sergey A. Fomenkov and Sergey G. Kolesnikov

Abstract The chapter presents software for extracting predicate-argument constructions that characterizing the composition of the structural elements of the inventions from cyber-physics domain and the relationships between them. The extracted structures reconstruct the component structure of the invention in the form of a net. Such data is further converted into a domain ontology and used in the field of information support of automated invention. A new method for extracting structured data from patents has been proposed taking into account the specificity of the text of patents and is based on the shallow parsing and segmentation of sentences. The ontology scheme includes the structural elements of technical objects as the concepts and the relationship between them, as well as supporting information on the invention. The results suggest that the proposed approach is promising. A further direction of research is seen by the authors in improving the existing method for extracting data and expanding ontology.

Keywords Patents · Information extraction · SAO · CAI-systems · Ontology

---

S. S. Vasiliev · D. M. Korobkin (✉) · A. G. Kravets · S. A. Fomenkov · S. G. Kolesnikov
Volgograd State Technical University, Lenin Avenue, 28, Volgograd, Russia
e-mail: dkorobkin80@mail.ru

S. S. Vasiliev
e-mail: svasilev2012@yandex.ru

A. G. Kravets
e-mail: agk@gde.ru

S. A. Fomenkov
e-mail: saf@vstu.ru

S. G. Kolesnikov
e-mail: sk375@bk.ru

A. G. Kravets et al. (eds.), *Cyber-Physical Systems: Advances in Design & Modelling*, Studies in Systems, Decision and Control 259,
https://doi.org/10.1007/978-3-030-32579-4_5

1 Introduction

In recent years software and hardware systems to support the simulation and the development of complex and interconnected Cyber-Physical Systems have been gaining momentum. Automated support systems are used to search for new technical solutions—CAI (Computer-Aided Invention) systems. The success of such systems largely depends on the completeness of the ontologies of the subject areas and the fullness of the various knowledge bases that allow generating new technical solutions. However, data enrichment is often a very laborious process. Therefore, an active search continues for an effective means of extracting structured data for this purpose.

The sources of technical information may be a patent array. However, current approaches to extracting structured data and natural language processing (NLP) tools are poorly oriented to work on an array of patents. In this regard, it is necessary to develop new efficient methods for extracting data from patents.

This chapter analyses the application of the current NLP tools and data extraction approaches in the context of the current task. The chapter offers a new method of extracting SAO structures and constructing a network of structural elements as part of solving the problem of information support for the synthesis of new technical solutions in the cyber-physics domain.

2 Research Background

Various systems [1–4] are known for processing English-language patents, including the use of the SAO formalism to extract various concepts. In [4], a tree syntax analysis was applied with a separate identification of the subject, the action and the object on the base of GATE system, which is poorly suitable for the Russian language. In [3], the SAO structures are extracted based on the rules using the Stanford parser software, but there are no ready-made models for the Russian language. Chapters [2, 5] process patents using linguistic markers (specific verbs and nouns) and lexical-syntactic patterns. The emphasis is mainly on rule-based systems since for statistical analysis systems a lot of marked-up data is obviously needed.

Considering the system for extracting structured data from Russian-language texts, the development of Yandex Tomita-parser [6] is most often distinguished. Despite the expressive power of context-free grammars and the Tomita-parser tool, the question of organizing the effective extraction of SAO structures from the claims remains open. Of the available systems of syntactic and semantic-syntactic analysis, for which there is an opportunity to work with the Russian language, we can distinguish Link Grammar Parser, MaltParser and UFAL UDPipe [7].

Among the available Russian language analyzers can be identified: TreeTagger, MyStem, TnT, pymorphy2 [8], FreeLing. A number of works [9–11] are devoted to the comparison of morphological analyzers of the Russian language: TreeTagger and Yandex MyStem tool. The full use of the analyzer MyStem is difficult due to the issuance of grammatical signs in the form of an unordered list of alternatives. The TreeTagger morphoanalyzer is devoid of such a feature—the grammar is issued without alternatives. The disadvantage of TreeTagger is that its speed is somewhat lower than the speed of processing MyStem on large texts.

Having considered certain aspects of the processing of Russian-language texts, we summarize the problems that arise when extracting structured data (SAO structures) from texts of Russian-language patents: (a) the claims are written in a specific sub-language of patents that impede text analysis (specific terminology); (b) the structure of the claims is poorly suited for parsing by ready-made syntactic analyzers; (c) the structure of the claims is not fully decomposed by simple heuristics; (d) the available NLP tools will certainly be mistaken in the analysis, so relying on one tool in the processing of complex structures is not always rational.

The presented limitations make it difficult to create an efficient data extraction system from patent claims using publicly available NLP tools (without considering complex commercial or closed systems).

However, given that the texts of patents are written in a template (not only in the structure of sections, but also in the structure of phrases), and it is possible to combine NLP tools for mutual correction of analysis errors; the authors suggest the following approach to processing the text of the claims, including the steps:

- segmentation of the patent claims based on shallow parsing;
- organization of the extraction of SAO structures based on the valence bond theory;
- post-processing of SAO structures with linking concepts to a common repository.

The main stages of processing and data flows of the proposed system are presented in Fig. 1.

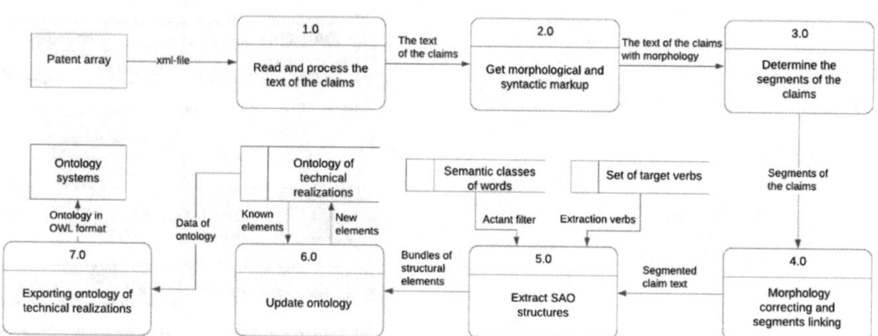

Fig. 1 DFD chart of the proposed system

3 Method Description

The unit of extraction is the predicate-argument construction in the form of the SAO structure, which semantically describes the structural elements of the device and the connections between them.

The ultimate goal of extracting data is to build a network of structural elements of the invention, reflecting the structure of the invention itself. Let us consider in more detail each of the stages.

3.1 *Segmentation of Patent Claims*

The purpose of segmentation is to identify the linear structure of the proposal of the claims for the subsequent extraction of SAO-objects based on the analysis of the morphological and linear-combinatorial characteristics of the text.

The overall sequence of operations for the segmentation of the claims is represented by the activity diagram in Fig. 2.

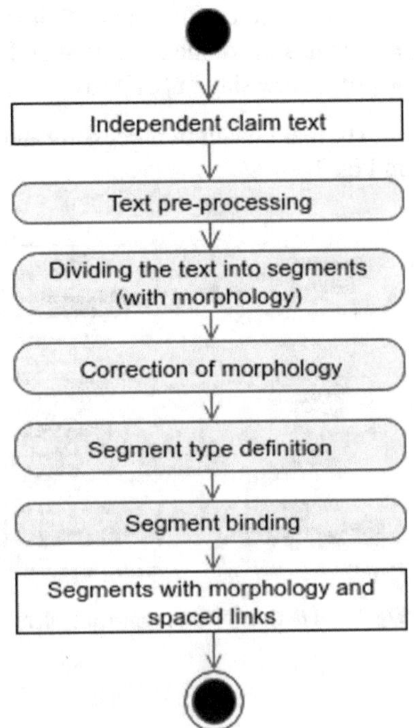

Fig. 2 The segmentation process of the claims

1. Text pre-processing includes the following steps:

   - deletion of introductory words: "for example", "at least", "however", etc.;
   - removal of brackets with all contents: for example, references to figures "(1), (12)";
   - deletion of the claim number at the beginning of the text ("1. Magnetic gear …");
   - removal of HTML-tags (<. *?>) in case of reading from XML-documents;
   - adding spaces to slashes (to correct the morphology to "and/or");
   - remove multiple spaces and duplicate commas.

2. The process of obtaining initial segments is associated with sequential fragmentation of the text by punctuation marks with auxiliary treatments. A specific feature of the stage is the final fragmentation based on the results of the syntactic analysis obtained after using the UDPipe tool. Tokenization is also performed by this tool.

   The division of the text into segments includes the steps:

- dividing the text according to the pattern, "characterized in that", while the segment is marked with a conditional part number: (0—all segments belong to one part (claim without division); 1—a segment of the restrictive part; 2—a segment of the distinctive part).
- division of the text by a semicolon;
- insertion of surrogate references to formulas and complex tokens (for example, "S = W = 0.00042λ") to reduce segmentation errors and morphological analysis;
- obtaining the primary morphology from the UDPipe tool and segmentation by commas.

3. The next step is the process of correcting morphology. If a segment token refers to parts of a verb, adjective, or adverb speech, then a part of speech is checked using MyStem. Moreover, if the result does not coincide with the initial one, then the morphological information of the token is updated taking into account the conversion of tags from MyStem into UDPipe format (CoNLL-U).
4. One of the fundamental stages of segmentation is the determination of the type of segment. The type of segment allows you to introduce certainty in the use of certain heuristics processing in the future. This problem is solved in two steps:

   - Determining of a segment pattern;
   - Determining the type of segment by pattern using a finite state machine (FSM).

The definition of a segment type by the template is implemented using a FSM. The principle is to search for elements of the segment template and search for specific sequences of Part-Of-Speech corresponding to the types of segments.

5. The main binding mechanism in the Russian language is coordination according to gender, number, and case. Therefore, the segmentation is carried out on the data of the extracted morphology, and the task is reduced to finding this agreement between the words in the sentence. At the same time, segment types limit the set of binding rules.

3.2 Extracting SAO Primary Structures

The extraction of predicate-argument constructions is based on the results of segmentation and valency of the target verbs. The valence of the verb determines the number of possible arguments. It is enough to define the "subject" and "object" of the structure being extracted.

The process of predicate-argument structures extraction in a generalized form is presented in the activity diagram (see Fig. 3).

When detected in the segment of the target verbs ("contain", "equip"; "include", "have", "supply", "enter", "consist"), the sequence of tokens to the right and left of the verb extends to the left and right parts, respectively.

The key stage of extraction: the search for the structures of the object and subject in the selected parts of the segment according to the predicate valencies.

Since the extraction of the subject and the object was carried out successfully, the formation of the triplet as an SAO object follows; noun groups of the subject and object are distinguished by labeling without taking into account additional valences.

Let's explain the presented methodology of SAO extraction on the example of a phrase: "[gasket] contains [at the other end one element]" ("[прокладка] содержит на [другом конце один элемент]" in Russian):

Subject	"gasket"
Action	"contains"
Object	"at the other end **one element**" → "one element"

This structure is used as an intermediate form of data processing. In the future, the primary SAO are subjected to processing aimed at clarifying the structural elements and linking them together.

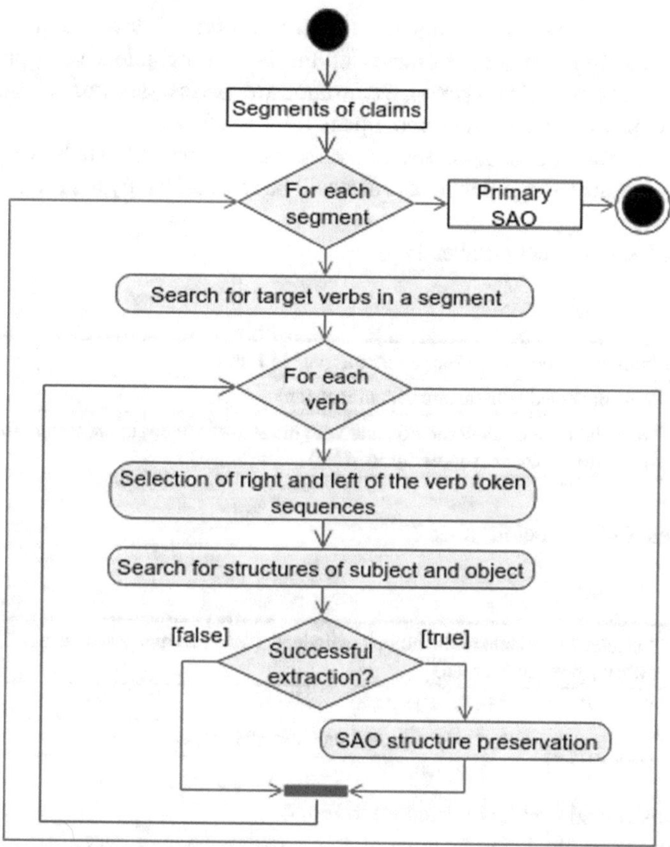

Fig. 3 The general algorithm for extracting SAO structures

3.3 *Creation a Net of Relations of Elements of Structures*

Creation a net of relationships requires determining the final vertices of the graph, taking into account the presence of identical concepts, as well as identifying implicit relations of generic terms between elements.

For further binding of the primary SAO objects, preprocessing is necessary:

- the separation of homogeneous members in the description of structural elements along the border of the composing union "and" provided it is located between the labels of the vertices of the noun group.
- generation of the normal form of the actants (in the nominative case) and the genitive case.
- after preprocessing, the primary SAO object is considered prepared.

To consider the complete structure of the output, take a small fragment of the text modeling the description of the claims: "Magnetic gearbox, characterized in that the hollow cylinders are connected with the rotor of slow rotation and with the stator, and the magnetic elements of the hollow cylinders have an angular position"/"Магнитный редуктор, отличающийся тем, что полые цилиндры связаны с ротором медленного вращения и со статором, а магнитные элементы полых цилиндров имеют угловое положение" (in Russian).

Primary predicate-argument constructs will be represented by the following SAO objects.

Prepared SAO-object number 1:

Subject	"hollow cylinders" ("полые цилиндры" in Russian)
Action	"are connected" ("связанный" in Russian)
Object	"with the rotor of slow rotation and with the stator" ("с ротором медленного вращения и со статором" in Russian)

Prepared SAO-object number 2:

Subject	"magnetic elements of the hollow cylinders" ("магнитные элементы полых цилиндров" in Russian)
Action	"have" ("имеют" in Russian)
Object	"an angular position" ("угловое положение" in Russian)

The constructed graph is presented in Fig. 4.

From the example, it can be noted that homogeneous members are distinguished from the object of the bundle SAO No. 1, and the parent relationship of "magnetic elements" and "hollow cylinders" is expressed by the pseudo-relationship "have". In this case, the actant "an angular position" was not added to the output set, since it falls into the forbidden semantic class of words.

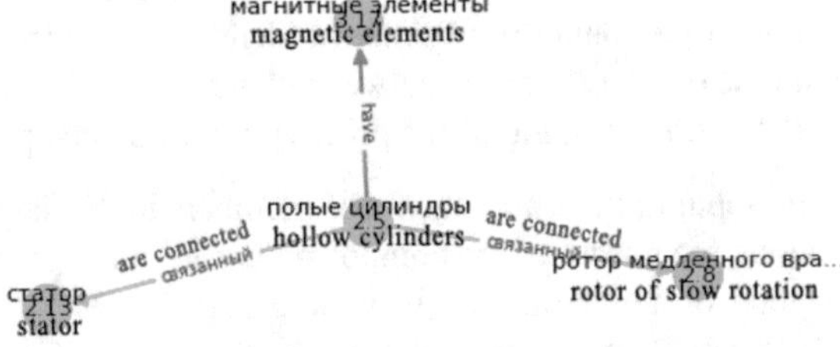

Fig. 4 Constructed graph of structural elements

4 Construction of Domain Ontology

The ontology-based patent processing technologies are developing more and more actively. Thus, in [12], ontology extracted concepts and relationships are used to improve patent search. In [13] the information extracted from the claims of the device is stored in an ontological representation and is used for visualization and processing. In [14] patent information in ontology is formalized with reference to technological areas. Thus, the ontological representation provides ample opportunities for the implementation of the description, linking and searching for patent information.

In this chapter, at the initial stage, ontology is considered to a greater extent as a storage medium. Inventions and connections between them act as concepts. The designed scheme of the domain ontology is presented in Fig. 5.

The invention (patent document) is assigned the name of the invention, the patent number, the owner organization and the IPC codes. Additional concepts of the technical function and the problem solved by the invention are introduced for the subsequent development of the system.

Relationships between components are specified through the following properties:

- connected_to—connection between elements (verbs "set", "connect", "connect", etc.);
- contains—indicating the presence of a component (verbs "contain", "have", etc.);

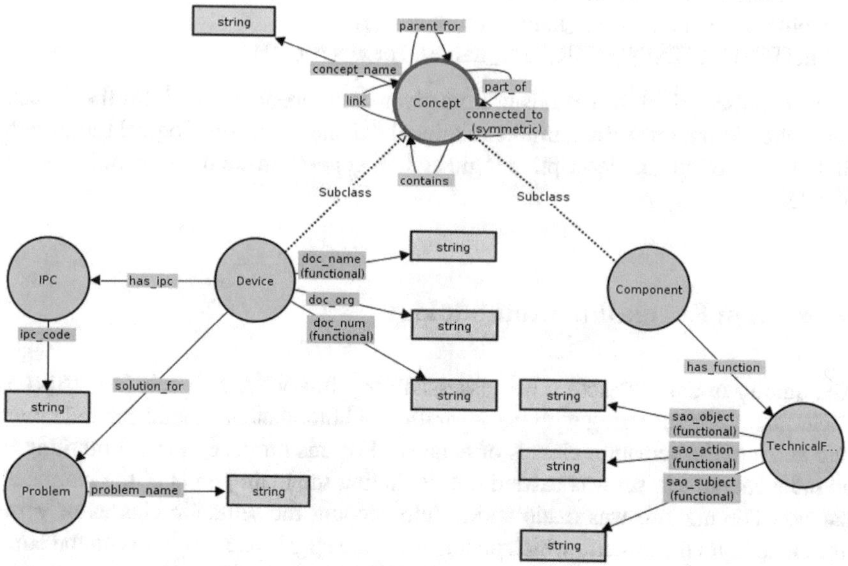

Fig. 5 Ontology scheme for storing patent information

- part_of—an implicit indication of the relationship to the invention;
- parent_for—an indication of an implicit parental relationship between the elements (for example, "hollow stator cylinders").

The ontology graph is described in RDF/XML format. The extracted semantic structures are embedded in the description of ontologies and are uploaded as an owl-file. The resulting data can then be loaded into the RDF storage and make the appropriate SPARQL queries. The data processing processed by the system amounted to 11,200 patent documents.

Further, it is possible to make various requests for information retrieval. An example of a simple request for the conclusion of all the components of a given invention (by the entry of the word "reducer" in the name):

```
PREFIX rdf: <http://www.w3.org/1999/02/22-rdf-syntax-ns#>
PREFIX rdfs: <http://www.w3.org/2000/01/rdf-schema#>
PREFIX owl: <http://www.w3.org/2002/07/owl#>
PREFIX cad: <http://www.vstu.ru/cad/ontology-of-devices#>
SELECT?device?dev_name?component_name
WHERE {

  ?device rdf:type cad:Device.
  ?device cad:concept_name?dev_name.
  ?device cad:doc_name?doc_name.
  ?device cad:doc_num?doc_num.
  ?component cad:part_of?device.
  ?component cad:concept_name?component_name
  FILTER(CONTAINS(STR(?dev_name), "редуктор"))}
```

You can see which inventions use the given component or search for the descendants, etc. At the same time, unlike relational databases, the ontological representation allows you to add descriptive logic and, in general, make more flexible queries to data.

5 System Evaluation Methodology

The quality of data extraction was assessed for primary SAO objects [14, 15] (i.e., excluding post-processing with the separation of homogeneous members and taking into account the semantic classes of actants). For this purpose, manual marking of an independent test set was carried out, including the main points of the claims 30 patents. The markup was made taking into account the semantic classes of verbs introduced into the system, which characterize the target relations between the components of the technical object. In this case, it was allowed to add into the system descriptions (valencies) of previously not encountered target verbs.

Since the unit of extraction is a predicate-argument construct [16], there may be cases of incomplete extraction of one (or both) of its arguments (due to the erroneous identification of the noun group and other reasons). Therefore, two types of evaluation were introduced: (a) with a rigorous entry of arguments (there must be a complete occurrence of the selected words); (b) with a lax entry of arguments (only the hit of the vertex of the noun group is significant).

Example of mild occurrence of arguments: "the second node is connected to *another* output of the primary winding": S—"the second node", A:—"is connected", O—"to output of the primary winding".

In the noun group of the object, the definition of "another" is omitted, and the vertex of the nominal group "output" is present. Therefore, according to the evaluation with a rigorous entry, the structure is considered to be incorrectly recognized. However, by a lax evaluation, the extraction can be considered successful.

The following metrics were used as evaluation metrics: precision, recall, and F1-measure. A test data set of 30 documents on the basis of manual marking included 318 SAO objects. The results of counting the number of extracted test data set objects are presented in Table 1 (first row). The design data set (that is, on which the system was designed) of 14 documents contained 136 recoverable structures. The results of counting the number of extracted SAO objects of the design data set are presented in Table 1 (second row).

where Rf—all SAO structures extracted by the system; Rrel_1—relevant SAO structures with a rigorous entry of the entire noun group in the argument; Rrel_2—relevant SAO structures taking into account that the vertex of the noun group is in the argument; Srel—manually selected SAO structures.

The result of the counting of the extraction quality metrics (precision, recall, and F1-measure) for the test and design data set is presented in Table 2.

According to the results of the evaluation on the test data set, the proposed method allows extracting data with an accuracy of 63%.

The performance indicator was the average patent processing time (reading from an XML document). The speed of work was determined by the arithmetic average

Table 1 Extracted test data set objects

Parameter	Rf	Rrel_1	Rrel_2	Srel
Amount	305	198	248	318
Amount	142	118	129	136

Table 2 Metric counting results

Metrics	Test data set		Design data set	
	Lax evaluation	Rigorous evaluation	Lax evaluation	Rigorous evaluation
Precision	0.81	0.65	0.91	0.83
Recall	0.78	0.62	0.95	0.87
F1-measure	0.79	0.63	0.93	0.85

Table 3 The results of the evaluation of the system speed

Program pass	Hardware configuration #1 (cluster)	Hardware configuration #2 (PC)
Average processing time (s)	2180	5606.94
Average processing time per document (s)	0.581	1.49

of the time of three runs on a test data set of 3755 documents without marking for each hardware configuration (see Table 3).

Patent processing speed is on average less than 1.5 s; which is not difficult for batch processing.

6 Conclusion

The general task of this research was providing information for new technical solutions synthesis based on the analysis of Russian-language patents. The data source was the main claim of the device from the patent text. And the object of extraction was the SAO semantic structures, which characterize the structural elements of a technical object and the relations between them. The ultimate goal was to build a graph of structural elements of the device on the selected structures.

The developed method of extracting SAO structures involves two main stages: segmentation of the sentences of the claims and the extraction of technical implementations in the form of predicate-argument structures. Segmentation is based on the derivation of typical structures in the sentence and coordination of dependent words on cases. The extraction of structural elements of technical objects is based on a verbs set corresponding to certain semantic classes. After that, the extracted structural elements are linked to a net, taking into account homogeneous members and parental relations.

The effectiveness of the method was evaluated with an independent test data set of patents with a total number of marked SAO structures of 318 elements. The value of the F1 metric with a rigorous evaluation (full comparison of arguments) and a lax evaluation (the presence of vertexes of noun groups suffices) was 63% and 79%, respectively. The average document processing speed was 1.49 s on a laptop with an average configuration.

The direct analogs of the system for comparison are unknown to the author. However, given that the accuracy of extracting structured data of similar systems is 75–85%, the results can be considered satisfactory.

The construction of the network of structural elements of a technical object allows going on to compile the ontology of the subject area and come closer to solving the problem of information support for new technical solutions synthesis, which is a further direction of the research.

The ontology scheme includes the structural elements of technical objects as concepts and the relationship between them, as well as supporting information on the invention. The initial content of the ontology is based on the processing of 11,200 patent documents for inventions. The existing scheme already allows to retrieve useful information about alternatives of structural components and communications between them. For example, searching for all elements of a structure in a given invention or tracking relationships. The results suggest that the proposed approach is promising. A further direction of research is seen by the authors in improving the existing method for extracting data and expanding ontology.

Acknowledgements The reported study was funded by RFBR according to the research projects 18-07-01086, 19-07-01200; and was funded by RFBR and Administration of the Volgograd region according to the research projects 19-47-340007, 19-41-340016.

References

1. Choi, S., et al.: SAO network analysis of patents for technology trends identification: a case study of polymer electrolyte membrane technology in proton exchange membrane fuel cells. Scientometrics, 863–883 (2011). https://doi.org/10.1007/s11192-011-0420-z
2. Guo, J., et al.: Subject–action–object-based morphology analysis for determining the direction of technological change. Technol. Forecast. Soc. Change (105), 27–40 (2016)
3. Wang, X., et al.: Identification of technology development trends based on subject–action–object analysis: the case of dye-sensitized solar cells. Technol. Forecast. Soc. Change (98), 24–46 (2015)
4. Yang, C., et al.: SAO semantic information identification for text mining. Int. J. Comput. Intell. Syst. **10**, 593–604 (2017). https://doi.org/10.2991/ijcis.2017.10.1.40
5. Souili, A., et al.: Starting from patents to find inputs to the problem graph model of IDM-TRIZ. Procedia Eng. **131**, 150–161 (2015). https://doi.org/10.1016/j.proeng.2015.12.365
6. Tomita-parser: Developer guide. https://tech.yandex.ru/tomita/doc/dg/concept/about-docpage/ (2019)
7. UFAL UDPipe: http://ufal.mff.cuni.cz/udpipe (2019)
8. Korobov, M.: Morphological Analyzer and Generator for Russian and Ukrainian Languages. Analysis of Images, Social Networks and Texts, pp. 320–332. arXiv:1503.07283v1 (2015)
9. Asiryan, A.K.: Morphological tagging tools comparison. Chapter presented at the Intellectual Potential of the XXI Century '2017, November. https://www.sworld.com.ua/konferu7-317/27.pdf (2017)
10. Blazhievskaya, A., et al.: Morphological Analysis for Russian: Integration and Comparison of Taggers. Analysis of Images, Social Networks and Texts, vol. 661, pp. 162–171. Springer, Berlin (2016)
11. Dereza, O.V., et al.: Automatic morphological analysis for Russian: a comparative study. In: Computational Linguistics and Intellectual Technologies: Proceedings of the International Conference «Dialogue» (2016)
12. Reis, S.R.N., Reis, A., Carrabina, J., Casanovas, P.: Contributions to Modeling Patent Claims When Representing Patent Knowledge. Lecture Notes in Computer Science, vol. 10791, pp. 140–156 (2018). https://doi.org/10.1007/978-3-030-00178-0_9
13. Ulmschneider, K., Glimm, B.: Semantic exploitation of implicit patent information. In: Proceedings of the 2016 IEEE Symposium Series on Computational Intelligence (SSCI 2016), Athens, Greece, December (2016). https://doi.org/10.1109/ssci.2016.7849943

14. Korobkin, D., Fomenkov, S., Golovanchikov, A.: Method of identification of patent trends based on descriptions of technical functions. J. Phys. Conf. Ser. **1015**, 7 (2018)
15. Korobkin, D., Fomenkov, S., Kolesnikov, S.: The method for detecting the dependencies between technical functions and physical effects. In: Proceedings of the MCCSIS 2018, Madrid, pp. 225–228 (2018)
16. Phan, C.-P., Nguyen, H.-Q., Nguyen, T.-T.: Ontology-based heuristic patent search. Int. J. Web Inf. Syst. (2018). https://doi.org/10.1108/IJWIS-06-2018-0053

Conceptual Approach to Designing Efficient Cyber-Physical Systems in the Presence of Uncertainty

A. P. Alekseev

Abstract Cyber-physical systems are one of the most advanced areas of research technology. However, being distributed systems with a large number of interconnected elements, they often pose a challenge for their designers concerning intra-network control and interaction. The chapter suggests a way to present a cyber-physical system as a multicommodity network model that demonstrates the dual character of connections between the elements of the system. The efficiency of such systems is studied regarding their ability to fulfill the requirements of their elements in the presence of uncertainty. Using the concept of the difficulty of achieving the goal, the authors developed an algorithm for analyzing the efficiency of the multicommodity network. The algorithm can be used to assess the efficiency of the system functioning in different conditions with different parameters. The suggested tools help to determine the most efficient version of the system, which can eventually broaden the scope of application domains of such systems.

Keywords Cyber-physical systems · Efficiency · Difficulty of achieving the goal · Multicommodity network

1 Introduction

Cyber-physical systems (CPS) are highly complex mechanisms which involve transdisciplinary approaches and effect various aspects of our lives depending on their application domain. A CPS is characterized by tight integration between physical and computation processes within it [1, 2]. Examples of CPS include a smart grid, autonomous automobile systems, automated industrial control systems (Industry 4.0) [3, 4], process control systems [5], robotics systems, and automatic pilot avionics [6].

A cyber-physical system is a complex distributed system controlled or monitored by computer-based algorithms and tightly integrated with the Internet and its users

A. P. Alekseev (✉)
Voronezh State University, Universitetskaya Ploshchad, 1, Voronezh 394018, Russia
e-mail: evil-emperor@mail.ru

A. G. Kravets et al. (eds.), *Cyber-Physical Systems: Advances in Design & Modelling*, Studies in Systems, Decision and Control 259,
https://doi.org/10.1007/978-3-030-32579-4_6

[7]. We can thus say that cyber-physical systems have a network structure. Due to several factors, such as a large number of elements and connections between them, the need for real-time processing of large amounts of data [8, 9], and the environmental influence, it becomes necessary to address the problem of the communication network for such complex distributed systems characterized by uncertainty not common for uniform networks. By now, there have been few studies focusing on intra-network modeling of such systems, as the emphasis tends to be more on the computational elements, and less on an intense link between the computational and physical elements. However, intra-network optimization modeling can significantly enhance the efficiency of such systems and broaden the scope of their application domains.

2 Modeling a Cyber-Physical System as a Multicommodity Network

An important thing in designing cyber-physical systems is formalization. When designing a CPS, structural modeling techniques should be used. Such techniques involve using graph representation models of complex systems. A CPS model can be presented as an aggregation of the CPS's algorithms and structure represented as graphs with the same vertex set [10]. Let the structure graph be defined as the physical graph, as it represents the conditional physical infrastructure allowing for the information flows. The algorithm graph should be then defined as the logical graph since it represents the structure of the connections between the system's elements, namely their mutual requirements for the information flow. The edges of the graph connect the elements of the system, which pass a flow with specific characteristics from one to the other. Such pairs of elements are called source-sink pairs. The information flows between a source-sink pair of the logical graph can only go through the channels of the physical infrastructure of the network, i.e. the edges of the physical graph. The kind of network described above is called a multicommodity network [11], because flows of different source-sink pairs are not interchangeable since every information flow is aimed at a specific addressee and cannot be substituted with any other flow. In fact, information flows between the nods of the logical graph correspond to different types of products, which go along the edges of the physical graph without interacting.

The requirements set by the source-sink pairs for the flows are estimated using specific units of measure for each parameter, such as the value of flow, its cost, etc. The edges of the logical graph are assigned corresponding values in the units adopted for the flow of a particular source-sink pair. The edges of the physical graph limit the flows within any source-sink pair that uses this communication channel. Therefore, every edge is assigned a characteristic measured in the same units as the requirements of the source-sink pairs. The challenge is to allocate the flows of the network so that the paths between the source-sink pairs going along the edges of the physical graph were optimal for each pair of nods of the logical graph [12].

Optimal allocation will account for the restrictions of both the physical graph (flow capacity or other parameters) and the logical graph (requirements of the components of the cyber-physical system).

Since we do not know which pairs will be exchanging flows at any specific moment in time, we are not able to forecast which flows will be going along each of the edges of the physical graph at any specific moment in time. We can thus consider two different situations.

1. Flows of every source-sink pair pass along the network at any moment in time. This is the maximum flow capacity of the network.
2. Only one flow passes along the network at any moment of time. This helps to determine the maximum degree to which the requirements of the corresponding source-sink pair can be met.

Both situations are rare, as it is more likely that a different number of single-product flows pass through the network at different moments. However, their analysis demonstrates the ability of the system to meet the requirements of the source-sink pairs, i.e. its ability to function efficiently. Analyzing the first situation, we can assess the efficiency of the whole system at maximum load, although it does not allow us to evaluate its ability to meet the requirements of a specific pair. Analyzing the second situation, we can assess the ability of the system to meet the requirements of every pair of elements and determine the safety margin in case the requirements or the system's capacity change. In this chapter, we will focus primarily on the second situation.

Limitations and requirements for the flows depend on the characteristics of the system. Let us consider a basic situation when it is necessary to maximize the flows between the source-sink pairs. In this case, there are certain requirements for the flow volume. It is obvious that for each source-sink pair the maximum flow should be determined, taking into account the network's flow capacity [13]. The system, however, may also require to minimize the cost of the flows, as well as to find the shortest paths, or the minimum-cost maximum flow, etc. In this case, the network's edges are assigned other parameters, such as the cost or the length of the path. We will further refer to these parameters as the characteristics of the edges of the network. The efficiency assessment procedure remains the same. If all the flows meet the logical requirements of the network's elements, it is considered acceptable, as it is able to function efficiently. If the opposite happens, it is either necessary to elaborate on the network (by improving the parameters of the existing edges or adding new edges to the physical graph) or to reconsider the conditions for the pairs whose requirements cannot be fulfilled.

The cyber-physical system represented by the above network model is characterized by uncertainties of three types.

The first type concerns the requirements of the source-sink pairs. In this case, either the decision-making agent is not fully aware of the requirements of the system's elements, or there is an objective necessity to increase the requirements (e.g. due to external factors), which the decision-making agent does not know about beforehand.

The second type concerns the characteristics matrix of the channels of the physical infrastructure, i.e. the physical graph. We assume that these values will be lower than those calculated while designing the network. Such uncertainty may result from the channels of the physical graph being damaged by external factors.

Uncertainty of the third type is caused by factors that are practically impossible to formalize. Although they do not affect either the characteristics of the edges or the requirements of the components, the system's ability to fulfill these requirements deteriorates.

Uncertainties of all the three types may be either internal or external. However, external influence is less predictable, which is why we will now focus on this type. We will further refer to any unpredicted or undesirable event (or a series of events) that may result in the system's malfunction as an incident. The degree of influence of an incident on the system should be referred to as the incident's gravity. Incident impacts may vary and effect the characteristics of the edges and the requirements for the flows between the source-sink pairs. They may also include non-formalized factors.

A decrease in the characteristics of the edges of the physical graph may be rather significant and difficult to compensate for. This means that a posteriori reallocation of flows will be required. Therefore, we will consider the problem taking into account the possibility of optimal allocation of flows after a damaging impact.

In case of uncertainty, when we do not have complete information, we must determine the guaranteed result, which means that we should expect the worst possible outcome. The localization of the impact resulting from the incident (edges and/or pairs subject to the damaging effect) and the way this impact is distributed between the edges and pairs of the network are considered to be unknown. To assess the efficiency of the network after the incident, it is necessary to determine the worst outcome of the incident, i.e. to determine the situation when the characteristics of the edges deteriorate, the requirements increase, or there are other factors that cause maximum damage to the network's functioning. The efficiency of the network is defined as its ability to fulfill the maximum flow requirements of the source-sink pairs.

Thus, analyzing the efficiency of a distributed cyber-physical system in the presence of uncertainty, we can say that a system which is not capable of performing its functions is not efficient. The most effective way to analyze the efficiency is to assess the efficiency of the system multiple times changing various parameters of uncertainty. This will help to establish the dependency between the system's efficiency and the uncertainty factors. Therefore, modeling different versions of the cyber-physical system functioning under various conditions enables us to analyze the system's efficiency in each situation and compare the results. For illustrative purposes, it is advisable to make dependency graphs for each version of the system and compare them afterward.

3 Evaluation of the Efficiency of the Multicommodity Network

By efficiency of the multicommodity network representing the cyber-physical system, we mean a complex parameter demonstrating how well the network can fulfill the requirements of the source-sink pairs. In other words, the degree to which the flows passing through the channels of the physical graph fulfil the requirements of the elements of the logical graph. The difficulty in fulfilling the requirements grows parallel to the increase in the requirements and the decrease in the quality of the flow between the nods of the pair. By the quality of the flow, we mean the degree to which the flow complies with the required characteristics specific to the network. If the quality of even a single flow is lower than necessary, it is not possible to meet the requirements. This dependency may be described by the concept of "difficulty of achieving the goal" introduced by Russman in [14]. The parameter "difficulty of achieving the goal" is an integrated characteristic of the quality of an object based on the ratio of the object's properties and the requirements for this object set by the system. These requirements most often depend on the requirements for the whole system.

Given below is a brief mathematical description of the "difficulty of achieving the goal" parameter. A particular estimate of the difficulty d_k depends on the requirements ε_k for the quality of the k-th object and the value μ_k of the quality of the k-th object. In order to determine the function d_k we need to determine its properties [15]:

1. If $\mu_k > 0$ and $\varepsilon_k = 0$, then $d_k = 0$, i.e. when there are no requirements for the quality, the difficulty level is minimal.
2. If $\mu_k = 1$ and $\mu_k > \varepsilon_k$, then $d_k = 0$, i.e. when the quality of the object is maximal, the difficulty level is minimal.
3. If $\varepsilon_k = \mu_k$, then $d_k = 1$, i.e. when the quality of the object complies with the requirement for the quality, the difficulty level is maximal.

Using the three conditions, we obtained the following formula for assessing the difficulty of achieving the goal [16]:

$$d_k = \frac{\varepsilon_k(1-\mu_k)}{\mu_k(1-\varepsilon_k)}, \tag{1}$$

where $d_k = 0$, when $\varepsilon_k = \mu_k = 0$, and $d_k = 1$, when $\varepsilon_k = \mu_k = 1$.

Since the quality of any object is a hierarchy of its characteristics, an integrated estimate of the difficulty of achieving the goal should be the function of scores d_k of separate parameters. Let us assume that there is an object with two characteristics whose scores the difficulty of achieving the goal are d_1 and d_2. The overall difficulty estimate will be determined as $D = f(d_1, d_2)$.

Russman demonstrated [17] that only one function of two variables meets the requirements:

$$D = d_1 + d_2 - d_1 d_2 = 1 - (1-d_1)(1-d_2) \tag{2}$$

When n components of the integrated resource are present, the following formula is used to calculate the integrated estimate [17, 18]:

$$D = 1 - \prod_{k=1}^{n} (1 - d_k) \tag{3}$$

Flow characteristics of the studied network may vary (the flow value, the cost of the flow, etc.) both in the measurement units and in application domains. The flow may also be characterized by several parameters (e.g. minimum-cost maximum flow), which makes the parameter of the difficulty of achieving the goal a very useful and flexible tool for assessing the degree to which flow in a multicommodity network complies with the requirements set by the corresponding source-sink pair.

We shall thus consider the efficiency of the multicommodity network as an integrated estimate of the difficulty of achieving the goal. This value is calculated using (3), where particular difficulty estimates are determined as the degree of fulfillment of the requirements of each source-sink pair. The quality of the flow between the components of the source-sink pair is a certain parameter of the flow (the value, the cost, etc.), while the requirement for the quality of the flow is the requirement set by the source-sink pair.

4 An Algorithm for Evaluating the Efficiency of a Cyber-Physical System in the Presence of Uncertainty

Before we describe the algorithm for evaluating the efficiency of the cyber-physical system modeled as a multicommodity flow network, let us introduce a number of designations. The multicommodity network $S = (V, P)$ is determined by a set $V = \{v_1, \ldots, v_n\}$ of nods and $P = \{p_1, \ldots, p_m\} \in V \times V$ of source-sink pairs or edges of the logical graph. Let the corresponding index sets be $N = \{1, \ldots, n\}$ and $M = \{1, \ldots, m\}$, with $V = \{v_i\}_{i \in N}$ and $P = \{p_k\}_{k \in M}$.

For any vertex $v \in V$ let $S(v)$ denote the set of indices of its outgoing edges, and $T(v)$—the set of indices of its incoming edges [11]. For each k-th source-sink pair let us introduce the designation $p_k = (v_{sk}, v_{tk})$, where $s_k < t_k$ is the vertex, v_{sk} is the source, and v_{tk} is the sink of the source-sink pair. g_k is the flow going from the source to the sink in every source-sink pair $p_k \in P$.

The network has quantitative restrictions determined by the edges of the physical graph. Let us assume that each edge (v_i, v_j) of the network has a certain value $c_{ij} \geq 0$, called the characteristic of the edge (the flow capacity, the cost of the flow, the length of the path, etc.) and measured in measurement units of the flow the network is created for. All the edges of the logical graph are assigned values $y_k \geq 0$, measured in the measurement units of the flow. These values also pass along the logical edge of the multicommodity network.

In order to determine whether the system is acceptable, it is not necessary to model all the possible allocations of the flows of the physical network. It is enough to determine the allocations that ensure best possible flows between all the source-sink pairs. Let us use z_k to denote the best of all the possible flows g_k. A set of such flows will then be denoted as $Z(c) = \{z_k\}$. This flow matrix ensures the maximal efficiency of the network [19].

Let us assuming that the incident's gravity is the vector of three variables $W = \{\beta, \gamma, \delta\}$.

Parameter β denotes the expected increase in the requirements of any source-sink pair.

Parameter γ denotes the expected deterioration of the characteristics of any edge of the network.

Parameter δ denotes non-formalized factors and the expected increase in the difficulty in fulfilling the requirements of any source-sink pair.

Using these designations, we can develop an algorithm for assessing the effectiveness of the described network. The algorithm is uniform for all the characteristics of the edges of the physical graph and differs only in the way the matrix $Z(c)$ is determined.

1. Make the physical and logical graphs of the multicomponent cyber-physical system based on the model of a multicommodity network using the parameters of the system and the requirements for them.
2. Assess the best flows between the source-sink pairs, i.e. matrix $Z(c)$. To determine the matrix, graph theory algorithms corresponding to the flow's parameters should be used. Thus, if we need to determine the maximum flow, the maximum flow computation methods are used, such as the Ford–Fulkerson algorithm, the Dinic's algorithm, the Gomory–Hu algorithm, etc.
3. Construct a matrix of the estimates of the difficulty of achieving the goal for all source-sink pairs. The quality of the obtained flows is evaluated using the following formula:

$$\mu_k = \frac{z_k}{\overline{Z} + Z_{corr}}, \tag{4}$$

where $\overline{Z} = \max_{1 \le k \le m} z_k$.
Z_{corr} is a special parameter for potential adjustments of the optimal quality (if no serious adjustments are required, the recommended value is $\overline{Z}/100$).

The next step is to evaluate the requirements for the quality of the flows:

$$\varepsilon_k = \frac{y_k}{\overline{Z} + Z_{corr}} \tag{5}$$

We should point out that both μ and ε are measured in the interval [0, 1], with $\varepsilon_k \le \mu_k \forall k$ for any source-sink par. If this condition is not fulfilled, the combination does not conform with the minimal quality requirements. In all the other cases, the difficulty value is:

$$d_k = \varepsilon_k(1 - \mu_k)/\mu_k(1 - \varepsilon_k) \tag{6}$$

Let us also introduce weighing coefficients in the range of $0 < \alpha_k \leq 0.1$. The final set of difficulties in fulfilling the requirements of the source-sink pairs is then determined as

$$D = \{d_k{}^{\alpha_k} \mid d_k{}^{\alpha_k} = 1 - (1 - d_k)^{\alpha_k}\} \tag{7}$$

The integrated difficulty is determined using the formula:

$$D_{\text{int}} = \sum_{k=1}^{m} d_k{}^{\alpha_k} \tag{8}$$

This parameter demonstrates the integrated difficulty in fulfilling the requirements of all source-sink pairs of the network and serves as a criterion for assessing the system's efficiency. The higher the difficulty value, the harder it is to meet the mutual requirements of the system's elements at the given flow capacity of the network. When $D = 1$, the difficulty value is maximal, which means that the system is highly vulnerable. If any of the values d_k is more than 1 (in the case when $\varepsilon_k > \mu_k$), the integrated value is also $D > 1$, which means that the flow between the vertices in this pair does not meet the requirements and the system does not function efficiently.

4. Estimate the expected incident's gravity $W = \{\beta, \gamma, \delta\}$. The incident's gravity can be determined by solving a number of special problems [17] or using expert forecasting methods. If the incident's gravity cannot be determined precisely, or it is not necessary for the current problem, steps 4 and 5 can be omitted or performed using a preset gravity value.
5. Calculate the matrix C^γ of the expected characteristics of the edges of the physical graph and vector Y^β of the expected increased requirements. The incident's gravity was estimated in step 4.

Formulas for calculating new parameters of the network:

$$c_{ij}^{\gamma} = (1 - \gamma)c_{ij}^{0} \tag{9}$$

$$y_{ij}^{\beta} = (1 + \beta)y_{ij}^{0} \tag{10}$$

It is now necessary to once more determine the best flows (perform step 2 with new parameters) and the set of estimates of the difficulty in fulfilling the requirements of the source-sink pairs (repeat step 3 with a new matrix).

To evaluate the non-formalized factors another parameter δ of the incident's gravity is used and D^δ is calculated:

$$d_{ij}^{\delta} = (1 + \delta)d_{ij}^{0} = 1 - \left(1 - d_{ij}\right)^{(1+\delta)} \quad (11)$$

The obtained parameter of the integrated estimate reflects the efficiency of the system after an incident of particular gravity. If the network is still acceptable, the system is considered efficient enough to resist the impact of the incident of expected gravity and meet the requirements of all the system's components.

6. Repeat step 5 gradually increasing the incident's gravity until the network stops being acceptable. Thus, the maximum incident's gravity for the system can be determined. Efficiency indices and corresponding incident's gravity values are marked on a diagram.
7. For a more detailed analysis, repeat step 6 for various incident impacts (with uncertainties of all the three types). The obtained diagrams will demonstrate the parametric dependence between the system's efficiency and the uncertainty factors.
8. Repeat steps 1–7 for different versions of the network in order to compare several versions and select the most efficient one.

The described algorithm, therefore, determines the way to design the most efficient distributed cyber-physical system.

5 Simulation Experiment

The suggested algorithm was tested on the information network of the company Technopark-V (Voronezh), whose chart is presented in Fig. 1.

The following parameters were determined using the algorithm.

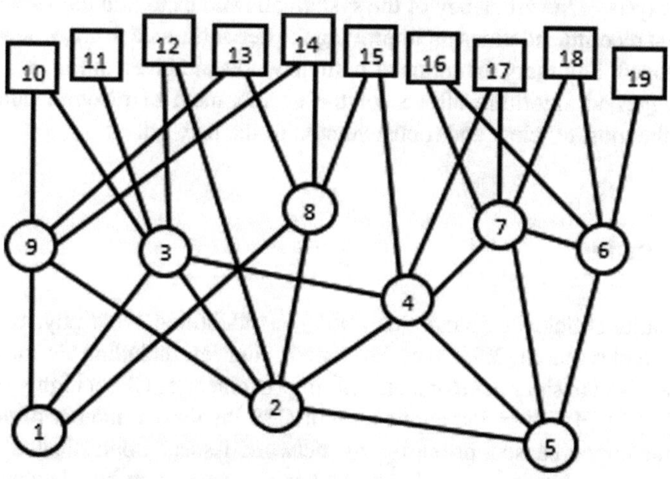

Fig. 1 Chart of information network

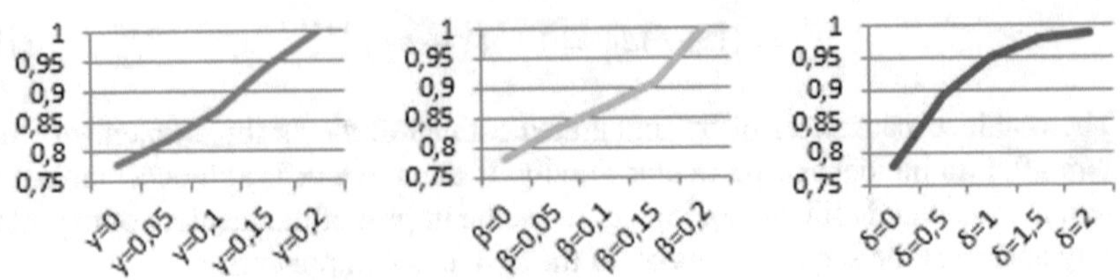

Fig. 2 Diagrams of the dependency of the system's efficiency on various uncertainty factors

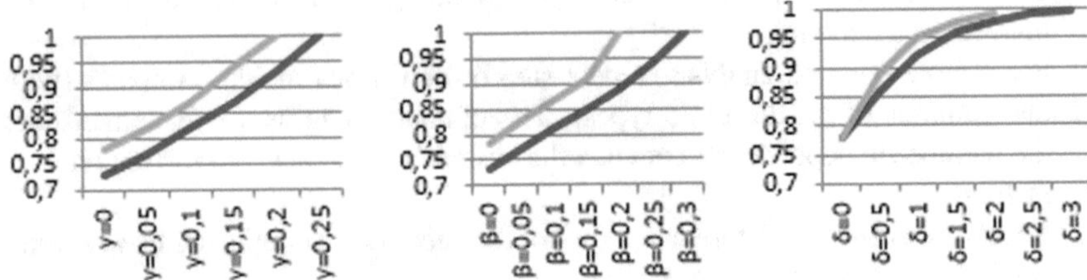

Fig. 3 Summary diagrams of efficiency for both networks

$D = 0.767$—the integrated efficiency estimation without uncertainties.
$D = 0.944$—the integrated efficiency estimation after an incident whose parameters were determined by experts.
$W_{lim}(\beta_{lim}; \gamma_{lim}; \delta_{lim}) = (0.18; 0.15; 0.19)$—vector of the maximal parameters of the gravity of the incident after which the system still functions.
(18; 6), (12; 3), (17; 2)—the most vulnerable source-sink pairs. Diagrams demonstrating the dependency of the system's efficiency on various uncertainty factors are presented in Fig. 2.

The analysis of the efficiency of the system allowed us to find the vulnerabilities and suggest recommendations on improving the network. As a result, a new network was developed. Summary diagrams of both networks are given in Fig. 3.

The suggested algorithm allows for the development of recommendations for selecting the most efficient and robust version of the network.

6 Conclusion

The US National Science Foundation (NSF) has identified cyber-physical systems as a key area of research [20]. Other developed countries, including Germany, Japan, and China, also consider development and improvement of CPS a highly promising sphere [21, 22]. However, the application of CPS involves a number of algorithm-related challenges caused primarily by network issues. Sophisticated tools for studying complex networks must be used when designing distributed cyber-physical systems, as they ensure the highest efficiency of the system. Modeling a cyber-physical system as a multicommodity network followed by the analysis of its efficiency in the presence of uncertainty allowed us to develop an algorithm for the analysis of efficiency of such systems based on the concept of difficulty of achieving the goal. The proposed techniques can be used when designing CPS under uncertainty. The simulation experiment carried out on an information network of the Technopark-V company demonstrated the effectiveness of the suggested method for determining the process of CPS design.

References

1. Rad, C.-R., Hancu, O., Takacs, I.-A., Olteanu, G.: Smart monitoring of potato crop: a cyber-physical system architecture model in the field of precision agriculture. In: Conference Agriculture for Life, Life for Agriculture, no. 6, pp. 73–79 (2015)
2. Wolf, W.: Cyber-physical systems. Computer **42**(3), 88–89 (2009)
3. Lee, J., Bagheri, B., Kao, H.: A cyber-physical systems architecture for Industry 4.0-based manufacturing systems. Manuf. Lett. **3**, 18–23 (2015). https://doi.org/10.1016/j.mfglet.2014.12.001
4. Lee, E.A., Seshia, S.A.: Introduction to embedded systems—a cyber-physical systems approach. LeeSeshia.org (2011)
5. Tarassov, V.B.: Enterprise total agentification as a way to Industry 4.0: forming artificial societies via goal-resource networks. In: Abraham, A., Kovalev, S., Tarassov, V., Snasel, V., Sukhanov, A. (eds.) Proceedings of the Third International Scientific Conference "Intelligent Information Technologies for Industry" (IITI'18). IITI'18 2018. Advances in Intelligent Systems and Computing, vol. 874. Springer, Cham (2019). http://dx.doi.org/10.1007/978-3-030-01818-4_3
6. Khaitan, S.K., et al.: Design techniques and applications of cyber-physical systems: a survey. IEEE Syst. J. **9**(2), 350–365 (2014)
7. Xia, F., et al.: Internet of things. Int. J. Commun. Syst. **25**(9), 1101 (2012)
8. Namiot, D.: On big data stream processing. Int. J. Open Inf. Technol. **4.3**(9.8), 48–51 (2015)
9. Lee, J., Bagheri, B., Kao, H.: Recent advances and trends of cyber-physical systems and big data analytics in industrial informatics. In: IEEE International Conference on Industrial Informatics (INDIN) (2014)
10. Tsvetkov, V.Ya.: Control with the use of cyber-physical systems. Int. Sci. Electron. J. (3), 55–60 (2017). ISSN 2307-2334
11. Malashenko, Yu.E., Novikova, N.M.: Models of Uncertainty in Multi-User Networks, p. 1999. Editorial URSS Publishing, Moscow (1999)
12. Malashenko, Yu.E.: About solving a multi-product problem with integer flows. J. Comput. Math. Math. Phys. **22**(3), 732–735 (2018)
13. Alekseev, A.P.: Vulnerability analysis of a multicomponent system as multicommodity network. Eur. Multisci. J. (Monthly international science journal) (9), 51–54 (2017)
14. Babunashvili, M.K., Bermant, M.A., Russman, I.B.: Analysis Methods for Graph of Targets. Research Planning and Information Support, Moscow (1972)
15. Kaplinskiy, A.I., Russman, I.B., Umyvakin, V.M.: Modeling and Algorithmization of Poorly Formalized Tasks of Choosing the Best System Options. VSU Publishing, Voronezh (1991)
16. Russman, I.B.: Complex system assessment and subsystem assessment. News USSR Acad. Sci. **2**, 201–204 (1978)
17. Russman, I.B.: Integral Quality Assessments in Organizational Systems. Structural Adaptation of Complex Control Systems. VPI Publishing, Voronezh (1977)

18. Alekseev, A.P., Abramov, G.V., Bulgakova, I.N.: Integral assessment of product quality as the difficulty of achieving the goal. J. Energy XXI Century (2), 79–86 (2017)
19. Alekseev, A.P., Abramov, G.V., Bulgakova, I.N.: The problem of reliability analysis of multi-component network systems under external disturbances. In: Applied Mathematics, Computational Science and Mechanics: Current Problems: Collection of the Works of International Science Conference. Voronezh, pp. 1557–1564 (2017)
20. Wolf, W.: The good news and the bad news (embedded computing column). IEEE Comput. **40**(11), 104–105 (2007). https://doi.org/10.1109/MC.2007.404
21. Suh, S.C., Carbone, J.N., Eroglu, A.E.: Applied Cyber-Physical Systems. Springer, Berlin (2014)
22. Fitzgerald, J., Larsen, P.G., Verhoef, M.: Collaborative Design for Embedded Systems: Co-modelling and Cosimulation. Springer, Berlin (2014). ISBN 978-3-642-54118-6

About Preparation of the Analytical Platform for Creation of a Cyber-Physical System of Industrial Mixture of Loose Components

A. B. Kapranova, I. I. Verloka and D. D. Bahaeva

Abstract The results of the preparation of an analytical platform for building a cyber-physical system (CPS) for industrial mixing of bulk components in a gravitational apparatus with additional mixing elements are presented. The latter includes rotating drums with brushes and inclined fender planes. To perform an analysis of the efficiency of the process of mixing bulk materials with different physical and mechanical properties in rarefied flows, the author's models are used. The above elements of the analytical platform for CPS contribute to the development of engineering methods for calculating gravity-type mixing equipment.

Keywords Cyber-physical system · Gravity mixer · Loose components · Brush elements · Distribution functions · Process parameters

1 Introduction

The problem of creation of the cyber-physical system (CPS) for the realization of an effective mixture of loose components can't be solved without the formation of the analytical platform, theoretical bases of the most specified technological process. Important stages, preceding model operation of the process of mixture of loose components, identification of its information parameters and the choice of basis (settlement) variables are. At the same time research of influence of input parameters of this technological operation as a mechanical process, on its output characteristics demands the description of mechanics of behavior of firm dispersion mediums in the displacement volume of the concrete mixer.

A. B. Kapranova (✉) · I. I. Verloka · D. D. Bahaeva
Yaroslavl State Technical University, Moskovskiy Prospect, 88, Yaroslavl 150023, Russia
e-mail: kapranova_anna@mail.ru

I. I. Verloka
e-mail: compvii@rambler.ru

D. D. Bahaeva
e-mail: bakhaevadd@mail.ru

A. G. Kravets et al. (eds.), *Cyber-Physical Systems: Advances in Design & Modelling*, Studies in Systems, Decision and Control 259,
https://doi.org/10.1007/978-3-030-32579-4_7

The research of behavior of loose components at their interfusing is of special interest at a development stage of theoretical bases of projection of mixing equipment for the needs of a wide range of industries and agriculture. One of the ways to overcome the prerequisites for the segregation effect, characteristic of the operation modes of many devices of the indicated purpose, is the constructive organization of crossing of rarefied flows of mixable bulk components in the working volume of the mixer [1, 2]. At the solution of specific technological objectives, it is required to consider a complex set of parameters of the studied process to which not only procedural indicators of operation of interfusing and its design parameters but also physicomechanical characteristics of loose materials belong. The analysis of the behavior of loose components in the described conditions often is complicated by the impossibility of realization of model experiments in connection with toxic properties of actuation mediums. In this case, it is expedient to select safe substitutes of compositions of the required mix with similar physicomechanical properties to perform «tunings» of an engineering methodic of calculation of key parameters of the mixer. In this time realization of theoretical researches on approbation of such replacement of toxic substances with model environments by means of the analysis of the corresponding results of stochastic modeling is required [1].

The significance of the creation of a cyber-physical system of an industrial mixture of loose components even more increases in the specified conditions.

2 Analysis of the Fundamental Scheme of the Mixer of Loose Materials

Briefly we will stop on the description of the key diagram of the device for mechanical mixture of the firm disperse components «1» and «2» with the continuous duty (Fig. 1) as the operation of hashing which is carried out before achievement of the specified coefficient of degree of inhomogeneity of the obtainable loose mix. Let the flow of the loose component «1» be «transporting», and a flow «2»—«key». Let's allocate the following main stages of mechanical technological process in the mixer with a set of design parameters $a = \left\{a_{j1} = cont\right\}, j_1 = \overline{1, u_1}$ flows «1» and «2».

- Realization of giving by means of the special device (a batcher or the feeder) of working materials with a set of input parameters $x(t) = \{x_i(t)\},\ i = \overline{1, 2}$ from area of their change of $X = \left\{x_1^- \le x_1 \le x_1^+; x_1^- \le x_1 \le x_1^+\right\}$. In particular, mass consumption $Q_i(t), Q_2(t)$ and mass fractions $\gamma_i(t), \gamma_2(t)$ of «1» and «2» components belong to components of entrance sets for each component.
- An increase in surfaces of shift S components is typical at their turbulent nature of hashing with a supply of energy of E during mixture. This energy is spent on the shift of particles with a rupture of intercommunications between them and also a macrodiffusion process. Therefore, a set of the regime parameters $b = \{S, D, E, \tau_S,\}$ includes the listed indicators of the studied process where D—is the macrodiffusion coefficient.

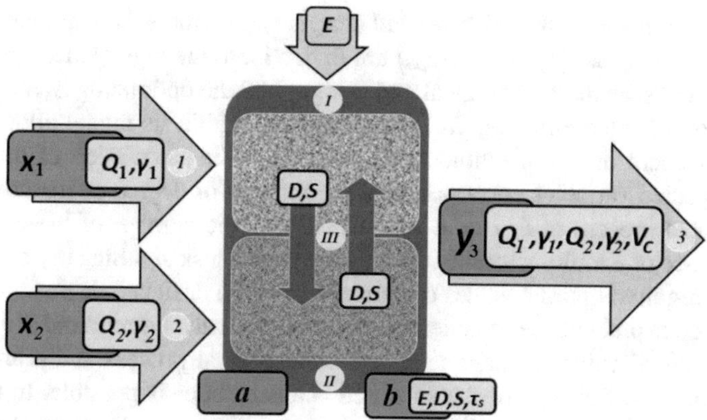

Fig. 1 The fundamental scheme of the device for the mechanical mixture of the firm disperse components «1» and «2» with the continuous duty: I and II—zones of driving of flows of «transporting» and «key» components and their mix, III—the diffusion surface of contact

- Removal (output) from displacement volume of the mixer of the obtained loose mix with coefficient in inhomogeneity of V_C which is characterized by the following set of output variables $y_3(t) = \{x_1(t), x_2(t), V_C\}$ or $y_3(t) = \{Q_i(t), Q_2(t), \gamma_i(t), \gamma_2(t), V_C\}$ of process of the given set of $Y = \{y_3^- \le y_3 \le y_3^+\}$.

Thus, the full set of z(t) = $\{x(t), y(t), a, b\}$ information variables with total number of w for the continuous mechanical mixing of two firm disperse components «1» and «2» with the continuous duty consists of several sets: input $x(t)$, output $y(t)$, constructive a, regime b. Further, there is a question of the division of information variables into two categories (Fig. 2): settlement z_m (basis, defined from the formed

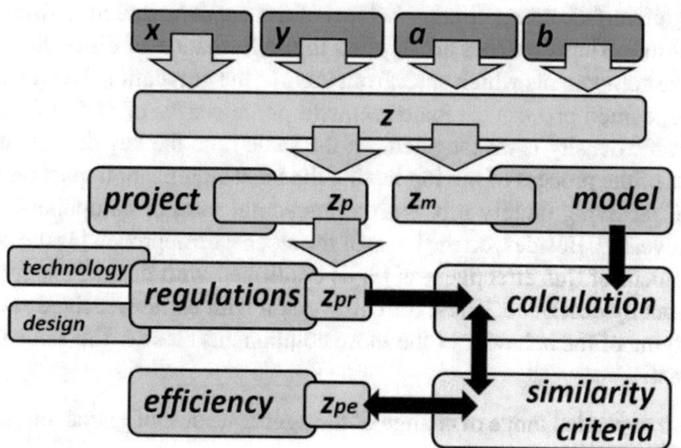

Fig. 2 The conditional scheme of calculation of the technological process by means of information variables

mathematical model of the mixture) and design z_p (all others from among informational). As a rule, last $\left(z_p = z_{pr} + z_{po}\right)$ are divided into the regulated z_{pr} parameters (which set consists of technological and design) and the optimizing z_{po} parameters.

Variation of the optimizing process parameters within its model allows formulating a problem of optimization. Let's notice that the calculation of the number of the projected variables corresponds to the number of degree of freedoms of the technological process of $S_{TP} = w - n$ where n is the number of basic variables. The existence of a multifactorality directed by such a task significantly complicates obtaining required optimal values of parameters of the designed device.

However, in practice the considerable accuracy isn't always required at the assessment of optimal values of parameters of technological process and searching for effective ranges of change of the specified characteristics is possible. In that case, restrictions for change of similitude criteria of a technological system (Fig. 2) can be used. In these conditions searching for limits of variation of effective values of parameters of the studied technological process within the analytical platform is carried out. Further, we will pass to the consideration of the formation of the corresponding mathematical model of the process of the mixture in the gravitational mixer with additional mixing elements.

3 Some Features of the Process of Interfusing of Loose Components in the Gravitational Apparatus

In the device of gravitational type for the increase in the effectiveness of the process of mixing of the not wetted loose components related to the I–III classes of looseness by Kerr's technique with particle sizes within 1.5–6.5 $\times$ 10^{-4} m it is offered to use additional mixing elements. They represent drums with the brush elements fixed on their surface along circular helixes of the opposite direction from both end faces of drums. These brushes are applied to the formation of dilute flow from the mixed loose components which enter from trays of the gravitational device. In works [3, 4] the specified process for materials with particle sizes of (1.5–4.0) $\times$ 10^{-4} m of comparable density is investigated. At the same time the key design and regime parameters of the process of mixing having the most significant impact on the effectiveness of receiving quality mix with a procedural ratio of components 1:10 and more are revealed. Besides, on the basis of the stochastic approach [1] the conditions of achievement of this effectiveness [3, 4] confirmed with data of experiments [5] are theoretically described. Thus, for comparison with earlier received results from [3, 4] studying of the behavior of the loose components close to the value of density is of interest:

- in a more expanded range of change of the aggregate size of grains, in that number up to 6.5 $\times$ 10^{-4} m;
- from toxic substances, the conduct of experimental studies which is difficult.

Table 1 Physico-mechanical characteristics of bulk components

Name of bulk material, model	Bulk density, $\times 10^3$ kg/m^3	Particle size average on fractions, $\times 10^{-4}$ m
Natural sand GOST 8736-93 [2, 3]	1.525	1.5
Semolina GOST 7022-97 [2, 3]	1.440	4.0
Soda ash granules GOST 5100-85 model A (OKP 21 3111 0200)	1.080	1.75
Elimination of the shallow crushed stone GOST 8267-93, model M300	1.600	5.75

Considering a wide range of purpose of the loose mixes received in the gravitational device (from food and mineral structures to structural and chemical), as actuation mediums elimination of the GOST 8267-93 shallow crushed stone model M300 and the GOST 5100-85 soda ash model A granules (OKP 21 3111 0200) in addition to earlier considered the GOST 8736-93 natural sand and the GOST 7022-97 semolina were chosen.

In particular, shallow crushed stone is applied to the production of concrete, glass, printing materials; soda ash (Na_2CO_3 sodium carbonate)—for receiving glass, paper, soap, cast iron, paint and varnish materials, etc. Let's notice that soda ash is a substance of the 3rd class of danger according to GOST 12.1.007. Bulk density and particle sizes of the corresponding loose materials are specified in Table 1.

4 Application of the Stochastic Approach to the Formation of the Analytical Platform of a Cyber-Physical System of the Process of Interfusing of Loose Components

The problem of modeling operation of the process of formation of dilute flows of loose components doesn't lose the relevance [6, 7] since this process is a component of many technological operations of chemical technologies [8]. According to the made analysis of literature sources, it is expedient to carry out the description of mechanics of behavior of particles of loose material in the formed dilute flows on the basis of stochastic approach [9, 10] owing to the unpredictability of trajectories of their driving. At the same time from all variety of the stochastic methods [8, 11–13] including involved at creation of informational management [14] and cell-like [15–17] models, descriptions with time series [18] and with the analysis of a power condition of a single microsystem [19] it is offered to allocate a power way of model operation.

This choice has a talk a possibility of account in the defined law of distribution of number of particles in the formed dilute flow in the given parameters of process of interfusing of a number of the characteristic factors of driving of particles, for example, of their rotations, interactions both with mixing elements, and at collisions among themselves, etc [8, 20, 21]. Application of the specified power way which was described in the monograph by Klimontovich [9], for the theory of technological processes is insufficiently developed [22, 23], however this way is successfully approved when studying shock interactions in dispersible and film systems [19] and technological operation of a refinement [24]. According to the stochastic model operation of process of formation of dilute flows of granular environments described in works [3, 4], the constructed differential distribution function of number of particles $\chi_{ij}(\alpha_j)$ for loose components $i = 1, 2$ on the angle α_j of spreading the drum with brushes depending on number of the deformed brush element $j = 1, 2, 3$ has the form:

$$\begin{aligned}\chi_{ij}(\alpha_j) = K_{ij}\Big\{\exp\Big[-k_4 k_0^2(\alpha_j+\varphi_{ij})^2\big[\mu_{ij}(\alpha_{sij})\big]^2/k_{1i}k_{2i}\Big]\\ \Big(\mathrm{erf}\Big\{\mu_{ij}(\alpha_{sij})\times\big[1+k_0 k_{3i}(\alpha_j+\varphi_{ij})\big]^2/k_{3i}\Big\}\\ -\mathrm{erf}\big[\mu_{ij}(\alpha_{sij})/k_{3i}\big]\Big)\Big\}/\big[\mu_{ij}(\alpha_{sij})\big]\end{aligned} \quad (1)$$

where $\mu_{ij}, K_{ij}, k_{vi}, v = \overline{1,4}$—the coefficients depending on a set of input data of model; $\varphi_{ij}, \alpha_{sij}$—are the characteristic angles of driving of particles in drum transverse sections for each brush element. Function $\chi_{ij}(\alpha_j)$ in (1) it is received at the adding E_{ij}—the energy of stochastic driving of a single spherical particle of component $i = 1, 2$ taking into account their headway, the accidental nature of their moments of impulses and elastic interaction with deformable brush element $j = 1, 2, 3$. At the same case, the element of a phase space $d\Omega_{ij}$ is defined by polar coordinates of this particle concerning a spin axis of a rotary drum mixer in the plane of its section for each deformed brush element

$$d\Omega_{if} = dv_{xij}dv_{yij} = \omega^2 r_{ij}dr_{ij}d\theta_{ij} \quad (2)$$

where ω—an angular velocity of rotation of the drum; r_{ij} and θ_{ij},—radial and angular coordinates for points of disruption of loose materials of $i = 1, 2$. Let's notice that (1)—the stationary solution of power representation of the Fokker-Planck equation, which, in particular, in the presence of macroscale fluctuations of conditions of particle systems of each component as collisions of their elements, has the following form

$$\chi_{ij} = A_{ij}\exp\Big[-E_{ij}/E_{0ij} + E_{ij}^2/\big(2E_{fij}^2\big)\Big] \quad (3)$$

where E_{0ij} and E_{fij}—energetic parameters of the model.

5 Model Results

The received results for function $\chi_{ij}(\alpha_j)$ from expression (1) for the GOST 5100-85 soda ash model A and elimination of the GOST 8267-93 shallow crushed stone model M300 (Table 1) are reflected in Figs. 3 and 4 respectively. Besides, the specified results allow approving this model [3] taking into account collisions between particles of the mixed components, which dropped by brush elements from screw spiralling of opposite directions, for a wider class of loose materials, than for considered in works [3, 4].

The analysis of these results is made depending on major factors revealed in works [3, 4] having the greatest impact on quality of mix at this stage of the studied process: angular velocity of rotation of the drum ω; extents of deformation of brush

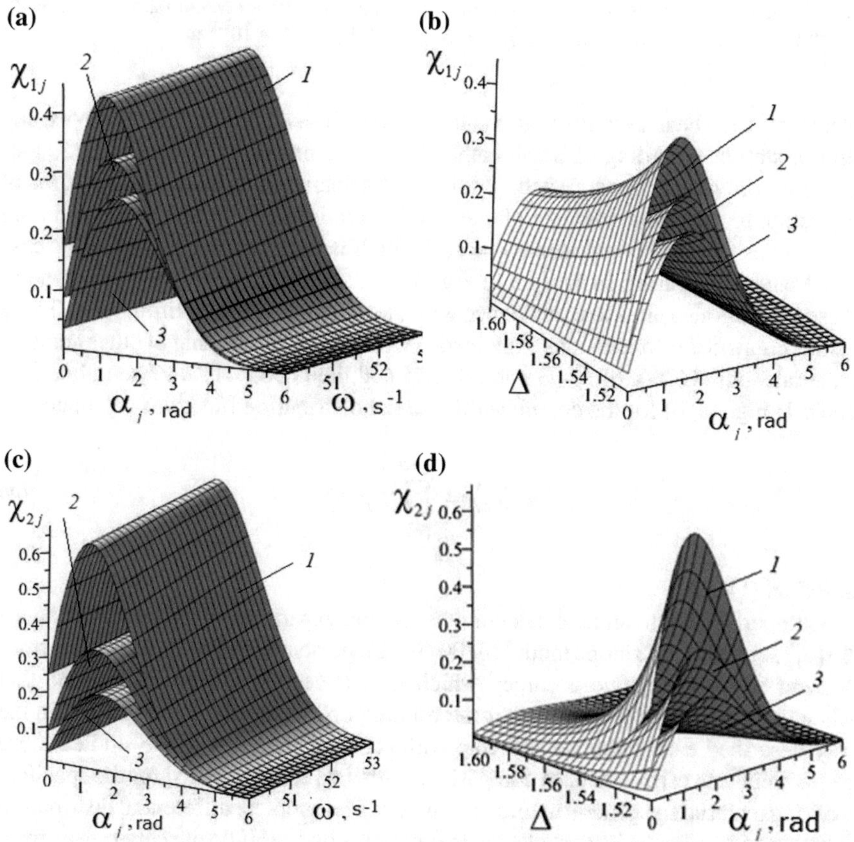

Fig. 3 Dependences $\chi_{ij} = (\alpha_j)$ for number j of the deformed brush element: **a**, **b**,—the GOST 5100-85 soda ash model A (i = 1); **c**, **d**,—elimination of the GOST 8267-93 shallow crushed stone model M300 (i = 2); **a**, **c**,—$\chi_{ij} = (\alpha_j, \omega)$, $\Delta = 1.5$; **b**, **d**—$\chi_{ij} = (\alpha_j, \Delta)$, $\omega = 52.36\ c^{-1}$; j = 1 (1); j = 2 (2); j = 3 (3); $h_s = 1.6 \times 10^{-2}$ m

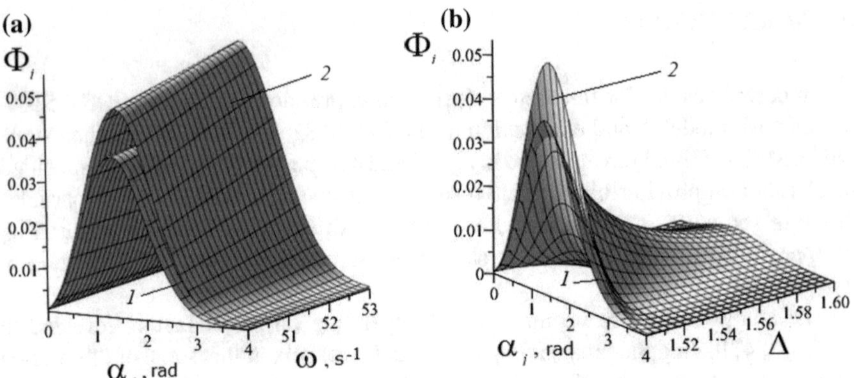

Fig. 4 Dependences $\Phi_i = (\alpha_j)$ for the GOST 5100-85 soda ash model A (a surface 1; i = 1) and elimination of the GOST 8267-93 shallow crushed stone model M300 (surface 2; i = 2): a – $\Phi_i(\alpha_j, \omega)$, $\Delta = 1, 5;$, b—$\Phi_i(\alpha_j, \Delta)$, $\omega = 52.36\ c^{-1}$; $h_s = 1.6 \times 10^{-2}$ m

elements Δ (as beater length relations to gap height between a tray of the device and drum); step of spiralling of brush elements h_s. The presented surfaces (Fig. 3a–d) illustrate providing an estimated condition of effective interfusing in the form of aspiration to the rapprochement of values of extremums for the angles of scattering of particles of loose components i = 1, 2 which is very marked observed for case j = 3 when comparing surfaces 3 in Fig. 3a, c or Fig. 3b, d. Moreover, the type of these surfaces also aims to coincidence with the mixed granular environments. These results confirm the conclusions explained in work [3] for interfusing of other granular materials—the GOST 8736-93 natural sand and the GOST 7022-97 semolina (see Table 1; Fig. 4a, b) for the complete differential distribution functions calculated by

$$\Phi_i(\alpha_j) = \prod_{j=1}^{n_b=3} \chi_{ij}(\alpha_j), \tag{4}$$

based on (1)–(3).

Comparison of theoretical calculations and the experimental data of dependence $\Phi_2(\alpha_j)$ screening of 8 stone model M300 (i = 2), presented in Fig. 5, has satisfactory consent with the relative accuracy which isn't exceeding 10%. Let's notice that when carrying out experiences near the rotating drum with the fixed brushes in the way described earlier the vertical trap with cells for particles of the studied loose component was established. In comparison with data from work [3] for the specified working material the general nature of change of the complete differential distribution function of number of its particles $\Phi_2(\alpha_j)$ remains on the angle of scattering α_j from (4) at formation of a dilute flow owing to interaction with brush elements both for theoretical experimental data. In Fig. 5 it is more evident, than in Fig. 4, the tendency to the rapprochement of values of extremums for α_{1jex} and α_{2jex}—the angles of

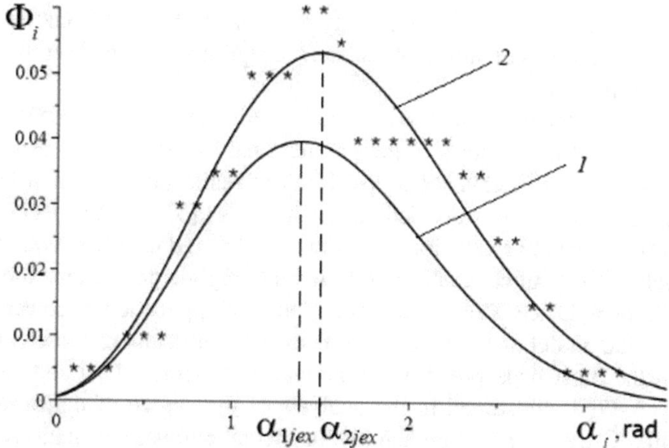

Fig. 5 Dependences $\Phi_i(\alpha_j)$: $\Delta = 1{,}5$; $\omega = 52.36\,c^{-1}$; $h_s = 1.6 \times 10^{-2}$ m; 1—GOST 5100-85 soda ash Brand A (i = 1); 2—elimination of the GOST 8267-93 shallow crushed stone model M300 (i = 2); lines—the theory; points are the experimental data for elimination of the GOST 8267-93 shallow crushed stone model M300 (i = 2)

scattering of particles of loose components i = 1, 2 is presented. This trend reflects a condition of effective interfusing of loose components.

6 Conclusion

So, in the work preparation of an analytical platform for the construction of a cyber-physical system of industrial mixing of bulk components was carried out. In the considered example, the following parameters can be chosen as the characteristic variables of the technological process of mixing bulk materials. Input variables are $x = \{x_i(t)\} = \{Q_{Vi}, n_{Vi}\}, i = 1, 2$ where Q_{Vi} volumetric costs of each component are; n_{Vi} are their volume fractions, specified by the technological features of the industrial process. Output variables are $y = \{y_i(t)\} = \left\{Q_{Vi}, n_{Vi}, V_{C\tau}^{tech}, \Delta V_{C\tau}\right\}, i = 1, 2$ where $V_{C\tau}^{tech}$ are routine values of the heterogeneity coefficient at the τ-stage of mixing, $\tau = \overline{1, 3}$; $\Delta V_{C\tau}$ are absolute parameter errors $V_{C\tau}^{tech}$. The set of regime parameters is $b = \{\omega_\tau, \psi_{1\tau}\}$ where ω_τ is drum rotation speed; $\psi_{1\tau}$ is the angle of the bump to horizontal. As the design parameters can be selected: length, width and angle of inclination of the tray; length and angle of the bump; width and radius of the roll; the length, radius and angle of the helical winding of the cylindrical brush element; the number of brush elements; the distance between the brush elements in the helical winding. Given that the input parameters are a subset of the output, the total number of information variables of the process of mixing bulk components in a gravity mixer is determined by the expression $N = N_{yi} + N_{b\tau} + N_{a\tau}$. For example, the basic variables can be the parameters $\Delta V_{C\tau} = \left|V_{C\tau}^{tech} - V_{C\tau}^{pr}\right|$ where $V_{C\tau}^{pr}$ are the calculated values of the inhomogeneity coefficient obtained from the proposed stochastic mixing model of bulk components and optimizing—total sets of structural and mode parameters.

The made the stochastic analysis of the behavior of various loose components at the formation of dilute flows in the gravitational mixer confirms expediency of application of brush elements and a possibility of realization with their help of the effective mixing conditions revealed earlier in works [3, 4]. The specified factors essentially influence the evaluation test of the mix [5] and the formation of an engineering methodology of calculation of design and regime parameters, the designed new gravitational device. Besides, the expediency of application of power approach at the stochastic model operation of the process of interfusing of loose components along with the most widespread informational management [14] and cell-like by methods [15–17] is illustrated in the analysis of the degree of uniformity of the received mix. Use of power approach [9] for special purpose calculations [5] in the field of engineering chemistries can be considered how the development of their theoretical bases.

References

1. Kapranova, A.B., Verloka, I.I., Lebedev, A.E., Zaitzev, A.I.: The model of dispersion of particles during their flow from chipping the surface. In: Czasopismo techniczne, Vol. 113, no. 2. pp. 145–150. Mechanika. Krakov, Poland (2016)
2. Kapranova, A.B., Bakin M.N., Verloka I.I.: Simulation of the quality criterion of a mixture in a drum-belt apparatu. Chem. Pet. Eng. **54**(5–6), 287–297 (2018). https://doi.org/10.1007/s10556-018-0477-0. More about changing the spelling of the author's name Vol. 54, no. 7–8, pp. 618 (2018). https://doi.org/10.1007/s10556-018-0524-x
3. Kapranova, A.B., Verloka, I.I.: Stochastic description of the formation of flows of particulate components in apparatuses with brush elements. Theor. Found. Chem. Eng. **52**(6), 1004–1018 (2018). https://doi.org/10.1134/S0040579518050330
4. Verloka I., Kapranova A., Tarshis M., Cherpitsky S.: Stochastic modeling of bulk components batch mixing process in gravity apparatus. Int. J. Mech. Eng. Technol. (IJMET) **9**(2), 438–444; Chapter ID: IJMET_09_02_045. Available online at http://www.iaeme.com/IJMET/issues.asp?JType=IJMET&VType=9&IType=2 (2018). ISSN Print: 0976-6340 and ISSN Online: 0976-6359
5. Kapranova A.B., Verloka I.I., Yakovlev P.A., Bahaeva D.D.: Investigation of the quality of mixture at the first stage of work of the gravitational type apparatus. Russ. J. Gen. Chem. **62**(4), 48–50 (2018)
6. Metzger, M.J., Remy, B., Glasser, B.J.: All the Brazil nuts are not on top vibration induced granular size segregation of binary, ternary and multi-sized mixtures. Powder Technol. (205), 42–51 (2011)
7. Anchal, J., Matthew, J.M., Benjamin, J.G.: Effect of particle size distribution on segregation in vibrated systems. Powder Technol. **237**, 543–553 (2013)
8. Dehling, H.G., Gottschalk, T., Hoffmann, A.C.: Stochastic Modeling in Process Technology, 279 p. Elsevier Science, London, 2007 (2017)
9. Klimontovich, Yu.L.: Turbulent Motion and the Structure of Chaos: A New Approach to the Statistical Theory of Open Systems, 328 p. LENAND, Moscow (2014)
10. Protodyakonov, N.O., Bogdanov, S.R.: The statistical theory of transport phenomenas in processes of engineering chemistry, p. 400. Chemistry, Leningrad (1983)

11. Sun, L., Xu, W., Lu, H., Liu, G., Zhang, Q., Tang, Q., Zhang, T.: Simulated configurational temperature of particles and a model of constitutive relations of rapid-intermediate-dense granular flow based on generalized granular temperature. Int. J. Multiph. Flow 77, 1–18 (2015)
12. Zhuang, Y., Chen, X., Liu, D.: Stochastic bubble developing model combined with Markov process of particles for bubbling fluidized beds. Chem. Eng. J. **291**, 206–214 (2016)
13. Almendros-Ibanez, J.A., Sobrino, C., de Vega, M., Santana, D.: A new model for ejected particle velocity from erupting bubbles in 2-D fluidized beds. Chem. Eng. Sci. **61**, 5981–5990 (2016)
14. Borodulin, D.M., Sablinskii, A.I., Sukhorukov, D.V., Andryushkov, A.A.: Study of the operation of the mixing unit, consisting of two successively installed centrifugal continuous mixer, to obtain a mixture with a 1: 1000 mixing ratio of the components to be mixed by serial dilution. Vestn. KrasGAU (6), 178–185 (2013)
15. Alsayyad, T., Pershin, V., Pasko, A., Pasko, T.: Virtual modeling of particles two-step feeding. J. Phys. Conf. Series **1084**(1). 5 September 2018. No. 012005 (2018)
16. Mizonov, V.A., Balagurov, I., Berthiaux, H.C.: Gatumel Markov chain model of mixing kinetics for ternary mixture of dissimilar particulate solids. Particuology **31**, 80–86 (2016)
17. Zhukov, V.P., Belyakov, A.N.: Simulation of combined heterogeneous processes based on discrete models of the Boltzmann equation. Theor. Found. Chem. Eng. **51**(1), 88–93 (2017). https://doi.org/10.7868/S0040357117010158
18. Kendall, M., Stewars, A.: Multidimensional Statistical Analysis and Time Series: Monograph, 736 p. Science, Moscow (1976)
19. Zaytsev, A.I., Bytev, D.O.: Shock Processes in Dispersible and Film Systems, 196 p. Moscow: Chemistry (1994)
20. Shaul, S., Rabinovich, E., Kalman, H.: Generalized flow regime diagram of fluidized beds based on the height to bed diameter ratio. Powder Technol. **228**, 264–271 (2012)
21. Lim, K.S., Zhu, J.X., Grace, J.R.: Hydrodynamics of gas-solid fluidization. Int. J. Multiph. Flow **21**, 141–193 (1995)
22. Lingineni, S., Srinivasaraghavan, S.: Stochastic analysis of transmission of diabetic threshold using particular distribution. Int. J. Civil Eng. Technol. **8**(11), 492–499 (2017)
23. Akhmadiev, F.G., Nazipov, I.T.: Stochastic modeling of the kinetics of processing of heterogeneous systems. Theor. Found. Chem. Eng. **47**(2), 136 (2013). https://doi.org/10.1134/S0040579513020012
24. Zemskov, E.P.: Time-dependent particle-size distribution in comminution. Powder Technol. (7), 71–74 (1998)

Development of an Automated System for Monitoring and Diagnostics a Guided Robotic Vehicle

Alexander Bazhanov, Roman Vashchenko, Vasily Rubanov and Olga Bazhanova

Abstract This research considers the development of an automated system for monitoring and diagnostics of a controlled robotic vehicle. This system is based on an algorithmic approach performed by creating models in the form of fuzzy behavior charts. A key feature is the ability to represent a continuous change in time as a set of modes. While decomposing the diagnostics object (a robotic vehicle), there were created models of its main nodes in the form of fuzzy behavior charts of the second rank. For the diagnostics of the selected nodes, there were considered all possible faults, which also was compared with those ones that can be traced using the created models. Based on the tables of possible malfunctions of the electric motor, the battery, and the infrared proximity sensor, there was obtained a list of malfunctions and abnormal situations that can be organized in the knowledge base structure. Analysis of fault tables allowed developing the structure of an automated system for monitoring and diagnostics of abnormal and emergency situations. The obtained results (models and algorithms) were used to create a software product.

Keywords Mobile robots · Diagnostics of abnormal situations · Fuzzy behavior charts · Knowledge base

A. Bazhanov (✉) · R. Vashchenko · V. Rubanov · O. Bazhanova
Belgorod State Technological University N.a. V.G. Shukhov, 46 Kostukova St, Belgorod 308012, Russia
e-mail: all_exe@mail.ru

R. Vashchenko
e-mail: madrid.59@mail.ru

V. Rubanov
e-mail: vgrubanov@gmail.com

O. Bazhanova
e-mail: olga_ryn@mail.ru

A. G. Kravets et al. (eds.), *Cyber-Physical Systems: Advances in Design & Modelling*, Studies in Systems, Decision and Control 259,
https://doi.org/10.1007/978-3-030-32579-4_8

1 Introduction

The autonomous robotic vehicle is a device that performs various operations associated with spatial movements with loads and is capable of effective behavior in a changing environment [1–7]. Considering a robotic vehicle as a diagnostics object, we can say that it is a complex system consisting of a large number of heterogeneous components and subsystems [8–12].

The state diagnostics of autonomous robots is necessary as it is complex technical objects, and such an approach is one of the main methods of increasing the efficiency of their use. The diagnosis objective of the robot state is to identify and alert abnormal situations and malfunctions in the operation of the main elements and subsystems. Among these problems, functional diagnostics tasks are of great importance when monitoring and malfunctions diagnosing are carried out in real-time directly during the robot operation [13–15].

It is possible to significantly improve the diagnosing quality for such objects as robots and various vehicles by creating embedded (onboard) diagnostic devices.

Essentially, any diagnosis is a classification: the definition of the object state from source data (signals from location sensors, control, video surveillance, etc.) can be considered as a task of classifying depending on the output information transformation, where the classes are the different states of the same object.

Another possible method for diagnosing the sensors of the mobile vehicle is based on the analysis of mismatch signals arising between physical and virtual sensors (models of sensor operation).

The most promising direction is the creation of methods based on artificial intelligence techniques, in particular, diagnostics methods based on artificial neural networks and fuzzy logic. Thus, in [16] is described the usage of neural network diagnostics of the robot state according to the integral parameter. At the same time, it is necessary to note the shortcomings of neural networks application: the lack of methods for determining the network structure, the number of neurons in the hidden layer and the parameters of the activation function, the complexity of network training. A method for diagnosing sensors of mobile vehicles using fuzzy logic methods is presented in [17].

Given the above, it is proposed to create a diagnostic system that will simplify a human's task of tracking the object state. To create such systems, it is advisable to use the algorithmic approaches, in particular, based on the construction of the models in the form of fuzzy behavior charts. A feature of this approach is the ability to represent a continuous in time change of a variable as a set of modes [18, 19].

2 The Selection of Nodes

The development of a system for monitoring and diagnosing abnormal and emergency situations based on fuzzy behavior charts for a controlled vehicle implies the sequential implementation of the following stages [20–22]:

1. Representation of the control and diagnostics object in the form of nodes (decomposition).
2. Determination of disturbing factors for each node.
3. Construction of membership functions for each disturbing factor.
4. Composing a rule that is based on fuzzy behavior charts of nodes (FBCN).
5. Composing the failures base on each node.
6. Determination of abnormal situation or malfunction of a specific node, taking into account the state of the FBCN and the decision-making control formation.

The robotic vehicle is a combination of the control object (robotic trolley) and an onboard control device. When decomposing a diagnostic object, let us single out: brushless DC motors built into the wheels, energy elements—batteries, and internal structure control devices—a controller with electronic interfaces to peripherals, for which monitoring and diagnostics are carried out by testing.

The node is a part of an automated vehicle. This part corresponds to a specific output variable and includes all disturbing factors affecting it [23, 24].

There are the following nodes with their output variables highlighted:

1. The wheel.
2. The rechargeable battery.
3. The inductive sensor.
4. The infrared proximity sensor.

For built-in brushless DC motors (Fig. 1), the input variables are a motor voltage (U), rotor current (A) and the speed of rotation (T). The output variable is the moment on the shaft—μ (N m).

For the battery (Fig. 2), the following variables were chosen as input: temperature (t), rotation speed (T), battery voltage (U). The output variable is the battery capacity—C (Ah).

For an analog inductive sensor (Fig. 3), the input variables are a supply voltage (U_{sp}), temperature (t), current (I), and the output variable is a distance—L (mm).

For the infrared proximity sensor (Fig. 4), the following input variables are taken:

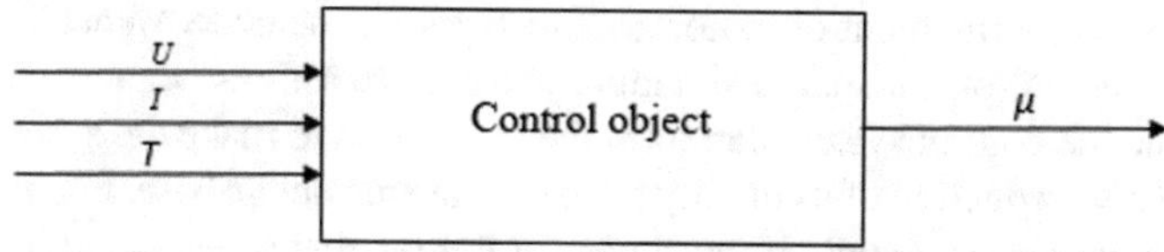

Fig. 1 The node "Wheel"

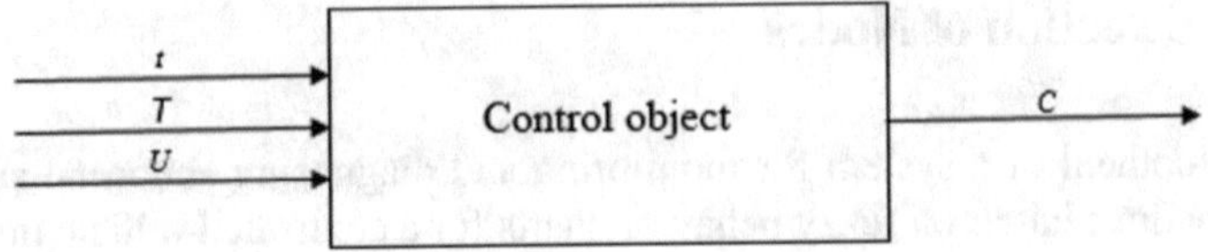

Fig. 2 The node “Rechargeable battery”

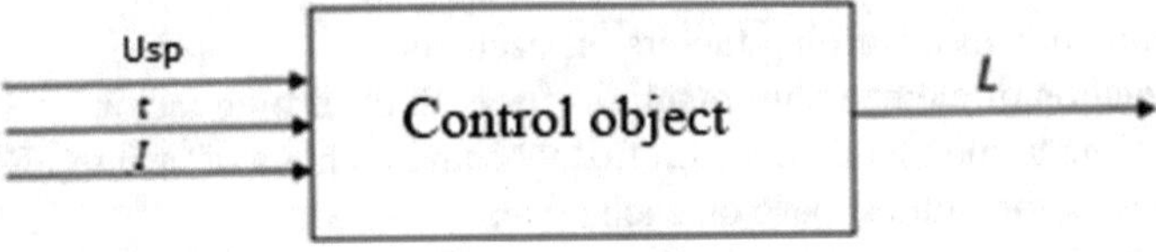

Fig. 3 The node “Inductive sensor”

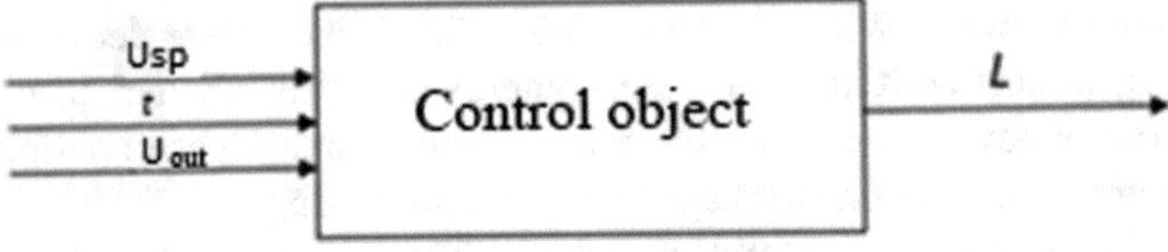

Fig. 4 The node “Infrared proximity sensor”

supply voltage (U_{sp}), temperature (t), output voltage (U_{out}). The output variable is a distance—L (mm).

3 Identification of Possible Faults for Each Node

There were created models in the form of fuzzy charts of the second rank for the selected nodes. In order to diagnose the malfunction of the presented nodes, it is necessary to consider all possible malfunctions, and also to compare them with those that can be traced using the created models in the form of the fuzzy behavior chart. The flexibility of this method lies in the fact that with the additional nodes and disturbing factors, a wider range of faults and abnormal situations can be diagnosed.

All faults for each node were considered and possible faults for diagnostics were selected.

Table 1 shows a fragment of motor malfunctions for the node “Wheel”.

Here are shown some standard situations and service actions when a malfunction happens, and the control system can form the suggestion to make the system able to work with some reduction of its quality by using the knowledge base. For example, if the motor could not rotate only if the power for it is on, that wheel could be powered off and let it operate in the freewheel mode with direction correction by the rest of wheels. Another example is when the motor couldn’t reach the set speed and heats.

Table 1 Drive malfunctions

Fault type	Cause of fault	Troubleshooting
When turning on the power, the rotor (anchor) is still fixed	There is no voltage at the input terminals of the motor or it is too low	Check the power supply line, repair the damage and ensure that the rated voltage is applied
When the power is on, the rotor is still fixed and intense heating	Bearing is destroyed; touching the rotor on the stator; the jammed shaft of the working mechanism	Disconnect the motor shaft from the shaft of the mechanism and turn on the engine again; if the motor shaft remains stationary, remove the motor
Motor stop	Voltage disconnected, motor protection was tripped	Find and eliminate the gap in the supply circuit, find out the cause of the protection operation (motor overload, the voltage in the network has changed significantly), eliminate it and turn on the motor
The motor does not reach the required speed, it is significantly overheated	Motor overloaded Bearing is out of service	Eliminate overload Replace bearing
The motor is overheating	Motor is overloaded The mains voltage is out of limits The outside temperature is higher than normal Motor ventilation is interrupted (air channels to the fan are blocked, engine surface is dirty)	Eliminate the overload Find out and eliminate the cause of the voltage deviation from the nominal Eliminate the cause and decrease the temperature to the acceptable value. Clean the air supply ventilating channels to the fan and clean the rotor surface
The motor is accompanied by a strong buzz, smoke appears	Turns of some stator winding coils have a shortcut	Send the motor to repair
The strong vibration of the motor	The fan wheel of the motor or another element mounted on the motor shaft is unbalanced	Eliminate unbalance of the fan or other element installed on the motor shaft
The bearing overheats, the noise is heard in it	Bearing and grease are dirty Bearing worn The alignment of the shaft of the engine and the working machine is broken	Remove the grease from the bearing, rinse it and lay down a new grease, replace the bearing, align the shafts

The suggestion could be to decrease the rotation speed to the not critical heating level or if it is not able then to power off and let it work in the freewheel mode.

3.1 Obtaining Knowledge Base and Identification of Abnormal and Emergency Situations with the Help of FBCN

When a table of all possible malfunctions is created for all major modes: the electric motor, the battery, and the infrared proximity sensor, it is created a list of malfunctions and abnormal situations that can be diagnosed using the suggested approach.

Each malfunction is associated with a state in a fuzzy behavior chart of a node and using indices, it indicates the terms to which the disturbing factors of a particular node in a certain state belong.

For the node "Wheel" (Fig. 5):

1. Motor stop—Status 1.
2. Slipping of one or several wheels—Condition 7.
3. The motor does not start—Status 2.
4. The motor does not reach the required speed—State 3.
5. The motor consumes high current during start-up—State 13.
6. The rotation frequency is higher than the nominal—State 6, 7.
7. The rotation frequency is less than the nominal—State 5, 9.

For the "Battery" node (Fig. 6):

1. Electrolyte freezing—Condition 1.
2. The internal circuit of the battery is open—State 2.
3. Fast discharging of the battery—Status 7.

For the "Infrared proximity sensor" node (Fig. 7):

1. The sensor is off—Status 1.
2. The termination of operation during normal operation of the power source—Status 9.
3. The inaccuracy of data—Status 7.

In this fuzzy behavior chart of the second rank, each position corresponds to a specific state that represents a specific fault and has a set of three values; this set is an index of the term belonging to one of *n*—disturbing influences.

Similar fuzzy behavior charts are constructed for the remaining nodes.

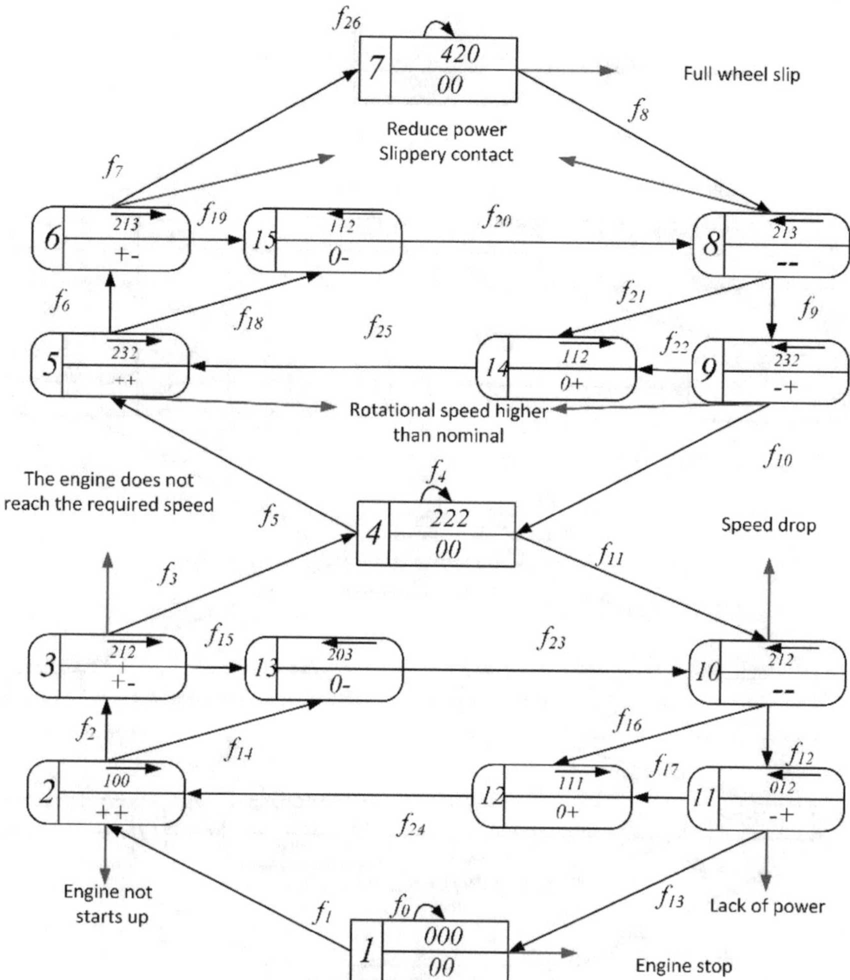

Fig. 5 FBCN mapped to the base of failures for the node "Wheel"

3.2 *The Development of the Diagnostic System Structure for Abnormal and Emergency Situations*

Analyzing the fault table, the structure of the system for diagnosing abnormal and emergency situations in the form can be shown in Fig. 8.

Wireless communication block. Responsible for transferring data from a robotic vehicle to the system of diagnostics and emergency situations, using LoRa modules.

Input block. Responsible for the processing of input data, that is, the PC records the received data into the array (the dimension corresponds to the number of disturbing factors) for further transfer to the main program.

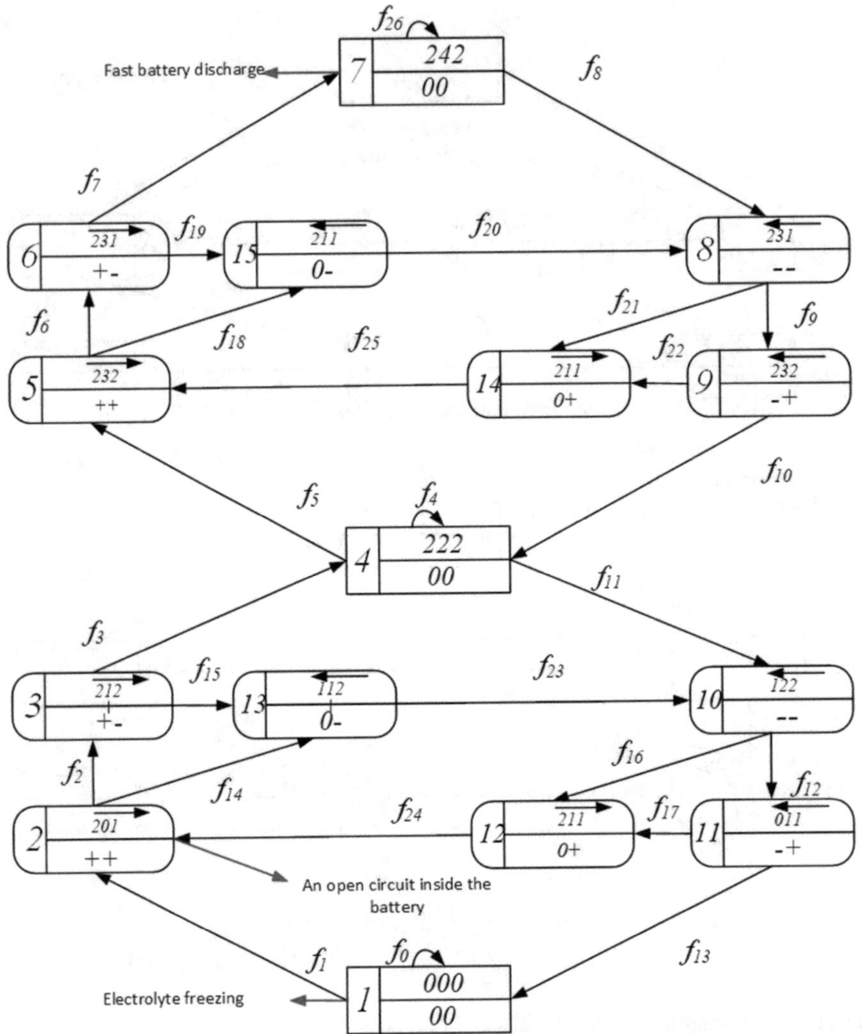

Fig. 6 FBCN mapped to the base of failures for the node "Battery"

Block of construction the membership functions. After entering the boundaries of the membership functions via the interface, they will be transferred to the "*trimf*" and "*trapmf*" methods together with the disturbing factor. These methods determine the numerical value of belonging to a particular term with given boundaries. These methods will be called in the next block.

Block for determining the degree of truth for each term. This block takes one of the disturbing factors as input, then passes it to membership functions, together with the boundaries for each of them, and returns the index of the term for which the degree of accepted membership value is greater than the rest.

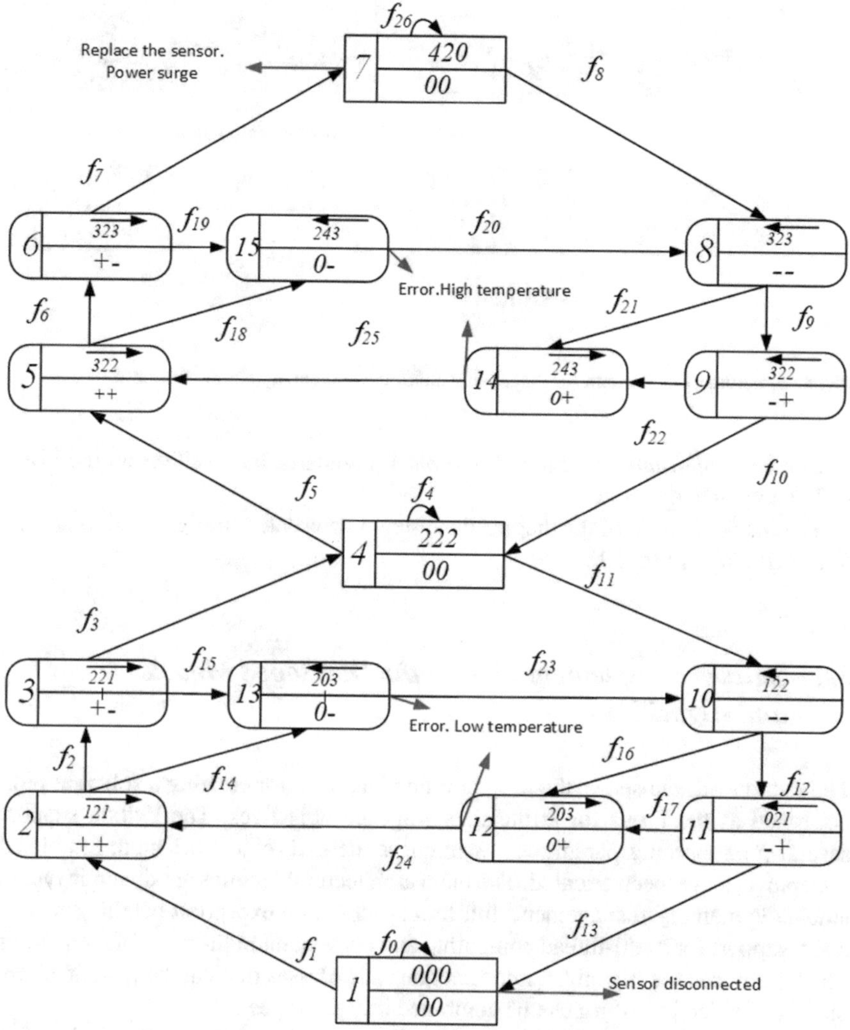

Fig. 7 FBCN mapped to the base of failures for the node "Infrared proximity sensor"

State definition block on FBCN. The block includes an array of Boolean states and rules for transition from one state to another. After determining the index for each disturbing factor, the combination of indexes that are used in the transition rules determines the state on the FBCN.

Knowledge base. It contains a list of faults and abnormal situations.

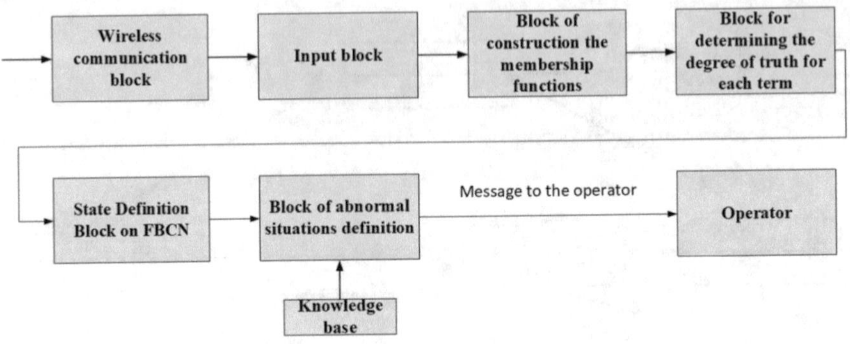

Fig. 8 Structure of the system for diagnosing abnormal and emergency situations

Block of abnormal situations definition. It compares the position on the FBCN with the knowledge base.

The block diagram of the diagnostic system is presented in Fig. 9 and its scheme of functioning in Fig. 10.

3.3 Software Implementation of the Developed Models and Algorithms

The Python was chosen as the programming language for creating a software product based as the fittest for artificial intelligence objectives. The Python supports several programming paradigms, including structural, object-oriented, functional, imperative, and aspect-oriented. The main architectural features are dynamic typing, automatic memory management, full introspection, an exception handling mechanism, support for multi-thread computing and convenient high-level data structures. The Python code is organized into functions and classes that can be combined into modules (which in its turn can be combined into packages).

Development and description of the software interface. The software interface consists of several windows. Figure 11 represents the main window in which could

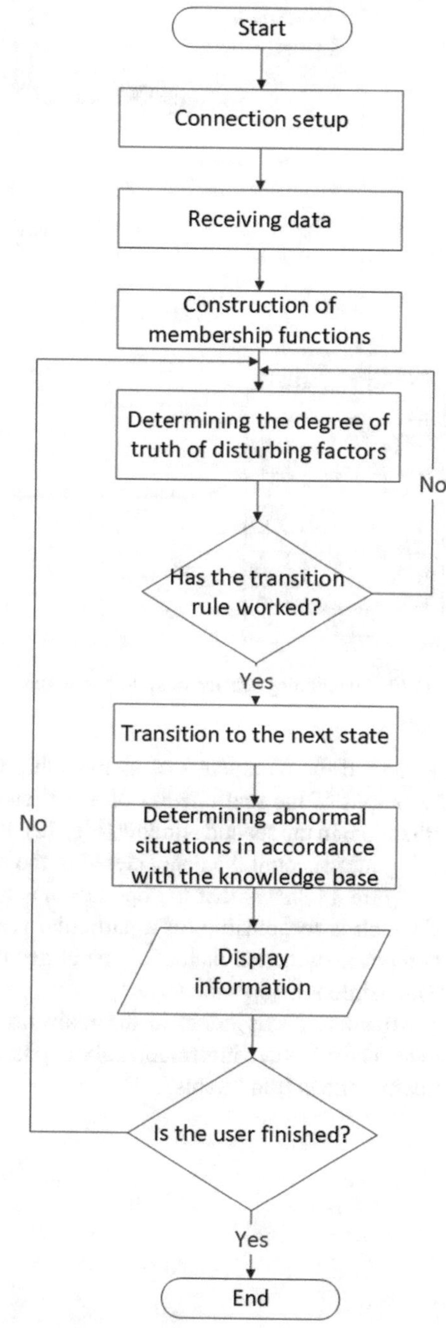

Fig. 9 Block diagram of functioning the automated system of monitoring and diagnosing abnormal and emergency situations

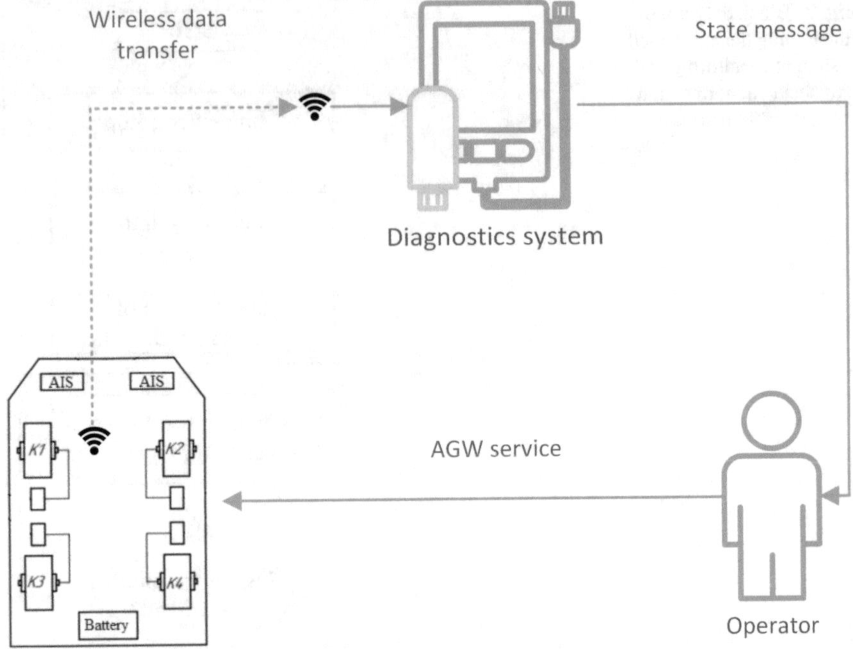

Fig. 10 Functioning scheme of system emergency situations diagnostics

be entered the parameters of membership functions for each disturbing factor and the output of the malfunction of a particular node, as well as possible emergency situations. In the second window (Fig. 12), the membership functions are constructed taking into account the data entered in the main window.

Figure 11 shows that the operator has a table for entering values, each element of which is responsible for a particular node (wheel, infrared sensor, battery) and its input disturbance factor (temperature, voltage, torque, etc.). Each input field is responsible for a specific term.

After transferring data to the main program, you can view the status of each node, as well as see information about possible abnormal or emergency situation by pressing the button "status".

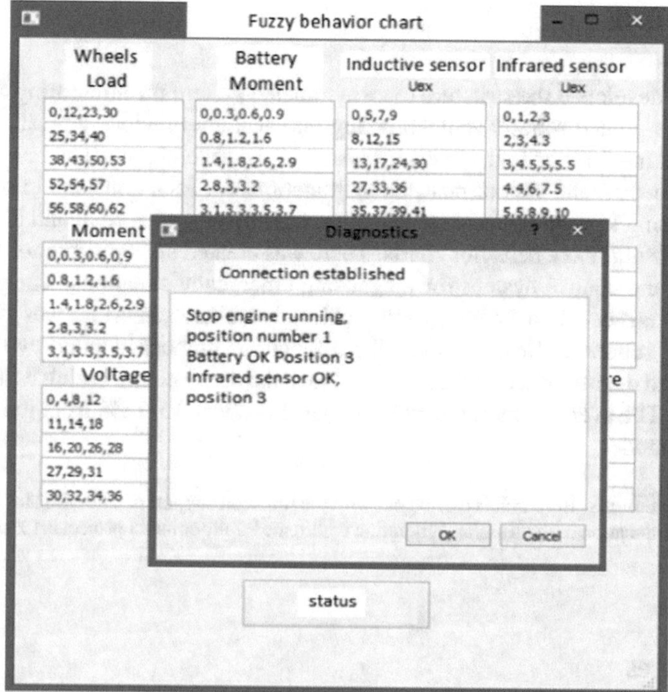

Fig. 11 Main software window

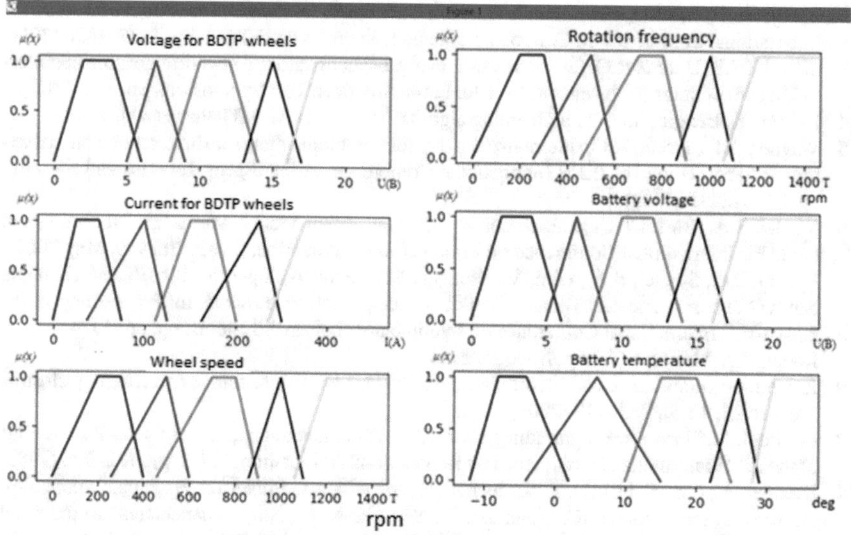

Fig. 12 Membership functions for each disturbing factor

4 Conclusion

There was developed the structure of an automated system for monitoring and diagnostics of a guided robotic vehicle, taking into account previously created models of nodes in the form of fuzzy behavior charts of the second rank. The research data allows to identify the area of possible malfunctions for each node and to obtain the corresponding knowledge base, which allows identifying abnormal and emergency situations using fuzzy behavior charts. There was created an algorithm for the functioning of the control system for diagnosing emergency situations. The proposed software was created on the basis of the achieved results. Its interface was also tested on the real robotic trolley. The overall malfunctions number is about two hundred and the third quarter of it could be resolved to make the system working without critical stops. The average quality decreasing level is about 15–18% from the faultless operation mode.

Acknowledgements Research is carried out with the financial support of The Ministry of Science and Higher Education of the Russian Federation within the Public contract project 2.1396.2017/4.6.

References

1. Khalastchi, E., Kalech, M.: On fault detection and diagnosis in robotic systems. ACM Comput. Surv. **51**(1), art. no. 9 (2018)
2. Yazdjerdi, P., Meskin, N.: Actuator fault tolerant control in a team of mobile robots. In: 15th International Conference on Control, Automation, Robotics and Vision, pp. 1885–1890 (2018)
3. Silva, G.N.P., Duarte, R.O. (2018) Towards evolvable hardware and genetic algorithm operators to fail safe systems achievement. In: 19th Latin-American Test Symposium, pp. 1–4 (2018)
4. LeCun, Y., Bengio, Y.: Scaling learning algorithms towards AI. MIT Press (2007)
5. Wagner, A.R., Arkin, R.C.: Internalized plans for communication-sensitive robot team behaviors. In: Proceedings of IEEE International Conference on Intelligent Robotics and Systems, pp. 2480–2487 (2003)
6. Stoeter, S.A., Burt, I.T, Papanikolopoulos N.: Scout Robot Motion Model. In: Proceedings of the IEEE International Conference on Robotics and Automation, Taipei, Taiwan, May 2003
7. Rybski, P.E., Stoeter, S.A., Gini, M., Hougen, D.F., Papanikolopoulos, N.: Effects of limited bandwidth communications channels of the control of multiple robots. In: Proceedings of the 2001 IEEE International Conference on Intelligent Robots and Systems, pp. 369–374
8. RoboCV X-Motion NG. http://robotrends.ru/r
9. Roland, L.: Dunbrack, Cyclic coordinate descent: a robotics algorithm for protein loop closure. Protein Sci. **12**(5), 963–972 (2003)
10. Baillieul, J.: Kinematic programming alternatives for redundant manipulators. In: Proceedings of the IEEE International Conference on Robotics and Automation, Vol. 2, pp. 722–728 (1985)
11. Drenner, A., Burt, I., Dahlin, T., Kratochvil, B., McMillen, C.P., Nelson, B., Papanikolopoulos, N., Rybski, P.E., Stubbs, K., Waletzko, D., Yesin K.B.: Mobility enhancements to the scout robot platform. In: Proceedings of the 2002 IEEE International Conference on Robotics and Automation, Washington, DC, May 2002
12. Lee, B., Ehsani, M.: Advanced simulation model for brushless DC motor drives. Electr. Power Compon. Syst. **31**(9), 841–868 (2003)

13. Abramov, I.V., Nikitin, Y.R., Abramov, A.I., Sosnovich, E.V., Bozek, P.: Control and diagnostic model of brushless DC motor. J. Electr. Eng. **65**(5), 277–282 (2014)
14. Lorincz, R.I., Basch, M.E., Bogdanov, I., Tiponut, V., Beschieru, A.: Hardware implementation of BLDC motor and control system diagnosis. Int. J. Circuits Syst. Signal Process **5**(6), 660–671
15. Harnefors, L.: Control of variable-speed drives. Applied Signal Processing and Control, Department of Electronics, Mälardalen University, Västerås (2002)
16. Ang, K., Chong, G., Li, Y.: PID control system analysis, design and technology. IEEE Trans. Control Syst. Technol. **13**, 559–576 (2005)
17. Steinbauer, G.: A survey about faults of robots used in RoboCup. In: RoboCup 2012: Robot Soccer World Cup XVI, pp. 344–355. Springer, Berlin (2013)
18. Christensen, A.L., O'Grady, R., Birattari, M., Dorigo, M.: Fault detection in autonomous robots based on fault injection and learning. Auton. Robots **24**, 49–67 (2008)
19. Timofeev, A.V.: Functional diagnostics and fault-stable control for mechatronic systems and robots. Tr. SPIIRAN **2**(1), 266–283 (2004)
20. Bazhanov, A.G., Vashchenko, R.A., Magergut, V.Z.: Fuzzy nodes behavior charts for complex technological objects, principles of its construction and usage. Instrum. Syst. Monit. Control Diagn. **9**, 26–34 (2014)
21. Magergut, V.Z., Ignatenko, V.A., Bazhanov, A.G., Shaptala, V.G.: Approaches to construction of discrete models of continuous technological processes for synthesis control machines. Bull. BSTU Named After V. G. Shukhov **2**, 100–102 (2013)
22. Magergut, V.Z., Bazhanov, A.G., Vashchenko, R.A.: Algorithmic approach to synthesis fuzzy control systems for objects with continuous technology. World Appl. Sci. J. **24**(10), 1291–1295 (2013)
23. Bazhanov, A.G., Magergut, V.Z., Vashchenko, R.A.: Operation model of the cement kiln node "material temperature in the drying zone" as a fuzzy behavior chart. In: Proceedings of the International Conference on Information and Digital Technologies, IEEE Xplorepp, pp. 35–38. Zilina, Slovakia (2015)
24. Vashchenko, R.A., Bazhanov, A.G., Magergut, V.Z., Stepovoy, A.A.: Application of the model based on fuzzy behavior charts in the advising control system of rotary cement kiln. In: 2016 International Conference on Proceedings of Information and Digital Technologies (IDT), Rzeszow, Poland, pp. 299–304 (2016)

About Formation of Elements of a Cyber-Physical System for Efficient Throttling of Fluid in an Axial Valve

A. B. Kapranova, A. E. Lebedev, A. M. Melzer and S. V. Neklyudov

Abstract An example of the calculation of an axial valve separator, as one of the stages of the formation of elements of a cyber-physical system designed for the effective throttling of a fluid, is performed. The engineering methodology of the authors for calculating the design parameters of the axial valve with the external location of the locking part was chosen as the basis. When obtaining the results, the author's models are used that describe hydrodynamic cavitation at its initial stage in a separator of an axial valve.

Keywords Cyber-physical system · Gravity mixer · Bulk components · Brush elements · Distribution functions · Process parameters

1 Introduction

Analysis of the current state of the problem of designing regulatory equipment for pipeline systems has revealed a number of difficulties related to the choice of appropriate engineering methods, the possibility of upgrading them when changing operating modes of regulatory devices and adapting to specific operating conditions. At the same time, the use of ready-made calculated hydrodynamic software products is insufficient. There is a need for the formation of a special cyber-physical system for calculating the parameters of the process of throttling the flow of liquid media, which ensures a decrease in pressure in the pressure pipeline systems due to the flow of fluid

A. B. Kapranova (✉) · A. E. Lebedev · A. M. Melzer · S. V. Neklyudov
Yaroslavl State Technical University, Moskovskiy Prospect, 88, Yaroslavl 150023, Russia
e-mail: kapranova_anna@mail.ru

A. E. Lebedev
e-mail: lae4444@gmail.ru

A. M. Melzer
e-mail: meltzer.a@mail.ru

S. V. Neklyudov
e-mail: neklydov.s@nporeg.ru

A. G. Kravets et al. (eds.), *Cyber-Physical Systems: Advances in Design & Modelling*, Studies in Systems, Decision and Control 259,
https://doi.org/10.1007/978-3-030-32579-4_9

through the constriction channel. The development of an engineering methodology for calculating the parameters of pipeline valves requires a preliminary analysis not only of the hydrodynamic picture of the process of throttling the fluid in the flow part of the valve, but also of individual elements of this cyber-physical system (CFS) associated with the choice of many parameters of the designed technological process. The decisive role in this is played by the system-structural analysis [1] of the results of theoretical and experimental studies [2–4]. One should not forget about taking into account the physical and chemical properties of the working fluid, the degree of its aggressiveness with respect to the internal surfaces of control valves, the presence or absence of dispersed inclusions. These issues require the study at each stage of the formation of the corresponding cyber-physical system for calculating the parameters of this technological process.

The urgency of designing new regulatory equipment for pipeline systems is confirmed by the need to comply with the national valve industry international standards in the face of fierce competition. At the same time, the problem of reducing the intensity of cavitation phenomena has several directions of the solution, one of which is the branching of the flow of the working medium [5]. Axial valves for this purpose use separators with throttling holes. At the same time, due to the design features of the axial valves, conditions are created for displacing the cavitation bubbles that form in the zone of the separator axis with their further withdrawal from the working volume. The description of the conditions of formation of the system of cavitation bubbles and their further behavior in the flow part of the valve being designed is of decisive character in the preparation of the analytical platform of the cyber-physical system. The description of the conditions of formation of the system of cavitation bubbles and their further behavior in the flow part of the valve being designed is of decisive character in the preparation of the analytical platform of the cyber-physical system. Note that the analysis of modern mathematical models for describing the evolution of cavitation bubbles [2] showed the possibility of using not only two classical approaches (deterministic [6] and stochastic for nucleation models: homogeneous [7, 8], with heterogeneity factor [9, 10], heterogeneous [11–14]), but also their combined variants [15–19]. In this work, preference is given to the stochastic method of forming a mathematical model of the formation of cavitation bubbles [20] in the indicated working volumes [21], which is associated with the necessity of evaluating the integral characteristics of the ensemble of this macro-cavity system [22, 23] and modeling the distribution functions of bubbles without postulating this law.

2 Features of the Structure of a Cyber-Physical System for Calculating the Parameters of the Process of Throttling Fluid Flows

Considering the operation of the regulating device from the standpoint of the process of throttling of the fluid flow, let us single out the main stages of this technological operation. At the valve inlet, the working fluid is supplied at a certain value of the medium velocity w in the pipeline with the specified maximum volumetric flow rate Q_{1max} and minimum pressure drop Δp_{min} at the fluid temperature t_1. It is believed that the valve should provide the required flow capacity K'_{vy}. It is believed that the valve should provide the required throughput. According to a special definition, the value K'_{vy} corresponds to the volume flow of water at $\Delta p_{min} = 10^5$ Pa.

In the process of reducing the pressure in the working volume of the valve, vacuum areas appear and cavities are formed, filled after the formation of spherical bubbles, subsequently with steam and gas dissolved in a liquid.

Hydrodynamic cavitation is observed, the evolution of which has negative consequences for the physical state of the valve and the mode of its operation. The presence of a system of cavitation bubbles that are formed requires consideration of this effect when designing a regulating device. Hydrodynamic cavitation is observed, the evolution of which has negative consequences for the physical state of the valve and the mode of its operation. The presence of a system of cavitation bubbles that are formed requires consideration of this effect when designing a regulating device.

Perform an example of calculating some set of projected parameters of an axial valve separator with an external location of the locking member [21]. According to the engineering methodology proposed by the authors for calculating the design parameters based on the performed theoretical and experimental studies [2–4], the following numerical values of the mode parameters. Perform an example of calculating some set of projected parameters of an axial valve separator with an external location of the locking member [21]. According to the engineering methodology proposed by the authors for calculating the design parameters based on the performed theoretical and experimental studies [2–4], the following numerical values of the mode parameters.

$$x = \{x_{k_2} = const\} = \{Q_{1max}, \delta_{\Delta P_{min}}, \delta_{\Delta P_{max}}, t_1, w\}, k_2 = \overline{1, \rho_2} \tag{1}$$

The output parameter is K'_{vy},

$$y = \{y_{k_5}\} = \{K'_{vy}\}, k_5 = \overline{1, \rho_5} \tag{2}$$

A set of additional characteristics covers the indicators of the physical and chemical properties of the working fluid.

$$\varphi = \{\varphi_{k_3}\}, k_3 = \overline{1, \rho_3}. \tag{3}$$

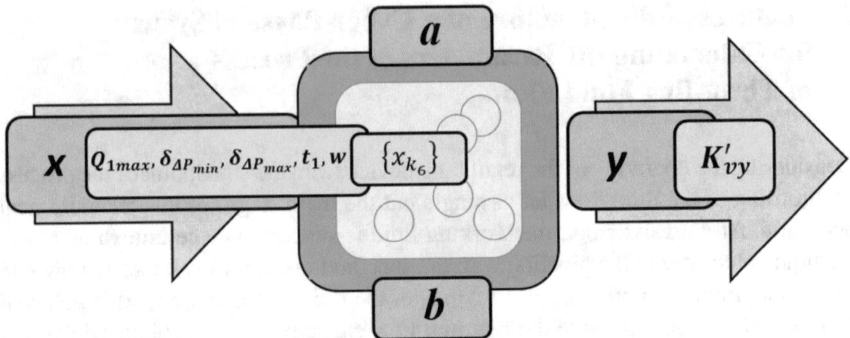

Fig. 1 Schematic diagram of the device for throttling fluid flow

The sets of design parameters of the process of throttling fluid flow in the valve $a = \{a_{k1} = cont\}$, $k_1 = \overline{1, \rho_1}$ cover the geometrical parameters of its design. The set of operational parameters includes $b = \{b_{k_4}\} = \{x_{k_2}, x_{k_6}\}$, $k_6 = \overline{1, \rho_6}$ where $\{x_{k_6}\}$ is a set of characteristic parameters of the cavitation system bubbles, for example, the minimum radius value; the pressure values of the bubble at its center are maximum and characteristic of the average radius value. The corresponding schematic diagram is shown in Fig. 1.

A schematic diagram of the calculation of the process of throttling in the valve using $\{x, y, a, b\}$, as a set of input, output, design, and regime parameters is shown in Fig. 2. Note that this set of z is divided into three conditional groups:

- basic z_m (calculated from models taking into account the characteristics of the properties of the working fluid φ from the expression (3));
- project regulated z_{pr} (constructive and technological);
- design optimizing z_{po} (replaced by rational values of z_{pe} within defined effective ranges of change).

3 Description of the Input Parameters of the Studied Process

Let the axial control valve with the external location of the locking part has a working fluid—water. For the flow of water, we specify:

- consumption parameters (the maximum achievable flow rate $Q_{1max} = 0.5\,\text{m}^3/\text{h}$);
- limit stresses (minimum pressure drop $\Delta p_{min} = 0.90\,\text{kPa}$);
- physical and mechanical properties (the density $\rho_L = 10^3\,\text{kg/m}^3$; the volume weight $\gamma = 0.995\,\text{g}\cdot\text{s/c}$ is the surface tension coefficient m^3; the temperature $t_1 = 30\,°\text{C}$; the kinematic viscosity $\nu_1 = 0.81 \times 10^{-2}\,\text{cm}^2/\text{s}$; $\sigma = 7.28 \times 10^{-4}\,\text{H/m}$ is the surface tension coefficient);
- kinematic indicators (the velocity of the fluid in the pipeline $w = 0.43\,\text{m/s}$).

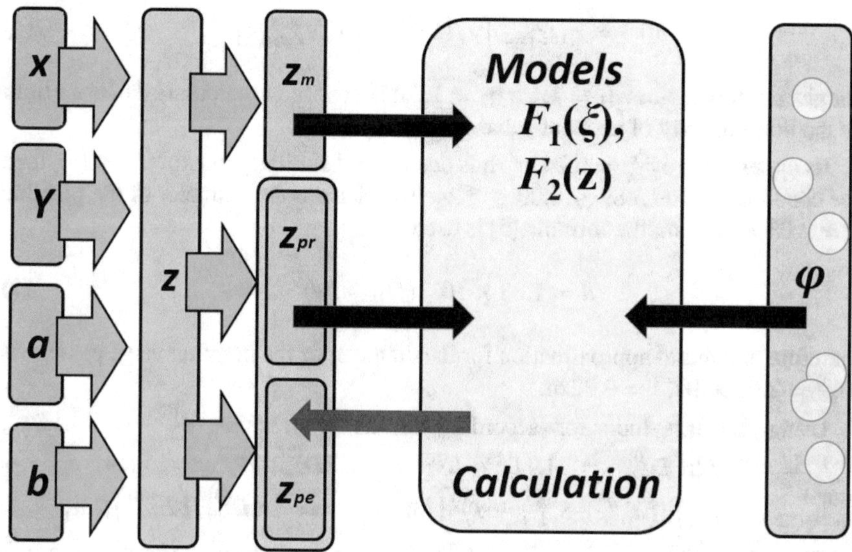

Fig. 2 Schematic diagram of the calculation of the process of throttling in the valve using information variables

For cavitation bubbles and their internal gas-vapor system, factors are considered known [2–4]:

- minimum radius $r_{min} = 10^{-3}$ m at the maximum pressure in its center $p_{max} = 1.3 \times 10^5$ кПа;
- pressure $p_s = 10^{-3}$ Pa in this center with average radius;
- air and steam density $\rho_g = 1.205\,\text{kg/m}^3$; $p_s = 1,44 \times 10^{-2}\,\text{kg/m}^3$;
- adiabatic index $k = 1.3$.

The output design parameters are the main design parameters of a cylindrical separator with choke holes [21], a shell in the form of a coaxial cylinder external to the separator, housing elements, as well as a routine value of valve flow capacity $K'_{vy} = 6\,\text{m}^3/\text{h}$.

4 An Example of the Calculation of the Projected Parameters of an Axial Valve Separator with an External Location of the Locking Part

Let at $t_1 = 30$ °C it is required to provide the following limit values for the differential pressure Δp_{min}:$\delta^{min}_{\Delta p_{min}} = 0.85$ kPa and $\delta^{max}_{\Delta p_{min}} = 0.92$ kPa.

Using the empirical formula from known literature sources [5]

$$\delta_{K_{vy1}}^{min} = \eta_1 Q_{1max}\left[\gamma/\left(1.02 \times 10^{-5}\delta_{\Delta p_{min}}^{max}\right)\right]^{1/2}, \tag{4}$$

for given safety factors $\eta_1 = 1.1$ и $\eta_2 = 1.2$ it is possible to calculate the lower limit of the flow capacity of an axial valve $\delta_{K_{vy1}}^{min} = 5.89\,\text{m}^3/\text{h}$.

Its upper value $\delta_{K_{vy1}}^{min} = 6.24\,\text{m}^3/\text{h}$ is determined similarly to expression (3), then the condition is true $5.89 \le 6.00 \le 6.24$. To calculate the diameter of the pipeline $d = 2.03 \times 10^{-2}\,\text{m}$, the formula [5] is used

$$d = 1.88 \times 10^{-2}(Q_{1max}/w)^{1/2}, \tag{5}$$

therefore, the initial approximation for the diameter of the nominal valve passage is $D_y^{(0)} = 2.03 \times 10^{-2} = 0.02\,\text{m}$.

Using the last indicator according to the expressions: $\delta_2^{(0)} = 0.1D_y^{(0)}$; $h^{(0)} = \delta_2^{(0)}/2$; $D_{out}^{(0)} = 1.1D_y^{(0)}$; $L^{(0)} = 2.2D_{out}^{(0)}$; $d_0^{(0)} = 0.14$; $S_{\varphi 1}^{(0)} = d_0^{(0)}$; $l_0^{(0)} = \left[3\left(d_0^{(0)} + S_{\varphi 1}^{(0)}\right) - d_0^{(0)}\right]/3$; $u_2^{(0)} = \pi D_{out}^{(0)}/\left(2d_0^{(0)}\right)$; $u_1^{(0)} = \left(L^{(0)} - l_0^{(0)}\right)/\left(l_0^{(0)} + d_0^{(0)}\right)$; $D_c^{(0)} = D_{out}^{(0)} + h^{(0)}$; $D_{ch2}^{(0)} = 3D_{out}^{(0)}/2$; $D_{cas2}^{(0)} = 2D_{out}^{(0)}$ the calculation of values for a set of the desired parameters in the 0-approximation is carried out.

These parameters include:

- the thickness of the body and separator $\delta_2^{(0)} = 3.0 \times 10^{-3}\,\text{m}$ and $h^{(0)} = 1.5 \times 10^{-3}\,\text{m}$;
- internal $D_{out}^{(0)} = 3.3 \times 10^{-2}\,\text{m}$ and external $D_c^{(0)} = 3.15 \times 10^{-2}\,\text{m}$ separator diameters;
- the separator length $L^{(0)} = 2.2D_{out}^{(0)} = 7.26 \times 10^{-2}\,\text{m}$;
- the diameter of the throttle holes $d_0^{(0)} = 1.14D_{out}^{(0)} = 4.62 \times 10^{-3}\,\text{m}$;
- the distance between them in the same row on the surface of the separator $S_{\varphi 1}^{(0)} = 4.62 \times 10^{-3}\,\text{m}$;
- the distances between these rows $l_0^{(0)} = 5.68 \times 10^{-3}\,\text{m}$;
- the total number of rows $u_2^{(0)} = [6.49] = 7$;
- the number of holes in one row $u_1^{(0)} = [11.23] = 12$;
- the external diameter of the internal valve body $D_{ch2}^{(0)} = 4.95 \times 10^{-2}\,\text{m}$;
- outer diameter for the cylindrical parts of the outer body of the axial valve $D_{cas2}^{(0)} = 6.6 \times 10^{-2}\,\text{m}$;
- the v angle of shell bevel (outer movable cylinder) $\alpha^{(0)} = 45°$.

The calculation of the critical value of the Reynolds number $\text{Re}_{kp} = 5209.58$ which is included in the condition of the valve-free cavitation mode $\text{Re}_y \le \text{Re}_{kp}$ is performed using the formula obtained using the stochastic model of cavitation bubble formation at the initial stage of hydrodynamic cavitation [2, 3, 20]

$$Re_{kp} = 32\rho_L k_{\zeta 1} \frac{u_1^{(0)} h_0^{(0)} \left(D_y^{(0)}\right)^2 \left[2\left(L^{(0)} - l_0^{(0)}\right) - \beta_0^{(0)}\right]}{\left\{\beta_0^{(0)} \left(d_0^{(0)}\right)^3 \left[4a_1 r_{min}^4 + \rho_L k_{\zeta 1}\left(B_1 + \beta_2^{(0)}\right)\right]\right\}} \quad (6)$$

where the designated coefficients $k_{\zeta 1}$, a_1, B_1, $\beta_0^{(0)}$, $\beta_2^{(0)}$ depend on the design parameters of the medium and its physical and mechanical properties.

Note that formula (6) is obtained from the extremum condition for the energy parameter $E_{02}(z)$, described in [3] by the degree of opening of the axial valve z at the moment of its full opening ($z = 1$). Then the calculation of the 1st approximation for the diameter of the conditional passage $D_y^{(1)} = 4.22 \times 10^{-2}$ m and the limits of its change $\delta_{D_y}^{min} = 4.19 \times 10^{-2}$ and $\delta_{D_y}^{max} = 4.28 \times 10^{-2}$ can be carried out respectively by the experimental formulas from [5]

$$D_y^{(1)} = 3.53 Q_{1max} / \left(v_1 Re_{kp}\right), \quad (7)$$

$$\delta_{D_y}^{max} = \left[4\delta_{K_{vy1}}^{max} \zeta_{12}^{(0)} \left[z\left(u_2^{(0)}\right)\right] / \left(5.04 \times 10^4 \pi\right)\right]^{1/2} \quad (8)$$

where the values of Q_{1max}, v_1 are input parameters of the calculation.

Here is marked $\zeta_{12}^{(0)}\left[z\left(u_2^{(0)}\right)\right]$—the value of the hydraulic resistance coefficient depending on the degree of opening of the valve z, as a function of $u_2^{(0)}$—values of 0 approximation for separator parameter (the number of rows of choke holes on its surface). Note that the calculation of the value $\delta_{D_y}^{min}$ is performed similarly to (4).

Due to the non-fulfillment of the condition $Re_{kp} \le 2 \times 10^3$ from [5], the calculation of correction factors ψ_1 and ψ_2 for the viscosity of the working medium for the limits of change in valve flow capacity $K_{vmax}^{(2)} \in \left[\delta_{K_{vy2}}^{min}; \delta_{K_{vy2}}^{max}\right]$ is not required and you can proceed to estimate the values of the 1st approximations for the limits of changing the following design parameters of the separator $d_{0j}^{(1)}$, $u_{2j}^{(1)}$, $L_j^{(1)}$, $l_{0j}^{(1)}$ the values $j = 1, 2$ correspond to $\delta_{D_y}^{min}$, $\delta_{D_y}^{max}$ to the limits of change of the value $D_y^{(1)}$ (m) according to (6).

The averaged values $\bar{u}_2^{(1)}$, $\bar{d}_0^{(1)}$, $\bar{l}_0^{(1)}$, $\bar{L}^{(1)}$ of the required parameters are determined according to [22]: $\bar{u}_2^{(1)} = [7.54] = 7$; $\bar{d}_0^{(1)} = 5.05 \times 10^{-3}$ m; $\bar{l}_0^{(1)} = 5.84 \times 10^{-3}$ m; $\bar{L}^{(1)} = 8.21 \times 10^{-2}$ m.

Search $\bar{D}_{out}^{(1)} = 4.24 \times 10^{-2}$ m; $S_{\varphi}^{(1)} = 6.03 \times 10^{-3}$ m the values of the 1st approximation for the internal movement of the separator $\bar{D}_{out}^{(1)}$ and the settings for $S_{\varphi 1}^{(1)}$ can be produce using a system of equations. One of them reflects the condition of the linear profiling of the valve's throughput characteristic of the degree of its opening, and the second, the extremum condition for the hydraulic resistance coefficient for the u_2 parameter according to [5].

The value of the 2nd approximation for the internal diameter of the separator $D_{out}^{(2)} = 4.19 \times 10^{-2}$ m and the 1st approximation of the bevel angle for the cylindrical part of the shell $\alpha^{(1)} = 44.84°$ are obtained from two consequences for the condition of the extremum of the energy parameter according to the degree of valve opening [4], for example, in the form

$$\alpha^{(1)} = \frac{\left\{\left[\beta_0^{(1)} - 2\left(\bar{L}^{(1)} - l_0^{(2)}\right)\right] + 2\beta_0^{(1)}[\theta_1 - \theta_4]\right\}}{\left[\beta_1^{(1)}\left(2\beta_0^{(1)}\theta_2\right)\right]} \tag{9}$$

where $\theta_1, \theta_2, \theta_4$ are coefficients depending on the design parameters of the environment and its physical and mechanical properties, as well as the values $\beta_0^{(0)}, \beta_1^{(0)}, \beta_2^{(0)}$ described earlier in (2).

According to [20], the calculation is performed for the 1st approximations of the following parameters: $\delta_2^{(1)} = 4.22 \times 10^{-3}$ m; $h^{(1)} = 2.11 \times 10^{-3}$ m; $D_c^{(1)} = 4.13 \times 10^{-2}$ m; $D_{ch2}^{(1)} = 6.25 \times 10^{-2}$ m; $D_{cas2}^{(1)} = 8.69 \times 10^{-2}$ m. Having calculated the value of the hydraulic resistance coefficient according to the formulas $\zeta_y\left[z\left(\bar{u}_2^{(1)}\right)\right] = 166.99$ proposed in [4], we can proceed to the determination of its critical value ζ_y^* from the condition of minimal r_{sb}^{min} for the average the ensemble values of the radius of the cavitation bubble according to [4, 22, 23,], then $\zeta_y^* = 1.19 \times 10^{-1}$. Checking the condition $\zeta_y^* \leq \zeta_y$, which reflects the possible manifestation of the cavitation effect in the flow part of the valve, allows you to evaluate the critical parameter $k_{C.max} = 0.06$ according to the graph $k_C(\zeta_y^*)$ of the cavitation number [5] dependence ζ_y^*. Consequently, for a given absolute pressure of the fluid up to the control valve at its maximum flow, the following values were obtained: critical (maximum allowable) pressure drop $\Delta p_{max}^* = 0.899$ kPa $= 9.179 \times 10^{-3}$ kg c/cm^2 from the classic formulas [5]

$$\Delta p_{max}^* = k_{C.max}(p_1 - p_{H1}), \tag{10}$$

where values are used for absolute pressures p_1 for liquids (up to the valve at maximum flow) and p_{H1} for saturated steam at a given temperature.

Using the formula of [5]

$$K_{vmax} = \eta_{12} Q_{1max}\left(\gamma / \Delta p_{max}^*\right)^{1/2}, \tag{11}$$

with a safety factor of η_{12}, you can set $K_{vmax} = 5.21$ m^3/h.

Thus, due to the fulfillment of the condition $K_{vmax} \leq K'_{vy}$, the following approximations of the axial valve parameters are the sought ones:

- for separator: $\bar{d}_0^{(1)} = 5.05 \times 10^{-3}$ m; $S_{\varphi 1}^{(1)} = 6.03 \times 10^{-3}$ m; $h^{(1)} = 2.11 \times 10^{-3}$ m; $\bar{u}_1^{(0)} = 12$; $\bar{u}_2^{(1)} = 7$; $\bar{l}_0^{(1)} = 5.84 \times 10^{-3}$ m; $D_{out}^{(2)} = 4.19 \times 10^{-2}$ m; $D_c^{(1)} = 4.55 \times 10^{-2}$ m; $\bar{L}^{(1)} = 8.21 \times 10^{-2}$ m;
- for shell: $\alpha^{(1)} = 44.84^{\circ}$;
- for housing: $D_{ch2}^{(1)} = 6.25 \times 10^{-2}$ m; $D_{cas2}^{(1)} = 8.69 \times 10^{-2}$ m; $\delta_2^{(1)} = 4.22 \times 10^{-3}$m.

5 Main Findings and Results

The choice of the numerical values of the input parameters of the calculation for the projected cyber-physical system is determined by the physicomechanical properties of the working fluid and the regulated value of the conditional valve capacity. In this chapter, water is chosen as the working fluid, which is traditional for testing regulating devices under production conditions. In the general case, a change in the working fluid is determined by the customer's regulations and entails a recalculation of the rational parameters of the regulating device according to the proposed engineering method.

In particular, according to the proposed method, the allowable ranges for changing the values of the main physicomechanical characteristics of a liquid medium are the following: $\nu_1 = (0.81–0.85) \times 10^{-2}$ cm/c for kinematic viscosity and $\rho = (1.0–1.2) \times 10^3$ kg/m^3 for the true density of the liquid. Note that for aggressive media in this engineering methodology, additional correction factors are required.

Changing the shape of the choke holes of the separator will lead to recalculation of the diameter of the nominal valve [2, 20]. In the future, this will be reflected in the assessment of the hydraulic resistance coefficient [3, 4, 23] and the results of the calculation using formulas (6)–(8) due to changes in the expressions for the differential distribution functions of the number of bubbles [2] and the opening degree of the valve [3].

Note that in the presented example, according to the calculations performed, a double check of the intervals of change in valve capacity is carried out. Initially, this was caused by the need to check the condition of compliance of the calculated intervals of change in valve throughput with its prescribed value. Further evaluation is carried out:

- the diameter of conditional passage in the initial;
- setting 0-approximations for a set of constructive and regime parameters;
- calculation of the critical value of the Reynolds number $Re_{кр}$ for the cavitation free flow regime;
- calculation of new limits of change in the diameter of the conditional passage;
- a new comparison of the Reynolds number values—calculated and characteristic.

If the calculated value of the Reynolds number is less than or equal to the characteristic value, then the calculation of correction factors for the viscosity of the working medium is required for the limits of change in valve flow capacity taking into account the first approximations for the desired parameters.

With the help of the indicated correction factors, new ranges of changes in valve flow capacity are set, which leads to the need for a new check of compliance of the obtained limits with the regulated value of flow capacity.

6 Conclusion

So, when building the engineering methodology for calculating the main constructive-mode parameters, we used the results of stochastic models of the formation of cavitation bubbles in the flow section of an axial valve proposed by the authors [2–4, 23]. In particular, we used expressions obtained taking into account the explicit form of the differential distribution functions of the number of bubbles F1 (ξ) according to their specific dimensions ξ from [2, 20] and the differential distribution functions of the number of bubbles F2 (z) according to the degree of valve opening from [3, 4, 23]. In particular, we used expressions obtained taking into account the explicit form of the differential distribution functions of the number of bubbles $F_1(\xi)$ according to their specific dimensions ξ from [2, 20] and the differential distribution functions of the number of bubbles $F_2(z)$ according to the degree of valve opening from [3, 4, 23].

The main distinctive features of the proposed calculation method from previously known ones are the application of the following provisions of the authors' stochastic models:

- relations between the critical value of the Reynolds number $Re_{кр}$ and the desired valve parameters when analyzing the cavitation free regime of the diameter of the conditional passage in the initial approximation [23];
- expressions for energy parameters of models [2, 4];
- expressions for calculating the average diameter over the ensemble of the macrosystem of cavitation bubbles in the separator region [22];
- expressions for the hydraulic resistance coefficient in the transition region of the fluid flow [3].

In addition, it is possible to find not only rational ranges of changes in the projected parameters of a cyber-physical system but also their optimal values with additional research into ways of constructing an objective function to solve an optimization problem.

References

1. Kafarov, V.V., Dorokhov, I.N., Koltsova, E.M.: System analysis of chemical technology processes. In: Entropy and Variational Methods of Nonequilibrium Thermodynamics in Problems of Chemical Technology, 367 p. Science, Moscow, (1988)
2. Kapranova, A., Lebedev, A., Melzer, A., Neklyudov, S.: Determination of the average parameters of cavitation bubbles in the flowing part of the control valves. Int. J. Mech. Eng. Technol. (IJMET) **9**(3), 25–31. Article ID: IJMET_09_03_003. Available online at http://www.iaeme.com/IJMET/issues.asp?JType=IJMET&VType=9&IType=3 (2018)
3. Kapranova, A., Neklyudov, S., Lebedev, A., Melzer, A.: Investigation of the energy of the stochastic motion of cavitation bubbles in the separator of the axial valve, depending on the degree of its opening. Int. J. Mech. Eng. Technol. (IJMET) **9**(8), 160–166. Article ID: IJMET_09_08_017. Available online at http://www.iaeme.com/ijmet/issues.asp?JType=IJMET&VType=9&IType=8 (2018)

4. Kapranova, A., Neklyudov, S., Lebedev, A., Melzer, A.: Qualitative evaluation of the coefficient of hydraulic resistance in the area of the divider of the fluid flow of the axial valve. Int. J. Mech. Eng. Technol. (IJMET) **9**(8), 153–159. Article ID: IJMET_09_08_016. Available online at http://www.iaeme.com/ijmet/issues.asp?JType=IJMET&VType=9&IType=8 (2018)
5. Arzumanov, E.S.: Calculation and Selection of the Regulatory Bodies of Automatic systems, 112 p. Energy, Moscow, (1971)
6. Christmas, V.V.: Cavitation, 148 p. Shipbuilding, Leningrad (1977)
7. Volmer, V., Weber, A.: Keimbildung in uebersaetigen Daempfen. Z. Phys. Chem. (119), 277–301 (1926)
8. Lienhard, J.H., Karimi, A.: Homogeneous nucleation and the spinodal line. J. Heat Transf. **103**(1), 61–64 (1981)
9. Ellas, E., Chambre, P.L.: Bubble transport in flashing flow. Int J. Multiph. Flow (26), 191–206 (2000)
10. Kwak, H.-Y., Kim, Y.W.: Homogeneous nucleation and macroscopic growth of gas bubble in organic solutions. Int. J. Heat Mass Transf. **41**(4–5), 757–767 (1998)
11. Kedrinskii, V.K.: Hydrodynamics of Explosion: Experiments and Models (Shock Wave and High Pressure Phenomena), Chap. 7, pp. 307–344. Springer, Berlin (2005)
12. Koch, S., Garen, W., Hegedűs, F., Neu, W., Reuter, R., Teubner, U.: Time-resolved measurements of shock induced cavitation bubbles in liquids. Appl. Phys. **108**, 345–351 (2012)
13. Shin, T.S., Jones, O.C.: Nucleation and flashing in nozzles-1. A distributed nucleation model. Int. J. Multiph. Flow. **19**(6), 943–964 (1993)
14. Hsu, Y.Y.: On the size range of active nucleation cavities on a heating surface. J. Heat Transf. **94**, 207–212 (1962)
15. Kumzerova. E.Yu., Schmidt, A.A.: Effect of bubble nucleation mechanisms on flashing flow structure (numerical simulation). Comput. Fluid Dyn. **11**(4), 507–512 (2003)
16. Petrov, N., Schmidt, A.: Effect of a bubble nucleation model on cavitating flow structure in rarefaction wave. Shock Waves. **27**(4), 635–639 (2017). https://doi.org/10.1007/s00193-016-0699-z (Springer)
17. Lin, H.: Inertially driven inhomogeneitiesin violently collapsing bubble: the validity of the Rayleigh-Plesset equation. J. Fluid Mech. **452**, 145–162 (2002)
18. Seung, S., Kwak, H.Y.: Shock wave propagation in bubbly liquids at small gas volume fractions. J. Mech. Sci. Technol. **31**, 1223–1231 (2017). https://doi.org/10.1007/s12206-017-0221-2 (Springer)
19. Sokolichin, A., Eigenberger, G., Lapin, A., Lubbert, A.: Dynamic numerical simulation of gas-liquid two-phase flows: Euler/Eler versus Euler/Lagrange. Chem. Eng. Sci. **52**, 611–626 (1997)
20. Kapranova, A.B., Miadonye, A. (2019) Stochastic simulation of cavitation bubbles formation in the axial valve separator influenced by degree of opening. J. Oil, Gas Petrochem. Sci. **2**(2), 70–75 (2019). https://doi.org/10.30881/jogps.00026/
21. Lebedev, A.E., Kapranova, A.B., Melzer, A.M., Solopov, S.A., Voronin, DV, Neklyudov, VS, Serov, E.M.: Utility Model Patent 175446 Russian Federation (2017), IPC F16K 1/12, F16K 47/14, F16K 3/24. Direct flow control valve. Publ. 05.12.2017, Bull. No. 34
22. Kapranova, A.V., Lebedev, A.E., Melzer, A.M.: The definition of the integral characteristics of the process of formation of cavitation bubbles when operating the control valve. J. Chem. Eng. Process Technol. **8**(5), 58. https://doi.org/10.4172/2157-7048-C1-009. https://www.omicsonline.org/ArchiveJCEPT/chemical-engineering-2017-proceedings-posters-accepted-abstracts.php (2017)
23. Kapranova, A.B.: On the influence of the degree of opening of the regulator valve separator on the process of formation of cavitation bubbles. J. Chem. Eng. Process Technol. **9**, 36. https://doi.org/10.4172/2157-7048-C3-016

A Study of a Trajectory Synthesis Method for a Cyclic Changeable Target in an Environment with Periodic Dynamics of Properties

Dmitrii Motorin, Serge Popov, Vadim Glazunov and Mikhail Chuvatov

Abstract Trajectory planning in a large dynamic environment is a computationally complex cyber-physical task. The chapter considers an environment with periodic dynamics that simulates the rhythm of the day and night. Robots move between two target points cyclically. To optimize the trajectory planning process, it is possible to use pre-calculated paths. The pre-calculated state space consists of the planned paths for environmental states that can be considered static for a given period of time. The planning of robot movement in such the state space is carried out using parts of the pre-calculated optimal trajectories for a certain time and criteria for the transition between them. The method and criteria are studied by simulating the robot movement on two fundamentally different realistic maps. The method allows to plan the trajectories asynchronously with the time of the beginning of the movement of the robot, as well as to estimate the energy costs of overcoming the route.

Keywords Trajectory synthesis · Robot · Dynamic environment · Dynamic targets · Control · Spatial-situational uncertainty · Cyber-physical system

1 Introduction

Trajectory planning for a robot in an environment with chaotic external effects is a computationally complex task, which cannot be solved in real time on existing hardware resources. However, the environment has basically predictable nature, the

---

D. Motorin (✉) · S. Popov · V. Glazunov · M. Chuvatov
Peter the Great St. Petersburg Polytechnic University, Saint Petersburg, Russia
e-mail: d.e.motorin@gmail.com

S. Popov
e-mail: popovserge@spbstu.ru

V. Glazunov
e-mail: neweagle@gmail.com

M. Chuvatov
e-mail: misha@iktp.spbstu.ru

A. G. Kravets et al. (eds.), *Cyber-Physical Systems: Advances in Design & Modelling*, Studies in Systems, Decision and Control 259,
https://doi.org/10.1007/978-3-030-32579-4_10

reaction on unplanned events is rarely required. Factory robots are used in the tasks of automating cyclic human actions that repeated periodically and slightly adjusted depending on the state of the environment. The real environment is periodic, for example, the rhythm of day and night, the seasons, the tides and the weather. This information limits the uncertainty of the environment state.

The trajectory in a dynamic environment can be planned by modifying classical static algorithms. For example, Dijkstra's algorithm applied for an unmanned surface vehicle in a real environment with static and dynamic obstacles, as well as flow on the surface of the water [1]. Comparison of the algorithm based on D* and the modified particle swarm optimization method in different dynamic environments shows that it is more efficient in terms of length and the computation time is the last one [2]. An improved artificial potential field path planning algorithm for unmanned aerial vehicle tracking the dynamic target in dynamic environment constructs a repulsion field and effectively solves a local minimum problem in optimization on a general potential field function, without introducing unexpected collisions with stochastically moving obstacles [3]. The considered algorithms work well in environments with low dynamics.

If we consider the environment as an urban landscape with uncontrolled moving obstacles, then the speed of the robot's reaction to external events should be increased. A trajectory planning algorithm for mobile robot navigation in crowded environments based on the informed optimal rapidly-exploring random tree method. This approach solves the problem of indecision behavior and saves computational resources, since the path is rarely computed from scratch, but is continuously updated while the environment changes [4]. A modified nearest neighbor based task allocation algorithm [5] allows distributing goals in a group of agents in a dynamic environment. The proposed approach can be used in real-time systems. The paper [6] considers two modules path planning method: the robot uses global information about his environment and plans the optimal path using genetic algorithms combined with Dijkstra algorithm through static obstacles. The adaptive formation control of a swarm of autonomous robots that pass through the obstacles of the dynamic environment to reach the target position [7] is built based on the change of the desired distance between the neighboring robots in the swarm.

The dynamic of the environment can have a specific nature, for example, in [8] collision-free trajectory planning for multiple mobile robots is considered in the environment with periodic motion obstacles. Cyclic motion takes place in patrolling and monitoring problems. Patrolling strategies [9] in the realistic static environment should take into account: uncertainty on action execution, characteristics of the environment, closed-loop coordinated behaviors, and realistic simulation environments. Periodic trajectory optimization method for a mobile sensor performing persistent monitoring to maintain the uncertainty in the environment at the minimum induces the periodic Riccati equation [10]. A study of periodic trajectories in simple n-sided regular polygons environments using a contact sensor and motion primitives is considered in [11].

The environment in the robot's control system can be simulated in fundamentally different mappings. The paper [12] deals with a novel approach to hybrid maps that are based on fixed-size interconnected occupancy grids organized in a topological graph. Building an occupancy grid from a monocular color camera can be realized by a method for fusion of camera data with data from a rangefinder [13]. Trajectories can be planned on the incomplete information about the environment, for example, a global path planning algorithm for an unmanned ground vehicle based on road map images is presented in [14]. Expansion of this task includes two-level planning of a global trajectory based on incomplete roadmap data and a local path based on information from sensors observing the environment [15].

The approach to model the probability of action success as a set of superimposed periodic processes allows the robot to predict action outcomes in a long-term data obtained in two real-life offices better than a static model [16]. Trajectory planning for both single and multiple sensing robots to best estimate a spatial-temporal field in a dynamic environment uses the Kalman filter and periodic Riccati recursions [17]. A distributed technique that allows a team of robots to plan the deformation of the boundary shape in order to escort the safe region from one place to a goal is proposed in [18].

Global motion planning for modular robots with local primitives based on a modification of rapidly exploring random tree algorithm implements trajectories on realistic maps with obstacles [19]. For teleportation of the mobile robot on rough terrain can be used a method to evaluate in advance the stability of robots on assumed routes, and to provide visual information of the stability to the operator [20]. Context-aware decision support for marine robot's mission planning within a dynamic environment [21] allows the setting of various parameters and incorporates them to dynamically allocate and develop search path planning strategies. The model [22] allows reducing the discrepancies in the forecasting data and in real returns from investment, as well as optimizing investment strategies by both sides of the investment process. Visual programming module [23] of the generator allows a user to construct a scheme of inter-agent collaboration with agents of component-based module set.

This chapter presents a method of trajectory synthesis for robot performing a cyclic changeable target in the environment with periodic dynamic. The main part of the method is using pre-calculated optimal trajectories for static states of the realistic dynamic environment, which planning is computationally difficult, and plan the transitions between these paths in the process of robot movement.

2 Problem Statement

The problem consists of path planning for robot R cyclic movement between two given target points A_R and Z_R in a realistic dynamic environment. The environment of movement E consists of m layers $E = \{\varepsilon_1, \ldots, \varepsilon_i \leftarrow [e_{xyi} \in \zeta_i], \ldots, \varepsilon_m\}$ marked up with regular grid with size axb, each cell e_{xyi} has a value $\zeta_i = [\min_i, \ldots, \max_i]$ from sets of physically possible values in grid coordinates (x, y). The state of the layers is static for Δt, and full time of events' period is equal to $\mathbb{K}1\Delta t$, and then it starts from the beginning.

The trajectory of the robot motion is $T_R = \{A_R, \ldots, Z_R\}$. The robot has a set of pre-calculated trajectories in the form of state space $\mathbb{T}=\{T_1, \ldots, T_{\mathbb{K}}\}$ which are optimal for the corresponding state $\mathbb{E}=\{E_1, \ldots, E_{\mathbb{K}}\}$. Multi-criteria algorithm uses for path planning [24], which is based on the assessment of energy cost to achieve the goal. The speed of robot moving $V = \{v_1, \ldots, v_k\}$ is assumed to be constant and independent of the state of the environment.

The aim of the method is to synthesize the robot's trajectory T_R for a given period of time with minimal energy costs based on a known set of optimal paths. The main complexity is the possibility of the asynchronous start of movement from target points. Thus, it is possible to plan one cyclic trajectory if the start time is known and robot motion is synchronized with the period of the environment dynamic. In this case of any external effects on the robot, its trajectory should be changed to optimize motion or avoid waiting on the start. Therefore, a distinctive feature of the proposed approach is the possibility of asynchronous trajectory planning with minimum additional calculations in case of any external or internal changes in the system or environment.

3 The Method of Trajectory Synthesis in Environment with Periodic Dynamic

The method of trajectory planning based on pre-calculated paths is considered to solve the stated problem. Moving along the paths $\mathbb{T}$is optimal for a limited period of time Δt. To overcome the entire route from initial to final point it is necessary to plan transitions between optimal trajectories from the state space $\mathbb{T}$. Transitions path planning requires fewer computation resources, then planning full trajectory in a dynamic environment and can be performed during of overcoming the planned segment of the route. The trajectory is synthesized as follows:

$$\mathrm{T_{i\to i+1} = T(E_i)(t_{i1} \ldots t_{i2}) + T(E_i \to E_{i+1})(t_{i2} \ldots t_{i3}) + T(E_{i+1})(t_{i3} \ldots t_{i4})}, \quad (1)$$

where $T_{i\to i+1}$ is the trajectory of the robot in the time interval from t_i to $t_{i+\Delta t}$; Δt is the time of conditional static state of the environment; $T(E_i) \in \mathbb{T}$is trajectory of movement along the optimal path in the state of the environment E_i; $T(E_i \to E_{i+1})$ is the transition path between the $T(E_i)$ and $T(E_{i+1})$ in states E_i and E_{i+1} respectively; $T(E_{i+1}) \in \mathbb{T}$is the robot's movement trajectory along a path that is not optimal at the given time, but the best for time t_{i+1}; t_{i1}, t_{i2}, t_{i3}, t_{i4} are the moments of the time determined by the transition criteria.

Multiplication of the robot's movement speed V and time of conditional static environment Δt allows taking into account the time in the form of the path's length overcome by the state of the trajectory of the robot. If the transition to the trajectory $T(E_{i+1})$ in the current state of the environment, or what is the same during the time Δt, is impossible, then the trajectory is ignored and made an attempt to go to the next path $T(E_{i+k})$ during the $k1\Delta t$, where k is the number of the trajectories impossible to go through.

Transitions between pre-calculated trajectories are planned according to the rules set forth in the criteria and corresponding to the constraints of the problem statement. Criteria are formed according to the following principles: conditional/unconditional transition between trajectories, a static/dynamic method of evaluating the movement of a robot. In view of the above, the following three criteria are formulated: a static unconditional transition criterion, a dynamic unconditional transition criterion, and a dynamic conditional transition criterion.

4 The Criterions of Trajectory Synthesis

4.1 The Static Unconditional Transition Criterion

Trajectory planning is carried out in accordance with formula 1 and Fig. 1, where the time of movement along the current optimal trajectory $T(E_i)$ is set as the parameter $\Delta t' = (t_{i1} - t_{i2})$, and the remaining $\Delta t' = (\Delta t - \Delta t')$ is spent on the transition and movement along the next trajectory: $(t_{i3} - t_{i2}) + (t_{i4} - t_{i3})$. Thus, the criterion is described as

$$(t_{i1} - t_{i2}) + (t_{i3} - t_{i2}) + (t_{i4} - t_{i3}) = \Delta t' + \Delta t'' = \Delta t.$$

A graphical representation of the synthesized trajectory is shown in Fig. 1, the green color indicates the optimal trajectory at the time of motion planning Δt_i, the blue color indicates the transition, and the orange color is the optimal trajectory in the next period of time Δt_{i+1}, but overcome in the current one.

The main advantages of the proposed criterion are the speed of decision-making and the construction of the route. The trajectories constructed on the basis of this criterion may not be optimal, however, the application of this approach will be appropriate in cases where the adoption of the wrong decision carries less loss than not making a decision at all. The transition is carried out without any estimates or conditions; the parameter is only the ratio of the desired movement along the current path, determined empirically in the formulation of the practical problem of robot

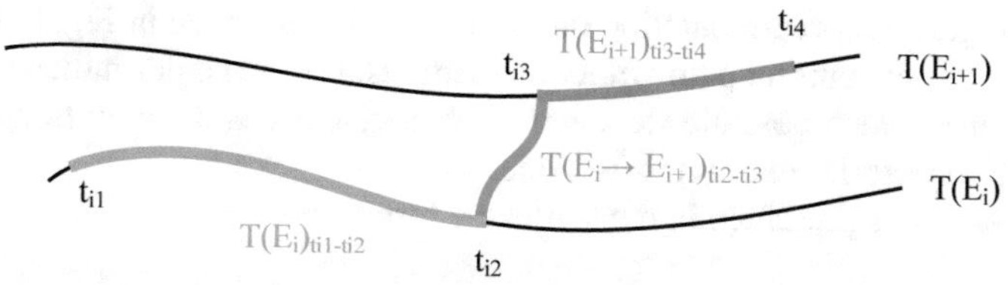

Fig. 1 Trajectory planning for unconditional transition

movement. The simplest division is the division by two, as in this case half of the route will surely be passing along the optimal trajectories.

4.2 The Dynamic Unconditional Transition Criterion

The weak point of the first criterion is the presence of a section $T(E_{i+1})t_{i3} - t_{i4}$ in each planned route. To optimize the trajectory, the minimization of this section is required. This is achieved by maximizing the travel time along the current optimal trajectory $T(E_i)$. The pathfinding algorithm provides the minimum of the transition path, to maximize and minimize the rest of the components requires the definition of the transition time. To determine this time, a distance estimate, such as Manhattan or Euclidean, is used for a possible transition to an adjacent trajectory from the last moment. The criterion is presented in the form of a system:

$$\begin{cases} T(E_i)_{ti1-ti2} \to \max \\ T(E_i \to E_{i+1})_{ti2-ti3} \to \min \\ T(E_{i+1})_{ti3-ti4} \to \min \end{cases}.$$

A feature of the dynamic criterion is the maximization of motion along the section $T(E_i) = t_{i1} - t_{i2}$, which provides an increase in the time of movement along the optimal trajectories. This increases the computational complexity used to estimate and minimize path segments.

4.3 The Dynamic Conditional Transition Criterion

Building routes on the basis of unconditional transition and local data does not allow to assess the prospect of movement and use all available information. The criterion of dynamic conditional transition is to estimate the robot's movement along the current path without transition and transition to each of the k nearest trajectories during $k1\Delta t$. A dynamic criterion is used to construct the estimated trajectories, but with the addition of the remaining path to the target point. The proposed trajectories take into account the periodic nature of the dynamics of the environment in assessing the movement, which allows to realistically predict the parameters of the planned paths. The graphical representation of the trajectories is shown in Fig. 2, the green color indicates the current path without transitions, the blue color indicates the tree of transitions at each possible step, and the orange color is the only transition that ignores all trajectories except the k-th one.

The criterion is presented in the following form:

$$\min\{T(E_i \to E_{i+i'} \to \ldots \to E_{i+i''})t_0\text{-}t_{i'}\text{-}\ldots\text{-}t_{i''\text{-}end} \mid i'=0..k,\ i''=0..k,\ i'\geq i'',\ k\in N\}. \quad (2)$$

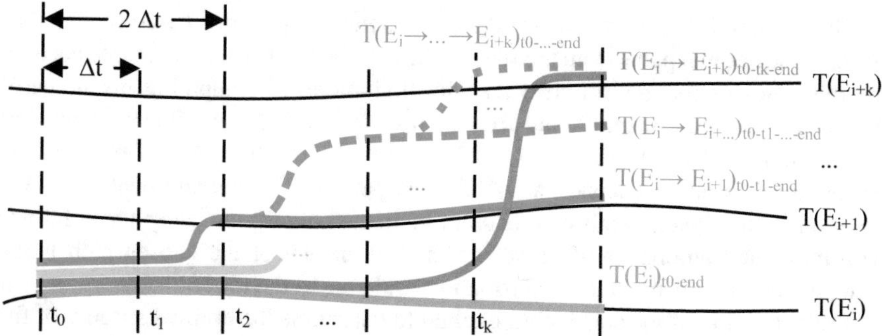

Fig. 2 Trajectory planning for conditional transition

The trajectories planned in this way can be represented as a triangular matrix of size $(k + 1)x(k + 1)$ filled with parameters describing the nature of the obtained paths. The rows of the generated matrix reflect the following k trajectories for the robot movement; the columns reflect the depth of the planning and evaluation.

$$O = \begin{bmatrix} T(E_i)_{t0-end} & 0 & \cdots & 0 \\ 0 & T(E_i \to E_{i+1})_{t0-t1-end} & \cdots & T(E_i \to \cdots \to E_{i+1})_{t0-\cdots-end} \\ \vdots & 0 & \ddots & \vdots \\ 0 & \cdots & 0 & T(E_i \to E_{i+k})_{t0-tk-end} \end{bmatrix}.$$

Searching for the minimum value in the resulting matrix allows choosing the minimum path of motion on the parameters of the robot autonomy and the number of pre-calculated paths used. The solution found will be Pareto-efficient if the maximum available amount of computing resources is used, determined experimentally depending on the task performed, and a route with minimal energy costs is chosen from the set of estimated trajectories.

The peculiarity of the criterion is the possibility of parametric adjustment of the depth of evaluation of the planned trajectories using the maximum amount of available information at the minimum cost of computing resources. In the course of trajectory planning, the actual route on $k1\Delta t$ is determined, which allows increasing the time of autonomous movement in the absence of chaotic changes in the environment.

In the proposed method, the calculations are minimized by reducing the planning area and minimizing trajectories based on the trajectory planning algorithm, as well as maximizing motion along the pre-calculated paths from the state space.

The periodicity of the task performed by the robots and the asynchrony with respect to the time of change of the initial and target position does not allow to build a trajectory for the entire life of the robot. Therefore, local suboptimal routes are constructed, in which there is no need to rebuild the entire trajectory with each change in the environment.

The proposed criteria for the synthesis of motion trajectories based on the state space of the optimal paths from the state of the environment allow taking into account the dynamics of the environment and the limitations of computing resources of the control object. Unconditional transition criteria use the minimum amount of information available to the robot, which limits the effectiveness of the solution but allows to make a decision as quickly as possible. The conditional transition criterion with a given depth of trajectory estimates allows choosing the solutions with maximum autonomy of movement and efficiency of the chosen path while using the maximum amount of information available to the robot.

The expansion of the considered method to determine the optimal speed v of the route, depending on the conditions of the robot position at the key points at which the change of the initial A and final Z points is made, can be carried out by simulating the dynamics of the environment at the preparatory stage. This is done by evaluating the movement of the robot at different speeds v for each initial time instant. The result of the simulation will be a set of realizable estimates of energy costs for movement in the environment at a given speed. The speed with minimal energy costs is optimal for the selected time of passing the key point. This approach allows choosing the speed of movement depending on the time of the task change when performing the task in real conditions.

5 The Study of the Method the Robot's Trajectories Synthesis in the Conditions Spatial-Situational Uncertainty

The feature research of the method the robot's trajectories synthesis in the conditions spatial-situational uncertainty is the time and computational complexity evaluation when performing cyclic tasks with the possibility of preliminary trajectories processing. The robot's tasks are performing the actions and movements of the same type between two given points. In case of significant and periodic changes of the environment, can be pre-calculate the optimal movement trajectory for the environment instantaneous state.

In research of the method, the robot's trajectories synthesis in the conditions spatial-situational uncertainty the maps of the dynamic environment are generated and the state spaces of the trajectories of the robots are calculated. The interaction the robot and the environment G is calculated as sum the energy costs of movement depending on the height between the points g_1 and the driving in the shadow g_2, in this case, $g_1 \ll g_2$, at that a fine on the cost of movement along the illuminated part of trajectory not impose.

The main method of the realistic landscape create is fractal midpoint displacement algorithm. The map is marked up a regular grid, so the dimension of the generated space is $2^n + 1$, where n is the depth of the fractal recursion. For creating models of the environment using the template that contains vertices and valleys. This template

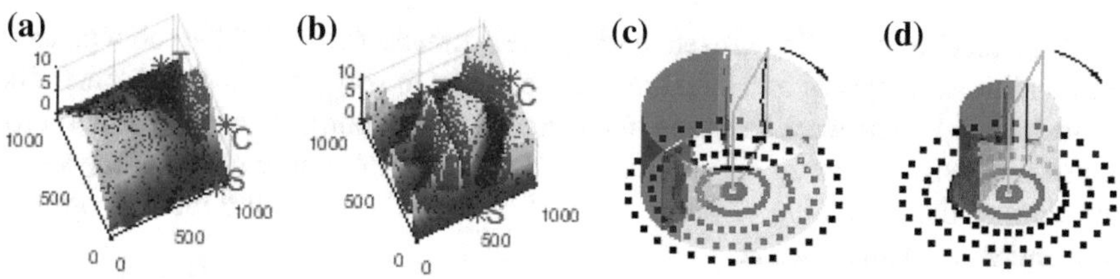

Fig. 3 Model of the realistic environment hill (**a**) and tunnels (**b**) and trajectories' state space (**c**) and (**d**) respectively

was used as input for the fractal algorithm thus determining the overall shape without disturbing the fractal structure. As a test two different type of terrestrial surface be generated: hills and tunnels. Each of the tests sites has several simple solutions for each environments state. On Fig. 3 shows the test sites of each type.

The map contains several layers. The first layer of the landscape is static and affects the robot energy only as a cumulative linear function of the height when the robot moving from one point to another. The second layer is the shadow layer, contains the illuminated and darkened parts of the map, changing over the day. The robot to avoid shadows as much as possible since it to remain without energy received from the sun, but it can nevertheless move in the shadows, but with a fine. Generated maps are 1025×1025 grid points.

The maps in Fig. 3a, b change their state cyclically, imitating the rhythm of day and night. The robot's task is to circulate between the points marked on the "S" and "T" maps, used as starting and target. The points "S" и "T" be used as starting and target points in calculating the space of robot solutions. The point "* C" denotes the side of the Sun. Figure 3a shows the map of the hill for trajectory planning. Solutions are shifted to the left as the shadows lengthen and to the right otherwise. Figure 3b shows a tunnel map consisting three of the tracks, from the shortest to the longest. The trajectory optimality depends on the position of the Sun above the horizon.

For each card under given initial and ending conditions the solution spaces are generated. To construct each trajectory in the solution spaces the algorithm for finding trajectories on multi-layered maps was used.

The state space is presented as cylinders consisting of trajectories at the corresponding moment of the state of the medium and the position of the light source. The colors of trajectory indicate the conditional of the environment: day (yellow), night (blue), twilight (red). Changes trajectories occur only during transitions between steady states. The choice of paths of movement are most relevant at the moment steady states change. Figure 3c depicts the state space corresponding to the environment in Fig. 3a, and the state space in Fig. 3d corresponds to Fig. 3b.

The movement over these paths optimally for a limited time Δt, the rest of the time, is not. To construct a route near to optimal without to waste computing resources, criteria for transitions between instantly optimal trajectories are proposed. From the calculated paths, one can compose the state space shown in Fig. 3c, d. The state space shows effectivity trajectory in the time.

Mathematical modeling is used to study the synthesis method of the robot paths in the conditions of spatial and situational uncertainty in solving cyclic problems. The pre-calculated paths were analyzed in the course of modeling the work of criteria for transitions between them. Figure 4 shows the graphs of energy costs on the speed and time of the beginning of the movement on the hill map along the first optimal path and when using criteria.

Figure 5 shows the graphs of energy costs on the speed and time of the beginning of the movement on the map of the tunnels along the first optimal path and using criteria.

Figures 4a and 5a show experiments without criteria are considered in the 600–800 and 1000–1200 ranges of visibility of the transition between steady states of the environment under the conditions of the existing change of position and shape of shadows on the map. Bursts on these bands are weakly dependent on speed. In Figs. 4 and 5 similar areas can be noted at high speeds at which it is possible to overcome the route before significant environmental changes occur. Figures 4b and 5b reflect the use of a static criterion of unconditional transition the effect of which is satisfactory at high speeds but worsens the results of planning at lower speeds. Figures 4c and 5c show use of a dynamic unconditional transition is implemented reduces the energy costs of movement at low speeds but does not effectively solve the problem of transitions between stable states of the environment. The criterion of conditional transition with an estimate of the movement along the paths to depth per unit presented in Figs. 4d and 5d show the best result and do not worsen the result of planning routes for all speeds and the time of the start of the movement.

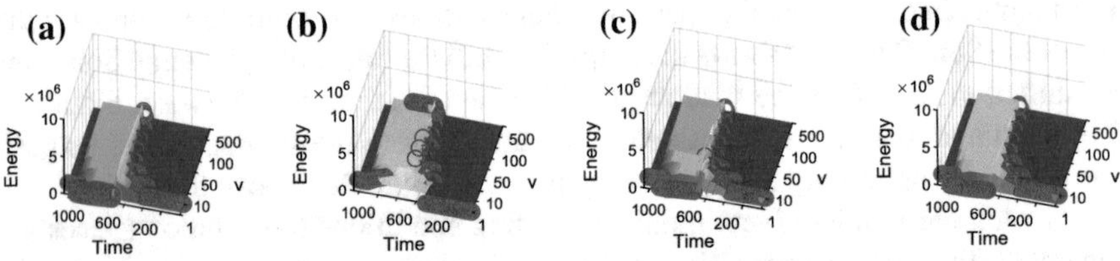

Fig. 4 The energy costs movement dependencies on the speed and time of the beginning of the movement for the hill map **a** without criteria **b** static criterion **c** dynamic criterion **d** evaluation criterion

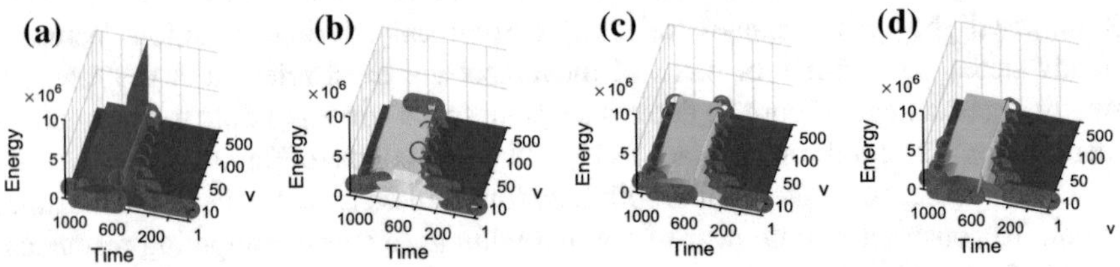

Fig. 5 The energy costs movement dependencies on the speed and time of the beginning of the movement for the map of tunnels **a** without criteria **b** static criterion **c** dynamic criterion **d** evaluation criterion

The use of the conditional transition criterion with a larger value of the valuation moves does not significantly affect the planning results and is similar to Figs. 4d and 5d respectively.

When comparing Figs. 4 and 5 it can be noted that the magnitude of the bursts at the transition in the range of 600–800 is much higher in the second case, and the change in the value of the energy costs of the movement depending on the speed and time is more smooth. This is due to the wide range of transitions on the map of the hills and the discrete choice of the route in the tunnels. In Figs. 4a–d and 5a–d red circles indicate the value of the minimum energy costs for the selected time of the beginning of the movement if the energy costs are the same for several speeds. The distribution of optimal speeds is similar in all cases. To minimize the energy costs of the task at a known time of the beginning of the movement of the robot, the speed of movement is selected.

The histogram of the distribution of deviations of the trajectory from the base paths calculated for the initial moment for the hill map is shown in Fig. 6a. The histogram compares the criteria of a static (blue) and dynamic (red) unconditional transition and a conditional transition with a rating of 1 (green), 5 (turquoise) and 10 (magenta) steps ahead.

The histogram of the distribution of the deviations of the trajectory from the base paths calculated for the initial moment for the tunnel map is shown in Fig. 6b. The color scheme is similar to Fig. 6a.

The histogram in Fig. 6a represents the result of applying static and dynamic un-conditional transition criteria that do not provide a steady improvement in trajectory planning results. The dynamic criterion of unconditional transition on average worsens the route less often and by a smaller value than the static one. The use of the criterion for estimating the energy expenditures of a movement with a conditional transition in the calculations one step forward significantly shifts the histogram to the left and, when planning 5 or 10 steps forward, improves the trajectory in each possible case. The conditional transition criterion improves the synthesized route due to the cut-off of the transitions to energy-intensive trajectories in a period of high environmental dynamics and low speed of the robot. The results in Fig. 6a confirm the conclusions made on the basis of Fig. 4 and supplement with the results of

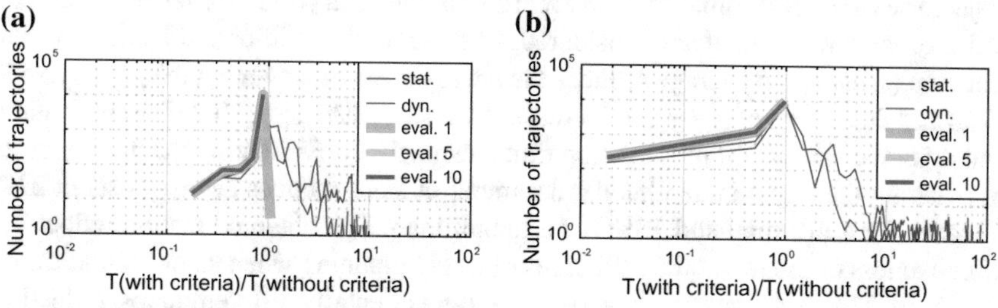

Fig. 6 Histogram of the distribution of the deviations of the routes from the basic paths calculated at the initial point for map hill (**a**) and tunnels (**b**)

experiments the study of the criterion of conditional transitions with an estimate of 5 and 10 steps.

Figure 6b shows using static and dynamic criteria, but when planning routes, the estimated conditional criterion does not matter the depth of calculations, the result differs by less than 1% and its use is always better than the initial paths. These results are based on a large difference in energy costs when moving along each of the tunnels of the map.

The study of the method of synthesis of the paths of the robot in the conditions of spatial and situational uncertainty was carried out on the basis of two different maps: hill (Figs. 3a, c, 4a–d and 6a) and the tunnels (Figs. 3b, d, 5a–d and 6b). Three criteria are analyzed: static unconditional transition, dynamic unconditional transition, and conditional transition with an estimate of a given number of steps. The use of unconditional transitions for the map of hills (tunnels) gives an average reduction in energy costs of achieving the goal by 25% (5%) and 30% (8%) of cases and an increase of 34% (23%) and 21% (10%) respectively with static and a dynamic estimate of transition time. When using the in-depth goal energy cost estimate, a fixed number of steps reduces energy costs by 22% of cases for a hill map and 13% for a tunnel map. The results are based on the geometry of the environment of robots using the proposed method allows reducing the energy costs of movement in the worst conditions.

6 Conclusions

The chapter considers the study of a trajectory synthesis method for a cyclic changeable target in an environment with periodic dynamics of properties.

Pre-calculated trajectories that are optimal for static states of the environment allow synthesizing routes that take into account the dynamics of the environment and respond to asynchronous random events in the process of movement. The proposed criteria for transitions between pre-calculated trajectories make it possible to plan paths that are close to optimal from the point of view of energy costs.

Three criteria for the synthesis of cyclic routes of movement along pre-calculated trajectories that are optimal for static representations of a periodic change in the state space of the environment are considered. The use of unconditional transition for the hill map (tunnels) on average reduces the energy costs of achieving the goal by 25% (5%) and 30% (8%) of cases and increases by 34% (23%) and 21% (10%) with static and dynamic estimation of transition time, respectively. Estimating the energy costs of reaching the target in depth a fixed number of steps reduces energy costs in 22% of cases for a hill map and 13% for a tunnel map. The obtained results reflect the effectiveness of the method for all cases of traffic planning when using a conditional transition. The study was conducted on two fundamentally different maps, reflecting both continuous and discrete transition between optimal trajectories.

In further work, the method is modernized and studied in case of trajectory planning with random external effects.

Acknowledgements The reported study was funded by RFBR according to the research project № 18-29-03250.

References

1. Singh, Y., Sharma, S., Sutton, R., Hatton, D., Khan, A.: Feasibility study of a constrained Dijkstra approach for optimal path planning of an unmanned surface vehicle in a dynamic maritime environment. In: IEEE International Conference on Autonomous Robot Systems and Competitions (ICARSC), pp. 117–122. Torres Vedras (2018). https://doi.org/10.1109/icarsc.2018.8374170
2. Sadiq, A., Hasan, A.: Robot path planning based on PSO and D∗ algorithms in dynamic environment. In: 2017 International Conference on Current Research in Computer Science and Information Technology (ICCIT), pp. 145–150. Slemani (2017). https://doi.org/10.1109/crcsit.2017.7965550
3. Chen, S., Yang, Z., Liu, Z., Jin, H.: An improved artificial potential field based path planning algorithm for unmanned aerial vehicle in dynamic environments. In: 2017 International Conference on Security, Pattern Analysis, and Cybernetics (SPAC), pp. 591–596. Shenzhen (2017). https://doi.org/10.1109/spac.2017.8304346
4. Primatesta, S., Russo, L., Bona, B.: Dynamic trajectory planning for mobile robot navigation in crowded environments. In: 2016 IEEE 21st International Conference on Emerging Technologies and Factory Automation (ETFA), pp. 1–8. Berlin (2016). https://doi.org/10.1109/etfa.2016.7733510
5. Biswas, S., Anavatti, S., Garratt, M.: Nearest neighbour based task allocation with multi-agent path planning in dynamic environments. In: 2017 International Conference on Advanced Mechatronics, Intelligent Manufacture, and Industrial Automation (ICAMIMIA), pp. 181–186. Surabaya (2017). https://doi.org/10.1109/icamimia.2017.8387582
6. Tazir, M., Azouaoui, O., Hazerchi, M., Brahimi, M.: Mobile robot path planning for complex dynamic environments. In: 2015 International Conference on Advanced Robotics (ICAR), pp. 200–206. Istanbul (2015). https://doi.org/10.1109/icar.2015.7251456
7. Dang, A., Horn, J.: Formation adaptation control of autonomous robots in a dynamic environment. In: 2015 IEEE International Conference on Industrial Technology (ICIT), pp. 3190–3195. Seville (2015). https://doi.org/10.1109/icit.2015.7125569
8. Mohri, A., Yamamoto, M., Fukuda, S.: Collision free trajectory planning for multiple mobile robots in environment with periodic motion obstacle. In: Proceedings of the 1996 IEEE IECON. 22nd International Conference on Industrial Electronics, Control, and Instrumentation, vol. 3, pp. 1572–1576. Taipei, Taiwan (1996). https://doi.org/10.1109/iecon.1996.570627
9. Iocchi, L., Marchetti, L., Nardi, D.: Multi-robot patrolling with coordinated behaviours in realistic environments, 2011 IEEE/RSJ International Conference on Intelligent Robots and Systems, pp. 2796–2801. San Francisco, CA (2011). https://doi.org/10.1109/iros.2011.6094844
10. Ha, J., Choi, H.: Periodic sensing trajectory generation for persistent monitoring. In: 53rd IEEE Conference on Decision and Control, pp. 1880–1886. Los Angeles, CA (2014). https://doi.org/10.1109/cdc.2014.7039672
11. Nilles, A., Becerra, I., LaValle, S.: Periodic trajectories of mobile robots. In: 2017 IEEE/RSJ International Conference on Intelligent Robots and Systems (IROS), pp. 3020–3026. Vancouver, BC (2017). https://doi.org/10.1109/iros.2017.8206140
12. Nitsche, M., de Cristóforis, P., Kulich, M., Košnar, K.: Hybrid mapping for autonomous mobile robot exploration. In: Proceedings of the 6th IEEE International Conference on Intelligent Data Acquisition and Advanced Computing Systems, pp. 299–304. Prague (2011). https://doi.org/10.1109/idaacs.2011.6072761

13. Stepan, P., Kulich, M., Preucil, L.: Robust data fusion with occupancy grid. In: IEEE Trans. Syst. Man Cybern. C (Applications and Reviews) **35**(1), 106–115 (2005). https://doi.org/10.1109/tsmcc.2004.840048
14. Hoang, V., Hernández, D., Hariyono, J., Jo, K.-H.: Global path planning for unmanned ground vehicle based on road map images. In: 2014 7th International Conference on Human System Interactions (HSI), pp. 82–87. Costa da Caparica (2014). https://doi.org/10.1109/hsi.2014.6860453
15. Ort, T., Paull, L., Rus, D.: Autonomous vehicle navigation in rural environments without detailed prior maps. In: 2018 IEEE International Conference on Robotics and Automation (ICRA), pp. 2040–2047. Brisbane, QLD (2018). https://doi.org/10.1109/icra.2018.8460519
16. Fentanes, J., Lacerda, B., Krajník, T., Hawes, N., Hanheide, M.: Now or later? Predicting and maximising success of navigation actions from long-term experience. In: 2015 IEEE International Conference on Robotics and Automation (ICRA), pp. 1112–1117. Seattle, WA (2015). https://doi.org/10.1109/icra.2015.7139315
17. Lan, X., Schwager, M.: Rapidly exploring random cycles: persistent estimation of spatiotemporal fields with multiple sensing robots. IEEE Trans. Rob. **32**(5), 1230–1244 (2016). https://doi.org/10.1109/tro.2016.2596772
18. Jahn, A., Alitappeh, R., Saldaña, D., Pimenta, L., Santos, A., Campos, M.F.: Distributed multi-robot coordination for dynamic perimeter surveillance in uncertain environments. In: 2017 IEEE International Conference on Robotics and Automation (ICRA), pp. 273–278. Singapore (2017). https://doi.org/10.1109/icra.2017.7989035
19. Vonásek, V., Saska, M., Košnar, K., Přeučil, L.: Global motion planning for modular robots with local motion primitives. In: 2013 IEEE International Conference on Robotics and Automation, pp. 2465–2470. Karlsruhe (2013). https://doi.org/10.1109/icra.2013.6630912
20. Awashima, Y., Fujii, H., Tamura, Y., Nagatani, K., Yamashita, A., Asama, H.: Safeness visualization of terrain for teleoperation of mobile robot using 3D environment map and dynamic simulator. In: 2017 IEEE/SICE International Symposium on System Integration (SII), pp. 194–200. Taipei (2017). https://doi.org/10.1109/sii.2017.8279211
21. Mishra, M. et al.: Context-aware decision support for Anti-Submarine Warfare mission planning within a dynamic environment. IEEE Trans. Syst. Man. Cybern. Syst. https://doi.org/10.1109/tsmc.2017.2731957
22. Akhmetov, B., Balgabayeva, L., Lakhno, V., Malyukov, V., Alenova, R., Tashimova, A.: Mobile platform for decision support system during mutual continuous investment in technology for smart city. In: Dolinina, O., Brovko, A., Pechenkin, V., Lvov, A., Zhmud, V., Kreinovich, V. (eds.) Recent Research in Control Engineering and Decision Making. ICIT 2019. Studies in Systems, Decision and Control, vol. 199. Springer, Cham (2019). https://doi.org/10.1007/978-3-030-12072-6_59
23. Kravets, A., Fomenkov, S., Kravets, A.: Component-based approach to multi-agent system generation. In: Kravets, A., Shcherbakov, M., Kultsova, M., Iijima, T. (eds.) Knowledge-Based Software Engineering. JCKBSE 2014. Communications in Computer and Information Science, vol. 466. Springer, Cham (2014). https://doi.org/10.1007/978-3-319-11854-3_42
24. Motorin, D., Popov, S.: Multi-criteria path planning algorithm for a robot on a multi-layer map. Informatsionno-upravliaiushchie sistemy [Inf. Control Syst.] (3), 45–53 (2018) (In Russian). https://doi.org/10.15217/issn1684-8853.2018.3.45

Cyber-Physical Systems Modeling

Intellectualization Methods of Population Algorithms of Global Optimization

Anatoly Karpenko, Taleh Agasiev and Maksim Sakharov

Abstract We consider constrained global optimization algorithms that are adaptive (self-adaptive) to the optimization problem being solved. We set tasks of parametric, structural and structural-parametric adaptation of these algorithms. We present the following methods for synthesis of adaptive algorithms for global optimization: tuning methods; control methods; self-control methods. We give some examples of adaptive algorithms and the results of research of their efficiency.

Keywords Global optimization · Meta-optimization · Adaptive algorithms · Hybrid algorithms

1 Introduction

Problems of continuous global optimization arise in the fundamental sciences—physics, chemistry, molecular biology, etc., as well as in many applied sciences. Frequently, features of such problems are: nonlinearity, non-differentiability, multi-extremity (multimodality), ravines, noisiness, lack of analytical expression (poor formalization) and high computational complexity of objective functions, high dimensionality of the search space, complex topology of the range of acceptable values, etc. As a rule, there is no a priori information about the specified features of global optimization problems.

Due to these circumstances, we observe a continuous increase in the complexity of algorithms for solving such problems, including intellectualization of these algorithms. Thus, modern global optimization algorithms widely use preliminary analysis

A. Karpenko (✉) · T. Agasiev · M. Sakharov
Bauman Moscow State Technical University, Moscow 105005, Russia
e-mail: apkarpenko@mail.ru

T. Agasiev
e-mail: taaalex@mail.ru

M. Sakharov
e-mail: max.sfn90@gmail.com

A. G. Kravets et al. (eds.), *Cyber-Physical Systems: Advances in Design & Modelling*, Studies in Systems, Decision and Control 259,
https://doi.org/10.1007/978-3-030-32579-4_11

and processing of problems to solve, including analysis of source data, dimensionality reduction of the search space, landscape analysis of the objective function, etc. [1]. To achieve high efficiency, composite algorithms that combine various "simple" optimization algorithms are used,—for example, the simulated annealing algorithm [2], evolutionary algorithms, including genetic ones [3, 4], meta-heuristic algorithms inspired by nature, such as the particle swarm algorithm [5], the bee swarm algorithm [6], ant colonies [7], etc. For the same reason, preliminary selection (using meta-optimization methods) of the most efficient optimization algorithm that allows for peculiarities of the initial optimization problem [8] is widely used. To reduce the number of objective function evaluations (the number of tests) during the process of solving the global optimization problem, approximating models of the objective function (meta-models, surrogate functions) are used [9]. Due to high computational complexity of objective functions, parallel computing systems of various types are used—systems with common and distributed memory, systems based on graphic processor devices, etc. In general, we believe that the general trend is transformation of global optimization algorithms into intelligent systems (intelligent decision support systems) [10].

Thus, currently, when developing global optimization algorithms, one should assume that they will work in uncertain and nonstationary conditions which are caused, firstly, by uncertainty of the optimization problem being solved, and, secondly, by uncertainty and nonstationarity of the computing system being used. It follows that these algorithms must be doubly adaptive: to the optimization problem and to the computing system. We would like to point out that, generally speaking, an adaptive system is a system that maintains efficiency when there are unforeseen changes in the properties of a controlled object, in control objectives or in the environment, and this property of the system is provided by changing the operation algorithm, the behavior program. Systems can be self-adjusting, self-learning and self-organizing according to the adaptation method used [11]. We believe that adaptive algorithms for global optimization are the first step in development of intelligent algorithms.

An algorithm that adapts to the computer system being used is called an algorithm that is self-consistent with the architecture of this system. Some aspects of construction of self-consistent algorithms are considered, for example, in the publication [12]. In this chapter, we limit ourselves to global optimization algorithms that are adaptive (self-adaptive) to the optimization problem being solved.

Intellectual algorithms for global optimization (I-algorithms) can use classical trajectory algorithms and population algorithms as basic ones (B-algorithms) when more than one candidate for solutions simultaneously evolves in the search space [13]. Unless otherwise specified, by B-algorithms we mean population algorithms for global optimization.

Traditionally, efficiency of B-algorithms is improved by meta-optimization methods [14]. If the result is an adaptive algorithm, then we believe it is correct to call this algorithm intellectualized, but not "meta-optimized". However, we use some terms from the theory of meta-optimization in this chapter.

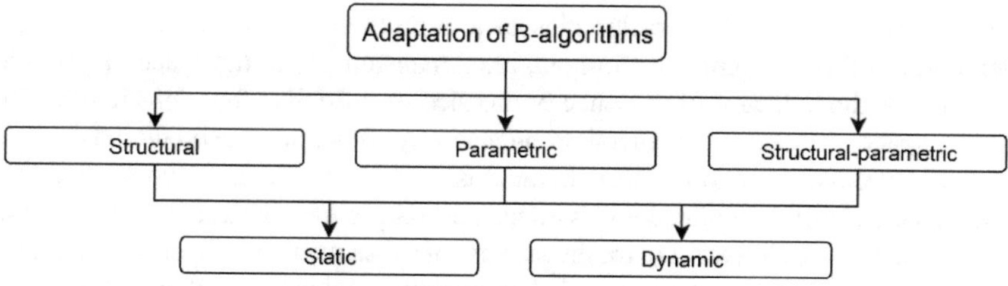

Fig. 1 Classification of adaptation methods of B-algorithms

Adaptation of the B-algorithm can be realized at two hierarchical levels: at the level of its free parameters; at the level of the algorithm structure (at the level of its operators). In accordance with this, we distinguish between parametric adaptation and structural adaptation of B-algorithms. Obviously, adaptation is also possible simultaneously at the levels of parameters and structure—structural-parametric adaptation (Fig. 1). The parameters and/or operators that are varied during the adaptation process are called the strategy of the B-algorithm.

We distinguish between static and dynamic adaptation. In the case of static adaptation, the found "optimal" strategy of the B-algorithm (B-strategy) does not change during its operation. Dynamic adaptation assumes that the B-strategy is some function of the iteration t number of this algorithm (program adaptation) or a function of its state $\left(v_1(t), \ldots, v_{|V|}(t)\right) = V(t) = V$ (positional adaptation).

We define the following classes of synthesis methods for adaptive B-algorithms:

- tuning methods;
- control methods;
- self-control methods.

2 Statement of Adaptation Problems

We consider the basic problem of global constrained optimization (B-problem) of the form

$$\min_{X \in D_X} f(X) = f\left(X^*\right) = f^*, \tag{1}$$

where X is the $|X|$-dimensional vector of variable parameters (dimension of the problem); $D_X \subset R^{|X|}$ is the search space; $f(X)$ is the objective function; X^*, f^* are the desired optimal vector X and the value of the objective function. The fitness function of the problem (1) is denoted as $\varphi(X)$.

Characteristics of the B-problem. The result of preliminary estimation of the properties of the problem $q = q(C)$ is a $|C|$-dimensional vector of characteristics $C = \left(c_1, c_2, \ldots, c_{|C|}\right)$ of this problem. We distinguish a priori and a posteriori characteristics. A priori characteristics of a B-problem directly follows from the definition of this problem, for example, the dimension $|X|$ of the space of variable parameters, the indicator of presence or absence of restrictive functions in defining the D_X space, the type of the objective and restrictive functions, the predicted number of local extrema, etc. A posteriori characteristics of B-problems, as contrasted to a priori characteristics, require computational cost for preliminary tests of the objective function in the search space D_X for the purpose of subsequent expert and/or automatic estimation of test results. A posteriori characteristics include, first of all, those of the objective function of the B-problem (1). Methods for estimating the values of a posteriori characteristics of the objective function without using expert estimates are called methods of landscape analysis [15].

Adaptation effectiveness indicators. Adaptation effectiveness criteria are called effectiveness indicators. Generally speaking, adaptation problems (M-problems) can be set as multi-indicator problems. We, however, limit ourselves to single-indicator setting.

The following groups of effectiveness indicators of global optimization algorithms are distinguished [16]:

1. performance (the number of arithmetic operations in the algorithm; the required RAM);
2. reliability of the algorithm (estimation of probability of achieving the specified accuracy of solution for a limited number of tests of the objective function—success rate; degree of constraint violation);
3. the quality of the computed solution (the best computed value of the objective function for a specified number of tests; the number of tests performed until the specified accuracy of the solution is reached).

When solving problems of high computational complexity, the maximum number of tests of the objective function is often fixed, and the best found value of the objective function is used as an effectiveness indicator.

To estimate the effectiveness of adaptation when solving B-problems of a certain class, integral estimates of the given indicators are used—for example, their average estimates. Productivity profiles can be used as integral estimates of effectiveness [17].

The parametric adaptation problem. The vector $(p_1, \ldots, p_{|B|}) = P$ of the free parameters of the B-algorithm $a = a(P)$ is called the B-strategy of the algorithm a. The set D_P of admissible B-strategies defines a set of algorithms $A(P) = \{a(P), P \in D_P\}$. The parametric adaptation problem is to find an "optimal" strategy P^* of the algorithm $a(P)$ for solving the given B-problem $q(C)$ of optimization. The effectiveness indicator of B-strategies for the B-problem $q(C)$ is denoted as $e = e(q(C), a(P)) = e(C, P)$. Thus, we set the (single-indicator) M-problem of static parametric adaptation in the form of

$$\underset{P \in D_P}{opt}\ e(C, P) = e\big(C, P^*\big) \tag{2}$$

Similarly, M-problems of program and positional parametric adaptation are, respectively,

$$\underset{P(t)\in D_P(t)}{opt} \quad e(C, P(t)) = e\big(C, P^*(t)\big) \tag{3}$$

$$\underset{P(V)\in D_P(V)}{opt} \quad e(C, P(V)) = e\big(C, P^*(V)\big) \tag{4}$$

We note that the M-problems (2)–(4), as well as the B-problem (1), are continuous constrained optimization problems.

The structural adaptation problem. Let the B-algorithm $a = a(O)$ include a set of operators $o_i, i \in [1 : |O|]$, for example, mutation operator, crossing operator, selection operator, etc., if the matter concerns a genetic algorithm. Suppose that $o_{i,j}, j \in [1 : |O_i|]$ is a set of possible realizations of the operator o_i. The set of admissible implementations of all the $|O|$-operators of the algorithm a forms a set of admissible strategies $D_O = \big(o_{i,j_i}, i \in [1 : |O|], j_i \in [1 : |O_i|]\big)$, which, in turn, determines the set of B-algorithms $A = \{a(O)|O \in D_O\}$. Using the introduced notations, by analogy with the M-problems of parametric adaptation (2)–(4) described above, static and dynamic M-problems of structural adaptation can be set. The only difference between these sets of problems is that the latter problems are discrete constrained optimization problems.

The structural-parametric adaptation problem. Suppose that each of the variants $o_{i,j}$ of the operator O_i has free parameters that form the vector $P_{i,j}; i \in [1 : |O|], j \in [1 : |O_i|]$. For simplicity, we assume that the dimensions and ranges of allowed values of all vectors $P_{i,j}$ are the same and equal $|P_i|$, D_{P_i} respectively, so $o_{i,j} = o_{i,j}(P_i)$, $P_i \in D_{P_i}$. In the introduced notations, the strategy of the B-algorithm turns out to be dependent on the vector of parameters $\mathbf{P} = (P_i, i \in [1 : |O|])$ and becomes a structural-parametric strategy of the form $O(\mathbf{P}), \mathbf{P} \in \mathbf{D}_P$, where $\mathbf{D}_P = D_{P_1} \oplus D_{P_2} \oplus \cdots \oplus D_{P_{|O|}}$; $\oplus$—is a symbol of the direct sum of sets. The set of B-algorithms, which is determined by the specified set of structural-parametric strategies, takes the form $A = \{a(O(\mathbf{P}))|O \in D_O, \mathbf{P} \in \mathbf{D}_P\}$. Using these notations, we can set M-problems of structural-parametric adaptation of the form (2)–(4) which are problems of mixed (continuous-integer) programming.

3 Tuning Methods

Tuning methods are based on preliminary analysis by a meta-program (M-program) of the characteristics $C = \big(c_1, c_2, \ldots, c_{|C|}\big)$ of the problem $q(C)$ using landscape analysis. The M-program uses a database containing a set of B-programs that implement a certain set of non-adaptive and/or adaptive algorithms $A = \{a_i, i \in [1 : |A|]\}$. During normal operation, the M-program stores the following data in this database:

the characteristics vector of the solved B-problem; the algorithm $a_{i_j} \in A$ by which the problem was solved; the corresponding estimation of the effectiveness indicator $e(C, a_{i_j})$.

When solving a new B-problem, the M-program (which implements the M-algorithm) performs the following actions:

1. using the landscape analysis of the problem, it finds its characteristic values C;
2. in the database it finds characteristics vectors $C'_1, \ldots, C'_{|C'|}$ close to C (in terms of a proximity metric);
3. as the "optimal" algorithm, it selects the algorithm $a_{i_j}(C'_k)$ which provides the best value of the effectiveness indicator $e' = e(a_{i_j}(C'))$.

It is important that parameter tuning methods allow predicting such effective strategies of B-algorithms that were not previously used for solving B-problems of optimization. On the basis of these methods the decision support systems are developed for optimization algorithms selection and tuning [18, 19].

4 Control Methods

We distinguish the following classes of control methods:

- methods of deterministic control;
- methods based on hybridization of M- and B-algorithms;
- coevolutionary methods.

Methods of deterministic parameter control. For example, in the case of parametric adaptation, the parameters of the B-algorithm are changed according to the a priori desired rule (e.g., according to the 1/5 rule by Rechenberg) [20]. Adaptive algorithms obtained in this way can hardly be classified as intelligent.

Methods of hybridization of M-and B-algorithms. In this case, the M-algorithm is built over the B-algorithm and uses the data obtained in the process of the latter's functioning. The I-algorithm obtained according to such a scheme is a high-level hybridization by embedding the M- and B-algorithms [21]. Most often, a genetic algorithm is used as an M-algorithm in this case. For example, there is a research in which a genetic algorithm is built over the particle swarm algorithm (PSO) which optimizes free parameters of the B-algorithm of the PSO in the process of solving the B-problem [22].

Let us explain the essence of hybridization methods by the example of the parametric adaptation problem. For the B-algorithm $a(P)$ variable parameters are components of the vector $X \in D_X$, and the fitness function is $\varphi(X)$. For the M-algorithm, in this case, the vector of the varied parameters is $P \in D_P$, and the fitness function is the effectiveness indicator $e(B)$.

Since the M-problem in hybridization methods is solved in the process of functioning of the B-algorithm, the indicator e should formalize the current estimation of the effectiveness of the B-algorithm. For example, the current convergence rate of this algorithm (to be maximized) could be such an estimate.

Coevolution methods are based on the use of coevolution of the rivalry type, when subpopulations differ in search strategies [23]. The idea is that in the process of coevolution, more resources get more successful subpopulations, thereby determining the optimal strategy of the B-algorithm, that is, the M-problem is automatically solved.

We also consider the essence of the coevolution control methods for the B-algorithm by the example of the parametric adaptation problem. Let the set D_P be discrete: $D_P = \{P_i, i \in [1 : |D_P|]\}$. Let us set up a correspondence of the algorithm $a(P)$ with a set of subpopulations $\mathbf{S} = \{S_i, i \in [1 : |\mathbf{S}|]\}$ in which the subpopulation S_i implements the B-algorithm $a(P_i)$ (that is, the algorithm with an admissible strategy P_i). The M-algorithm implements evolution of subpopulations $\mathbf{S}$. We use the best current value of the fitness function $\varphi(X)$ achieved by subpopulations $\mathbf{S}$ as an effectiveness indicator for this algorithm.

An important advantage of control methods in comparison with tuning methods is the fact that the former allow to automatically synthesize dynamic B-strategies.

5 Self-control Methods

For the parametric adaptation problem, the idea of self-control by the B-algorithm is as follows. The strategy parameters P are included in the vector of variable parameters of the B-problem, thus increasing the dimension of the search space. The same parameters are included in the fitness function of the I-algorithm, modifying the fitness function of the B-problem. As a result, in the process of the population evolution, the B-algorithm begins to perform the functions of the M-algorithm [24].

Let $a(P)$ be the considered B-algorithm with the strategy $P = \left(p_1, p_2, \ldots, p_{|P|}\right)$. The vector of variable parameters of the B-problem is $X = \left(x_1, x_2, \ldots, x_{|X|}\right)$, and the fitness function is $\varphi(X)$. We introduce the $(|X| + |P|)$-dimensional vector

$$\bar{X} = (X, B) = \left(x_1, x_2, \ldots, x_{|X|}, p_1, p_2, \ldots, p_{|P|}\right).$$

If the genetic algorithm is used as the B-algorithm, the first part of the corresponding multi-chromosome $\bar{X}$ is called exon, and the second part is called intron [25].

In the introduced notation, the essence of the self-control method is in the fact that the population $S = (s_i, i \in [1 : |S|])$ implements the B-algorithm $a(P)$, using the vector $\bar{X}$ as coordinates of its individuals. The first and second parts of this vector evolve in accordance with the fitness functions $\varphi(X)$ и $e(P)$ correspondingly.

6 Example 1. Co-evolution Algorithm of the Particle Swarm Co-PSO

The effectiveness of the particle swarm algorithm (PSO) largely depends on the values of its free parameters and on the topology of the neighborhood space which this algorithm uses [26]. The coevolution method can be used both to search for "optimal" values of free parameters (the parametric adaptation problem) and to find the "optimal" neighborhood topology (structural adaptation problem) [27, 28].

We consider the adaptive co-evolution algorithm Co-PSO [29] the basic algorithm of which is the canonical algorithm PSO. The algorithm Co-PSO uses $|\mathbf{S}|$ of co-evolving PSO sub-populations each of which, in general, has its own neighborhood topology of particles and differs from other sub-populations by values of its free parameters $\alpha,\ \beta,\ \gamma$ [27].

Free parameters of the Co-PSO algorithm are: the resource size n_f—the maximum allowable number of tests of the objective function; the adaptation interval Δ; the penalty value n_p; the minimum allowable size of the subpopulation $|S|_{\min}$.

The main steps of the Co-PSO algorithm:

- asynchronous independent evolution of subpopulations;
- estimation of effectiveness of subpopulations;
- resource rearrangement between subpopulations;
- migration of the best particles (individuals) of the population.

Let us consider the specific steps for the Co-PSO algorithm.

Estimate of the effectiveness of the subpopulation S_i is made on the basis of calculation of its fitness $e_i = e_i(t)$ which is a measure of adaptation of the subpopulation to the B-problem being solved. The fitness value is calculated by the formula

$$e_i(t) = \sum_{\tau=0}^{\Delta-1} \frac{\Delta-\tau}{\tau+1} d_i(\tau), \quad i \in [1:|\mathbf{S}|], \tag{5}$$

where $d_i(\tau) = 1$, if the subpopulation S_i at the iteration $t-\tau$ included the best particle in the entire population, and if not $d_i(\tau) = 0$.

Resource rearrangement is made on the basis of adaptation of subpopulations (5). If the current value of fitness e_i is maximal compared to the fitness of the remaining subpopulations, then this subpopulation is recognized as the winner, and the rest of the subpopulations—as losers. We rearrange the resource by reducing the size of each of the losing subpopulations by the penalty value determined by the value n_p, and by increasing the size of the winning subpopulation by a number equal to the sum of the losers' losses. Thus, the total population size does not change.

If the size of the losing subpopulation S_i turns out to be less than the size of the value $|S|_{\min}$, then on the next interval of adaptation Δ the size of this subpopulation will be equal to $\left|S_i'\right| = \max\left(n_p|S_i|, |S|_{\min}\right)$, $i \in [1:\ |\mathbf{S}|]$.

If the subpopulation S_i won on the current adaptation interval, then on the next interval of adaptation Δ, its size is determined by the expression

$$\left|S_i'\right| = |S_i| + \sum_j \max\left(n_p\left|S_j\right|, |S|_{\min}\right), i, j \in [1 : |\mathbf{S}|], i \neq j$$

Migration of the best particles is carried out according to the following scheme:

(1) particles of all subpopulations are combined into a single array and sorted in decreasing order of their fitness values;
(2) to each of the subpopulations we transfer from this array the number of particles proportional to the current size of this subpopulation.

In the paper [29] effectiveness of the Co-PSO algorithm was investigated, in particular, by the example of the problem of optimal control of a spacecraft during its descent in the Earth's atmosphere. In the original statement, the problem is a three-criterion problem of optimal control of a dynamic system, which is described by a non-rigid system of four ordinary differential equations. The problem is to determine the admissible control $u^*(\tau) \in D_u$, $\tau \in \left[0;\ \hat{\tau}\right]$ which provides the minimum to optimality criteria $f_1(u)$, $f_2(u)$, $f_3(u)$ having the meaning of maximum overload, deviation from a given point on the Earth's surface and maximum amplitude overload, respectively.

We use the method of solving the specified multicriteria optimization problem using additive scalar convolution of its particular optimality criteria (despite the known disadvantages of this method). The resulting single-criterion optimal control problem is reduced to the global optimization problem

$$\min_{U \in D_U} f(U) = f\left(U^*\right), D_U = \left\{U|u_j \in \left[u^-;\ u^+\right], j \in [1 : |U|]\right\}$$

as follows:

1. we cover the interval $\left[0;\ \hat{\tau}\right]$ with a uniform mesh with nodes τ_j, $j \in [1 : |U|]$;
2. at each of the intervals $\left[\tau_j;\ \tau_{j+1}\right]$ we approximate the control $u(\tau)$ with a segment of a straight line passing through the points $\left(\tau_j, u\left(\tau_j\right)\right)$, $\left(\tau_{j+1}, u\left(\tau_{j+1}\right)\right)$;
3. we assume that $U = \left(u_1, u_2, \ldots, u_{|U|}\right)$,

Computational experiments were performed for five subpopulations S_1–S_5, each of which has "click-type" neighborhood topology of particles and parameter values α, β, γ which are randomly uniformly distributed in a given interval. The following values of free parameters of the Co-PSO algorithm were adopted: population size $|S| = 100$; minimum allowable size of the subpopulation $|S|_{\min} = 4$; adaptation interval $\Delta = 9$; penalty value $n_p = 0.2$. In the process of initialization of the population all subpopulations get the number of particles equal to 20. The vector length $|U|$ is equal to 50.

Figure 2 shows dependences of the number $|S_i|$ of subpopulations of the Co-PSO algorithm in the function of the iteration number t. The graph of the function $|S_i|(t)$ of the winning subpopulation S_4 is highlighted in red.

Figure 2 shows that after the first adaptation interval, the subpopulation S_1 becomes the leader. The winning subpopulation S_4 takes the lead after the third

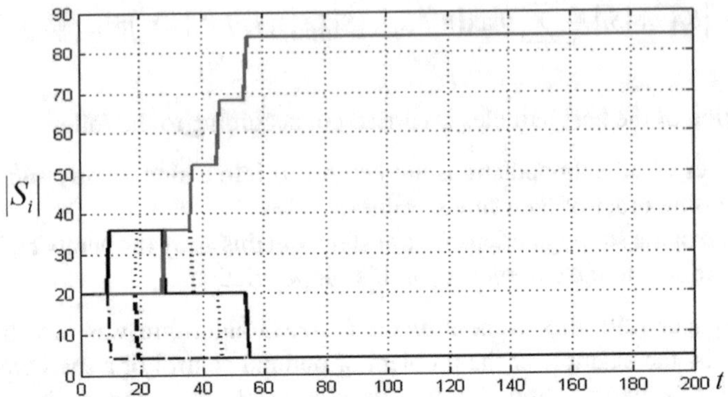

Fig. 2 The number $|S_i|$ of subpopulations in the function of the iteration number: ———, — — —, ..., ———, — · — · —subpopulations S_1–S_5, respectively

adaptation interval and maintains its leadership until the end of the iterative process when the number of particles in it becomes equal to 84.

The diagram of values of free parameters α, β, γ of subpopulations $S_1 - S_5$ is shown in Fig. 3. It is noteworthy that the values of these parameters of the winning subpopulation are closest to their recommended values [30] and are 0.9846, 1.13299, 1.6412, respectively.

It is known that singularity of the dynamic system under consideration is presence of optimality of so-called zero-overshoot responses in it, when, within the limit, control switches between its minimum and maximum allowed values with an infinite

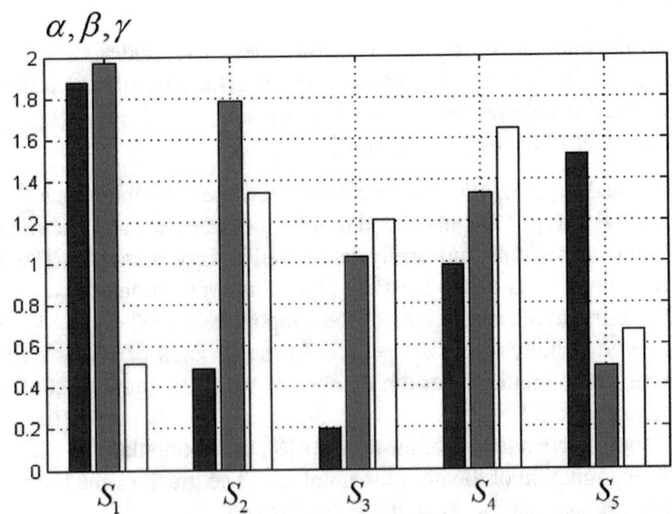

Fig. 3 Values of free parameters of subpopulations of the co-algorithm Co-PSO: ■—parameter α; ■—β; □—γ

frequency [30]. The results of the study show that the Co-PSO co-algorithm provides a much better approximation to the specified mode than the canonical PSO algorithm.

7 Example 2. Co-evolutionary Multi-memetic Algorithm M3MEC

The base for the adaptive co-evolutional algorithm M3MEC [31] is a hybrid algorithm based on the simple mind evolutionary algorithm (SMEC) [32] and the multi-memetic algorithm [33].

Adaptation in the M3MEC algorithm is performed at two hierarchical levels:

1. at the top level, adaptation is performed by tuning the parameters of the SMEC algorithms, using the data about the objective function of the B-problem obtained by landscape analysis methods;
2. at the lower level of the hierarchy, adaptation is performed on the basis of multi-memetic hybridization of the SMEC algorithm with several local and global search algorithms (memes).

The upper level of adaptation is static, which allows setting free parameters of SMEC algorithms before starting to solve the B-problem. The lower level of adaptation realizes dynamic adaptation in the process of solving the B-problem.

The main steps of the M3MEC algorithm are:

- landscape analysis;
- tuning the parameters of SMEC algorithms;
- solving the B-problem by hybridization of SMEC and multi-memetic algorithms.

Landscape analysis of the B-problem is performed as follows.

1. We generate N of vectors $X_r \in D_X, r \in [1 : N]$ whose components are uniformly randomly arranged in the area D_X.
2. For all vectors $X_r, r \in [1 : N]$ we calculate the corresponding value of the objective function $f_r = f(X_r)$.
3. We sort the vectors $X_r, r \in [1 : N]$ by increasing values f_r and uniformly split the final set of vectors into $|\mathbf{S}|$ of subsets (groups).
4. For each of the groups $G_i, i \in [1 : |\mathbf{S}|]$ we calculate the value of its diameter d_i—the maximum Euclidean distance between any two individuals of this group (Fig. 4b).
5. Using the least squares method, we approximate dependence of the diameter d_i on the number of the group i of linear function:

$$\tilde{d}(i) \approx b_{i,1}i + b_{i,0}, \quad i \in [1 : |\mathbf{S}|]. \tag{6}$$

6. The result of the landscape analysis of the B-problem are the diameters d_i of the groups and coefficients $b_{i,1}, \ b_{i,0}$ of the approximation (6); $i \in [1 : |\mathbf{S}|]$.

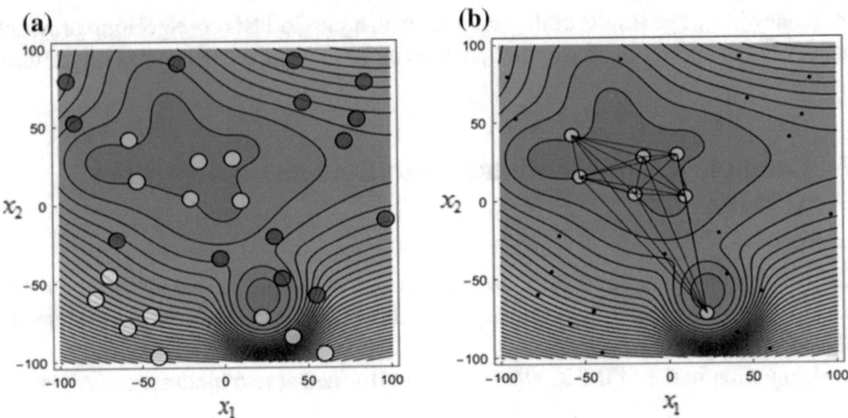

Fig. 4 Two-dimensional test composite function № 1 from the set CEC'14 [34]: **a** decomposition of domain D_X into subdomains, basing on landscape analysis; **b** determination of the group diameter

We emphasize that at this stage of the M3MEC algorithm, along with the landscape analysis, the domain D_X is decomposed into subdomains. The novelty and originality of this procedure lies in the fact that decomposition is carried out on the basis of proximity in the specified subdomains of values of the objective function. The content of the procedure is illustrated in Fig. 4 (it is assumed that the population **S** consists of four subpopulations $S_1 - S_4$).

Tuning parameters of SMEC algorithms. The groups G_i, $i \in [1 : |\mathbf{S}|]$ are assigned with subpopulations S_i, $i \in [1 : |\mathbf{S}|]$. The vectors forming the groups G_i, $i \in [1 : |\mathbf{S}|]$ are identified with the main individuals of these subpopulations. Basing on the values $d(D_X)$, d_i and values of the coefficients $b_{i,2},\ b_{i,1},\ b_{i,0}$, we determine the values of free parameters of the SMEC algorithm for the subpopulation S_i; $i \in [1 :\ |\mathbf{S}|]$. In the paper [31] a heuristic algorithm was proposed for determining values of the indicated free parameters. In the space of these values, the algorithm allocates six domains, each of which has its own set of values of free parameters of corresponding SMEC algorithms. Each of these domains, ultimately, is determined by the singularity of topology of the objective function in the part of the search space where the SMEC algorithm evolves.

Solving B-problem by a multi-memetic modification of the SMEC algorithm. The set of available memes is denoted as $M = \{m_j,\ j \in [1 : |M|]\}$.

To determine the meme that is the best for a given subpopulation S_i, $i \in [1 : |\mathbf{S}|]$ at a given step of its evolution t, the M3MEC algorithm uses hyper-heuristics based on the *selection function* ϕ. This function is an additive scalar convolution of three *particular* selection functions reflecting various aspects of meme quality:

$$\begin{aligned}\varphi_i(m_j) = \lambda_{i,1}\,\varphi_{i,1}(m_j) + \lambda_{i,2}\,\varphi_{i,2}(m_j)\\ + \lambda_{i,3}\,\varphi_{i,3}(m_j) \to \max,\ i \in [1 : |\mathbf{S}|],\ j \in [1 : |M|].\end{aligned} \quad (7)$$

Here λ_1, λ_2, λ_3 are weight coefficients (free parameters of hyperheuristics).

In the formula (6), the component $\phi_{i,1}(m_j)$ formalizes the value of relative improvement of the objective function of the B-problem at the current search point, reached by the meme m_j when used by its subpopulation S_i.

The component $\phi_{i,2}(m_j)$ has the meaning of probability of selecting the meme m_j by the subpopulation S_i, regardless of its success. During the initialization of the algorithm, all the probabilities $\phi_{i,2}(m_j)$ are identical and equal $1/|M|$. During the evolution of a subpopulation S_i, the value $\phi_{i,2}(m_j)$ increases in proportion to the number of iterations during which the meme $\phi_{i,2}(m_j)$ was not used by the given subpopulation. The component $\phi_{i,2}(m_j)$ is designed to prevent stagnation of the computational process with long-term use by the subpopulation of the same meme.

The component $\phi_{i,3}(m_j) \in [0;\ 1]$ allows to include expert knowledge in the selection process. The component formalizes a priori user's estimation of the meme m_j success.

Effectiveness of the M3MEC algorithm was investigated in a significant number of complex test problems [31]. With the help of this algorithm, according to the scheme p. 5, two practical significant optimal control problems [35] were also solved:

1. optimal control of vaccination;
2. optimal control of thermally stimulated luminescence of polyarylenephthalides.

8 Conclusion

At present, it is probably early to speak about the extensive use of intelligent algorithms and systems of global optimization. Such algorithms and programs are only at the stage of their formation. Therefore, the algorithms presented in the work should be called intellectualized rather than intellectual. The fundamental quality of such algorithms is their adaptation to peculiarities of the B-problem of global optimization without participation of the decision maker.

The chapter does not consider the important class of adaptive B-algorithms—the class of algorithms that are self-consistent with the architecture of the parallel computing system which is being used. However the algorithms of this class are known—for example, a modification of the SMEC algorithm discussed above, which is oriented to loosely coupled computing systems and allows the algorithm to adapt to this system [36]. There are publications considering automated synthesis of global optimization algorithms on the basis of PSO which are self-consistent with the architecture of parallel computing systems based on graphic processor devices.

References

1. Shan, S., Wang, G.G.: Survey of modeling and optimization strategies to solve high-dimensional design problems with computationally-expensive black-box functions. Struct. Multi. Optim. **41**(2), 219–241 (2010)
2. Van Laarhoven, P.J.M., Aarts, E.H.L.: Simulated annealing. Simulated Annealing: Theory and Applications, pp. 7–15. Springer, Dordrecht (1987)
3. Michalewicz, Z., Schoenauer, M.: Evolutionary algorithms for constrained parameter optimization problems. Evol. Comput. **4**(1), 1–32 (1996)
4. Wright, A.H.: Genetic algorithms for real parameter optimization. Foundations of Genetic Algorithms, vol. 1, pp. 205–218. Elsevier (1991)
5. Kennedy, J.: Particle swarm optimization. Encyclopedia of Machine Learning, pp. 760–766 (2010)
6. Karaboga, D., Basturk, B.: Artificial bee colony (ABC) optimization algorithm for solving constrained optimization problems. International Fuzzy Systems Association World Congress, pp. 789–798. Springer, Berlin, Heidelberg (2007)
7. Dorigo, M., Blum, C.: Ant colony optimization theory: a survey. Theor. Comput. Sci. **344**(2–3), 243–278 (2005)
8. Karpenko, A.P., Svianadze, Z.O.: Meta-optimization based on self-organizing map and genetic algorithm. Opt. Mem. Neural Netw. **20**(4), 279–283 (2011)
9. Forrester, A.I.J., Keane, A.J.: Recent advances in surrogate-based optimization. Prog. Aerosp. Sci. **45**(1–3), 50–79 (2009)
10. Kerschke, P., Trautmann, H.: Automated algorithm selection on continuous black-box problems by combining exploratory landscape analysis and machine learning. Evol. Comput. **27**(1), 99–127 (2019)
11. José Antonio Martín, H., de Lope, J., Maravall, D.: Adaptation, anticipation and rationality in natural and artificial systems: computational paradigms mimicking nature. Nat. Comput. **8**(4), 757–775 (2009)
12. Branke J., Elomari J.A.: Meta-optimization for parameter tuning with a flexible computing budget. In: Proceedings of the 14th Annual Conference on Genetic and Evolutionary Computation, pp. 1245–1252. ACM (2012)
13. Nobile, M.S. et al.: Fuzzy Self-Tuning PSO: A settings-free algorithm for global optimization. Swarm Evol. Comput. **39**, 70–85 (2018)
14. Neumüller, C. et al.: Parameter meta-optimization of metaheuristic optimization algorithms. In: International Conference on Computer Aided Systems Theory, pp. 367–374. Springer, Berlin, Heidelberg (2011)
15. Mersmann, O. et al.: Exploratory landscape analysis. In: Proceedings of the 13th Annual Conference on Genetic and Evolutionary Computation, pp. 829–836. ACM (2011)
16. Beiranvand, V., Hare, W., Lucet, Y.: Best practices for comparing optimization algorithms. Optim. Eng. **18**(4), 815–848 (2017)
17. Dolan, E.D., Moré, J.J.: Benchmarking optimization software with performance profiles. Math. Program. **91**(2), 201–213 (2002)
18. Polkovnikova, N.A., Kureichik, V.M.: Hybrid expert system development using computer-aided software engineering tools. In: Joint Conference on Knowledge-Based Software Engineering, pp. 433–445. Springer, Cham (2014)
19. Kosmacheva, I. et al.: Algorithms of ranking and classification of software systems elements. In: Joint Conference on Knowledge-Based Software Engineering, pp. 400–409. Springer, Cham (2014)
20. Hansen, N., Ostermeier, A.: Completely derandomized self-adaptation in evolution strategies. Evol. Comput. **9**(2), 159–195 (2001)
21. Eiben, Á.E., Hinterding, R., Michalewicz, Z.: Parameter control in evolutionary algorithms. IEEE Trans. Evol. Comput. **3**(2), 124–141 (1999)
22. Gong, Y.-J., Li, J.-J., Zhou, Y., Li, Y., Chung, H.S.-H., Shi, Y.-H. , Zhang, J.: Genetic learning particle swarm optimization. IEEE Trans. Cybern. **46**(10), 2277–2290 (2016)

23. Kavetha, Jeniefer: Coevolution evolutionary algorithm: a survey. Int. J. Adv. Res. Comput. Sci. **4**(4), 324–328 (2013)
24. Qin, A.K., Suganthan, P.N.: Self-adaptive differential evolution algorithm for numerical optimization. In: 2005 IEEE Congress on Evolutionary Computation, vol. 2, pp. 1785–1791. IEEE (2005)
25. Popov, V.: Genetic algorithms with exons and introns for the satisfiability problem. Adv. Stud. Theor. Phys. **7**(5–8), 355–358 (2013)
26. Xing, Bo, Gao, Wen-Jing: Innovative Computational Intelligence: A Rough Guide to 134 Clever Algorithms, p. 450. Springer International Publishing, Switzerland (2014)
27. Koua, X., Liua, S., Zhang, J., Zheng, W.: Co-evolutionary particle swarm optimization to solve constrained optimization problems. Comput. Math. Appl. **57**, 1776–1784 (2009)
28. Chen, Q., Jiao, B., Yan, S.: A cooperative co-evolutionary particle swarm optimization algorithm based on niche sharing scheme for function optimization. Advances in Computer Science, Intelligent System and Environment, pp 339–345. Springer Verlag, Berlin Heidelberg (2011)
29. Vorobeva, E.Y., Karpenko, A.P.: Co-evolutionary algorithm of global optimization based on particle swarm optimization. Science and Education of the Bauman MSTU, vol. 11, pp. 431–474 (2013)
30. Acary, V., Brogliato, B.: Numerical methods for nonsmooth dynamical systems. Applications in Mechanics and Electronics. Springer-Verlag, Heidelberg, LNACM 35, 519 p (2008)
31. Sakharov, M., Karpenko, A.: Multi-memetic mind evolutionary computation algorithm based on the landscape analysis. In: Theory and Practice of Natural Computing. Proceedings of 7th International Conference TPNC 2018, pp. 238–249. Springer, Dublin, Ireland, 12–14 Dec 2018 (2018)
32. Chengyi, S., Yan, S., Wanzhen, W.: A Survey of MEC: 1998–2001. In: Proceedings of 2002 IEEE International Conference on Systems, Man and Cybernetics IEEE SMC2002, vol. 6, pp. 445–453. Institute of Electrical and Electronics Engineers Inc., Hammamet, Tunisia, 6–9 Oct (2002)
33. Neri, F., Cotta, C., Moscato, P.: Handbook of Memetic Algorithms, p. 368. Springer, Berlin Heidelberg (2011)
34. Liang, J.J., Qu, B.Y., Suganthan, P.N.: Problem definitions and evaluation criteria for the CEC 2014 special session and competition on single objective real-parameter numerical optimization. Technical Report, Computational Intelligence Laboratory, Zhengzhou University, Zhengzhou, China and Technical Report, Nanyang Technological University, Singapore, 32 p (2013)
35. Sakharov, M.K.: Investigation of a disease monitoring model with pulse vaccination policy. Technologies and Systems 2018, pp. 116–120. Bauman MSTU Publ., Moscow (2018)
36. Sakharov, M.K., Karpenko, A.P.: Adaptive load balancing in the modified mind evolutionary computation algorithm. Supercomput. Front. Innovations **5**(4), 5–14 (2018)

Development of Models and Algorithms for Intellectual Support of Life Cycle of Chemical Production Equipment

Evgenii Moshev, Valeriy Meshalkin and Makar Romashkin

Abstract The article gives the statement of tasks necessary for automation of processes of adoption of intellectual decisions on the integrated logistic support of chemical production equipment, the solution of which is aimed at creating a cyber-physical system of the life cycle of the equipment. An example of formalization of this equipment life cycle with the help of functional modeling methods is given. There are given the models of knowledge representation about the equipment in a form of frames, and also—the processes of adoption of intellectual decisions on integrated logistic support of the equipment in a form of production rules. A heuristic-computational algorithm that allows automating the determination of classification characteristics of the equipment according to the degree of danger of the working substance is presented.

Keywords Cyber-physical system · Integrated logistic support · Frame · Production rules · Heuristic-computational algorithm · SADT-model · Functional model

1 Introduction

To ensure the process of continuous improvement of quality of integrated logistics support (ILP) of chemical production equipment under modern reality conditions and the level of technology development, it is necessary to process a significant amount of information arising at all stages of the life cycle of this type of equipment with the

E. Moshev (✉) · M. Romashkin
Perm National Research Polytechnic University, 29 Komsomol prosp., Perm 614990, Russia
e-mail: erm@pstu.ru

M. Romashkin
e-mail: t_romash_63@mail.ru

V. Meshalkin
D. Mendeleev, University of Chemical Technology of Russia, 9 Miusskaya pl., Moscow 125047, Russia
e-mail: clogist@muctr.ru

A. G. Kravets et al. (eds.), *Cyber-Physical Systems: Advances in Design & Modelling*, Studies in Systems, Decision and Control 259,
https://doi.org/10.1007/978-3-030-32579-4_12

usage of promising methods and mechanisms, namely, using cyber-physical systems. Creation of such a system of the life cycle of chemical production equipment for its further application in activities of chemical industry enterprises is inextricably linked with the need to develop models and algorithms with the help of which it is possible to solve actual ILP problems of this industrial equipment arising at various stages of its life cycle.

One of the most important tasks of integrated logistics support [1, 2] of industrial equipment is modeling of engineers', technical, organizational and technological processes for making intelligent decisions as a necessary condition for computerization. In accordance with State Standard of the Russian Federation GOST R 53394-2009, the ILP in the article refers to a set of types of engineering activities carried on through management, engineering and information technologies that provide a high level of performance readiness of industrial equipment (in particular by the features determining readiness of performance, that is, reliability, durability, ability for maintenance, serviceability and repair technological effectiveness, etc.) at a simultaneous reducing operating expenses [3, 4]. ILP of industrial equipment is a complex organizational and technological process that contains not only a large number of system interrelationships between different processes for making intelligent decisions but also special industrial requirements for nomenclature and methods of performing these ones. Among the ILP processes of chemical production equipment that require intellectual decision-making, we can distinguish the following:

- determination of features of equipment classification according to the degree of hazard of the working environment;
- selection and calculation of characteristics of structural elements of the equipment;
- selection and calculation of characteristics of welded and flange joints;
- determination of the characteristics of electric-arc welding and the weight of welding electrodes;
- formation of a structured list with the equipment data;
- calculation of the frequency spectrum of pressure pulsations generated by piston compressors in adjacent pipelines;
- determination of characteristics of structural elements of equipment, not included (missing) in the passport and technical documentation, on the basis of a set of introduced (known) characteristics;
- verification of the values introduced into the passport and technical documentation of equipment of chemical production, and their structural elements on conformity to the requirements of normative and technical documentation.

Currently, the implementation of above processes is carried out mainly manually or with the help of non-specialized software, which requires significant amounts of time and involves a large number of errors, presence of which decreases quality of ILP, and, consequently, leads to reducing profitability indicators and industrial safety of chemical production.

Preliminary analysis of intellectual decision-making processes for ILP equipment of chemical production showed that their formalization is impossible without the use of methods of artificial intellect theory [5, 6] and mathematical modeling.

Analysis of scientific and technical literature, particularly in the field of aircraft industry [7], building [8], studies of dynamic equipment [9], engineering [10], production planning [11, 12], design of systems and databases [13, 14], as well as analysis of current software [15–19] did not reveal the models, the algorithms and the modes of computer program operating that can be used to automation the above processes for making intelligent decisions on ILP of chemical production equipment.

Based on the above, the purpose of this study was to develop the models and the algorithms that allow you to automate the fulfillment of processes of making intelligent decisions on ILP for equipment of chemical production. This article presents only a small part of the results of this work.

To achieve this objective, it was necessary to solve the following tasks:

- to carry out analysis of scientific and technical literature, normative and technical and operational documentation, as well as the life cycle of chemical production equipment as an object of modeling;
- to formalize life cycle (LC) of chemical production equipment as a system of interrelated decision-making processes necessary for effective implementation of ILP;
- to create models of knowledge representation about the equipment of chemical production (digital counterparts of equipment);
- to develop models of knowledge representation about the processes by which decisions are made on ILP equipment of chemical production;
- to develop algorithms that formalize decision-making processes for ILP equipment of chemical production.

2 Analysis of Scientific and Technical Literature, Normative-Technical and Operational Documentation, and Life Cycle of Chemical Production Equipment as an Object of Modeling

In the process of analyzing a significant amount of scientific and technical literature, normative and technical and operational documentation, as well as life cycle processes of equipment of chemical production, the following provisions were established:

- presence of a large number of organizational and technological processes for implementation of which requires intellectual support during each stage of the life cycle of equipment of chemical production;
- containing a considerable amount of routine operations in many organizational and technological processes among which we can note a search of normative and technical data, an implementation of engineering calculations, a fulfillment of geometric representation of the elements that belong to the equipment of chemical production, a formation of technical documentation;

- existence of the dependencies between characteristics of structural elements of equipment of chemical production and working media, which in many cases are discrete in nature and in general can be described by Eq. (1):

$$\langle Y_f \rangle = (k_{f,1}, \ldots, k_{f,i}, \ldots, k_{f,N}) = FP_f, \tag{1}$$

where $\langle Y_f \rangle$—a subset of characteristics of structural elements of equipment of chemical production of *f*-type; $f = \overline{1, N_f}$), N_f—a number of types of structural elements of the equipment; $k_{f,i}$—*i*-characteristic of the *f*-type of the structural element; P_f—a subset of parameters of the equipment of chemical production and working media required to determine characteristics of the structural elements of the *f*-type;

- formalization of many procedures for making intelligent decisions on ILP of equipment of chemical production by means of models and algorithms used in the theory of artificial intellect;
- feasibility of the usage of SADT-models to represent knowledge about organizational and technological component of decision-making processes;
- the relevance of using frame models to represent knowledge about the equipment of chemical production and working environment;
- the advisability of the use of production models or of production rules to represent knowledge about the processes by which decisions on equipment of chemical products are made [20].

3 Formalization of Life Cycle of Chemical Production Equipment

Formalization of life cycle of chemical production equipment as a system of interrelated decision-making processes of ILP was carried out with the help of methodology SADT (Structured Analysis & Design Technique), described by David A. Mark and Clement McGowan in his book "Methodology of the Structures Strength Analysis and Design SADT". This methodology widely used in the development of complex systems is considered in the standards of the IDEF0 family, as an integral part of CALS technologies and is used in many countries, including Russia, where it is also known as methodology of functional modeling.

To simplify the overall formalization task, the complete SADT life cycle model of chemical production equipment was reduced to three conditionally independent models: a model of designed chemical production equipment, the one of the mounted one and that of an operated one (Fig. 1).

All the designed SADT-models differ with taking into account the complex relationships both within the stages of the life cycle of chemical production equipment and between them. With the help of specially developed algorithms, these models make it possible to carry out complex automation of ILP, as well as to ensure the

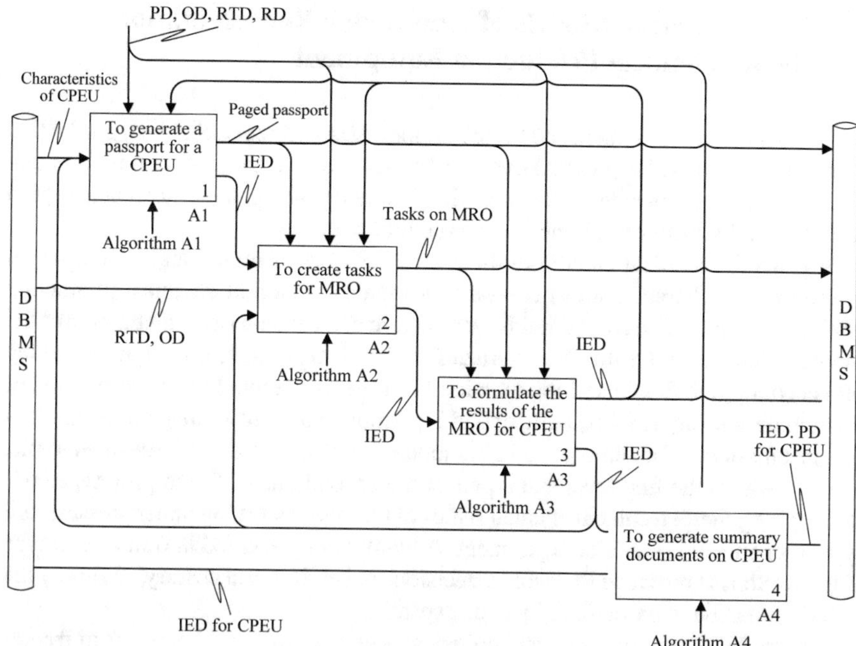

Fig. 1 The diagram of the zero level SADT-model "To execute ILP of the operated chemical production equipment unit" (CPEU): PD—project documentation; OD—operational documentation; RTD—regulatory and technical documentation; RD—repair documentation; IED—interactive electronic documentation; MRO—maintenance and repair; DBMS—database management system

interaction of all the subjects of the life cycle of chemical production equipment in a single information space. At these models usage the fulfillment of duplicate operations is eliminated and the speed of data exchange increases that promote quality improvement and reduction the cost of maintenance and repair of chemical production equipment. A complete description of the model shown in Fig. 1 contains 11 diagrams and about 60 functional blocks.

One of the most complex objects of ILP are piston compressors [21], this is due to physical peculiarities of their work, namely, to the pulsation of the pressure of compressed gas, that can become a source of increased vibration of pipelines adjacent to the compressors. Pressure pulsation is an integral feature of piston compressors' functioning and it is formed due to the periodicity of processes of inlet and outlet of \compressed gases from compressor cylinders. One of the ways to reduce the dynamic impact of gas pressure pulsations on pipelines and on technological equipment that is adjacent to compressors is the use of buffer tanks and smoothing diaphragms. Calculation of buffer tanks is a complex engineering and technical task that requires high qualification of specialists. Based on the above, the authors have also developed models and algorithms which permit to automate the calculation of buffer capacities and diaphragms.

4 Development of Models of Knowledge Representation About Chemical Production Equipment

An analysis of the equipment and the technological structures of chemical production shows that it is logical to present knowledge about the chemical production equipment in a form of interconnected frame models of three types: a general- technical type, a technical and technological one and a structural one.

A general- technical model is a three-level network of frames and contains a set of interrelated characteristics necessary for the description of chemical production equipment as an element of hierarchical structure of an enterprise. A distinction of the model is that it presents the characteristics of chemical production equipment, taking into account global points of the junction for a specific technological installation of coordinate with adjacent equipment and it permits to automate image of technological installations of an enterprise in 3D mode as a single structure. As an example, Fig. 2 presents the first level and a part of the second one of frame-prototype containing the general technical characteristics of the vessel working under pressure as a type of chemical production equipment. A blank field presented in frame-prototype (Fig. 2) (Slot) is provided to enable a decision-maker, if it is necessary, to supply the knowledge about a particular object or expand it.

The model of technical and technological type is a three-level network of frames and contains a set of interrelated characteristics, for example: pressure and temperature of working environment, test pressure, general passport and technical data, records of the results of the repairs and of the audits, chemical composition of the working environment, the group of chemical production equipment according to the class of hazard. A set of the characteristics contained in the model was determined by the analysis of scientific and technical literature, normative and technical and operational documentation, which is formed throughout the life cycle of chemical production equipment. The characteristics used in this model, as a rule, relate to the unit of equipment as a whole, are indicated without correlation to the elements of the structure of chemical production equipment and do not contain information about its structural design and dimensions. The model of technical and technological type differs with its taking into consideration the technological structure of chemical production equipment and also with the requirements of normative and technical documentation that permits to automate ILP of chemical production equipment, and, therefore, to increase its quality.

The model of the structural type is a three-level network of frames and contains a set of characteristics that describe chemical production equipment in a form of an interconnected and purposeful combination of connected together standard structural elements (shells, bottoms, flanges, supports, pipes, fittings, transitions, bends), which is intended to implement the technological processes occurring in chemical production equipment. In addition to topological and geometric characteristics of the structural elements, the model contains regulatory and technical requirements for manufacturing these elements and their weight. The values of the characteristics

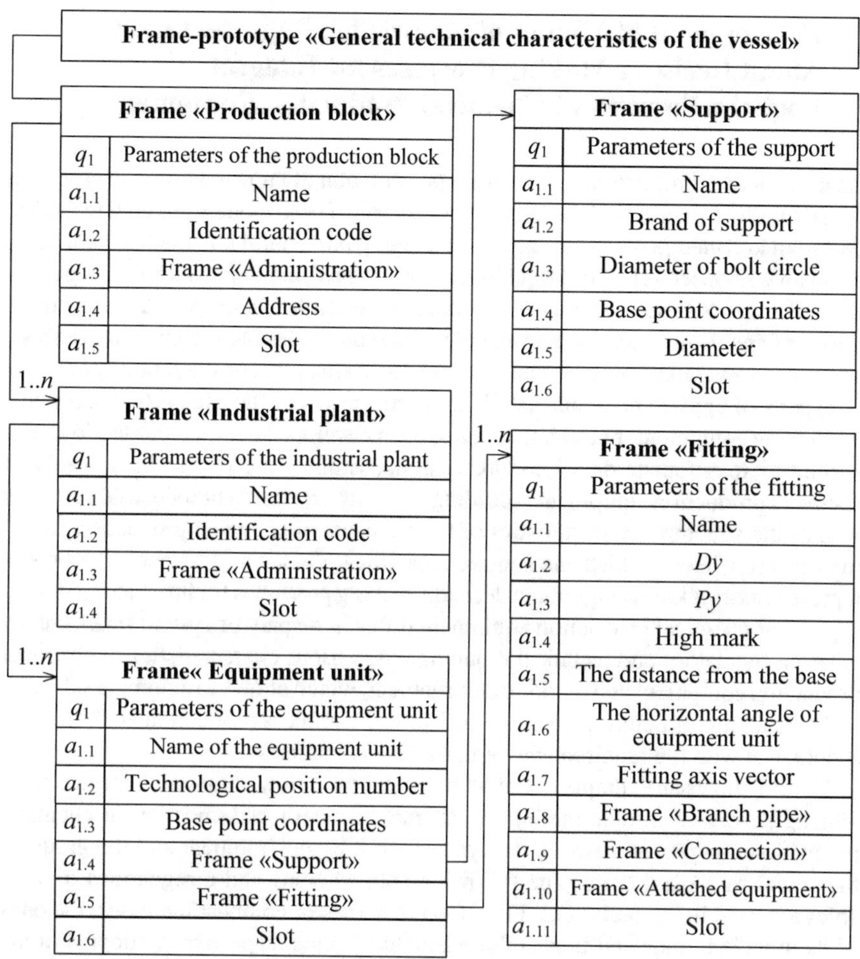

Fig. 2 Frame-prototype "General technical characteristics of the vessel": Dn—nominal diameter; Pn—nominal pressure; 1…n—the relationship between structural components of the model

given in the model are used in the formation of drawings, diagrams, forms and specifications of chemical production equipment throughout all the stages of the life cycle. A distinction of the structural type model is that it permits to automate both the image of the equipment in 3D and 2D modes and the formation of technical documentation throughout all the stages of the life cycle of chemical production equipment.

5 Development of Models of Knowledge Representation About Decision-Making Processes for Integrated Logistics Support of Chemical Production Equipment

As noted above, production models for representation of knowledge on the processes for adoption of intellectual decisions were chosen. These models are widely used to represent knowledge of various objects, including procedural knowledge (heuristics, normative knowledge) [20]. Development of production models of knowledge representation about the processes of intellectual decision-making was carried out by means of analyzing scientific and technical literature; normative-technical, passport, operational and repair documentation, as well as taking into consideration knowledge of experts of applied field and specificity of the stages of the life cycle of chemical production equipment. In total, there have been developed over ten production rules permitting to automate decision-making in determining a group and a category of chemical production equipment according to the requirements of industrial safety, as well as the structural characteristics of the elements of chemical production equipment and pipelines, welded and flange connections, elements of fasteners. Models of representation of knowledge about decision-making processes for integrated logistics support of chemical production equipment differ in display of system relationships between regulatory and technical requirements, various characteristics of chemical production equipment and working environment, which allows to automate selection and calculation of determined characteristics of chemical production equipment in accordance with the requirements of normative documents.

Table 1 shows an example of production rules that allow automating the decision-making processes in determining the characteristics of classification of chemical production equipment, namely, the type of working environment and the group of the vessel by hazard class. The following abbreviations and designations of variables are used in the table: CG, FL, CF—the names of combustible gases, the ones of flammable liquids and those of combustible liquids respectively, adopted in the

Table 1 An example of a production model of knowledge representation about the type of working medium and the group of vessels

I	Working environment: S_I	Type of work environment by hazard class: Ts_I	Conditions of applicability				Group of vessel: G_I
			Operating pressure range, MPa		Operating temperature range, °C		
			P_I^{min}	P_I^{max}	t_I^{min}	t_I^{max}	
1	Benzol	Second class	0.07	320.1	−273	700	1
2	Butane	CG	0.07	320.1	−273	700	1
3	Petrol	FL	0.07	320.1	−273	700	1
4	Fuel oil	CF	0.07	320.1	−273	700	1
5	Water	water	0.07	2.50	−40	400	3

regulatory and technical literature; $P_I^{\min}$, $P_I^{\max}$ и $t_I^{\min}$, $t_I^{\max}$—the boundary standard values of pressure and temperature of the working environment defining the sphere of existence of the characteristics sought in the process.

Determining the type of environment by hazard class is carried out by searching the array of regulatory attitude $R_I = (S_I, Ts_I)$ the relationship that meets the condition $(S_I, Ts_I) \cap S \neq \emptyset$ (2),

$$R_I = \left\{ \frac{1, (S = S_I)}{0}; if(R_I = 1)then(Ts = Ts_I), \right. \quad (2)$$

where $I = \overline{1, d}$—an identifier of string in an array of regulatory attitude; d—a number of strings in an array of regulatory attitude; S_I, Ts_I—the reference values of component of the working environment and the of working environment by hazard class in I-relation; S, Ts—a given name of the working environment and a sought value of its type.

The definition of the vessel group is carried out similarly by searching the string satisfying the condition in an appropriate array of normative reference relations (3),

$$(Ts_I, (P_I, R_2), (t_I, R_2), G_I) \cap \langle Ts, P, t \rangle \neq \emptyset, \quad (3)$$

where R_2—binary ratio, G_I—a group of a vessel on the hazard class of the working environment.

6 Development of Algorithms Formalizing Decision-Making Processes for Integrated Logistics Support of Chemical Production Equipment

In development of algorithms that formalize the decision-making processes for ILP life cycle of chemical production equipment, the methods of discrete, linear, logical and cyclic programming, the results of system analysis of scientific and technical literature, normative-technical and passport-technical documentation, as well as the knowledge of subject area experts [22] were used. In total, more than twenty necessary algorithms have been developed. The algorithms are differed in displaying the system relationships between heuristic and computational operations within the ILP procedures, as well as in use of frame and production models of knowledge representation, which allows automating intellectual decision-making processes for ILP for chemical industries equipment.

An example of a block diagram of a heuristic-computational algorithm for determining the characteristics of the classification of chemical production equipment by hazard class of the working environment is presented in Fig. 3. Following abbreviations and designations of variables are adopted in the flowchart: PKB—production

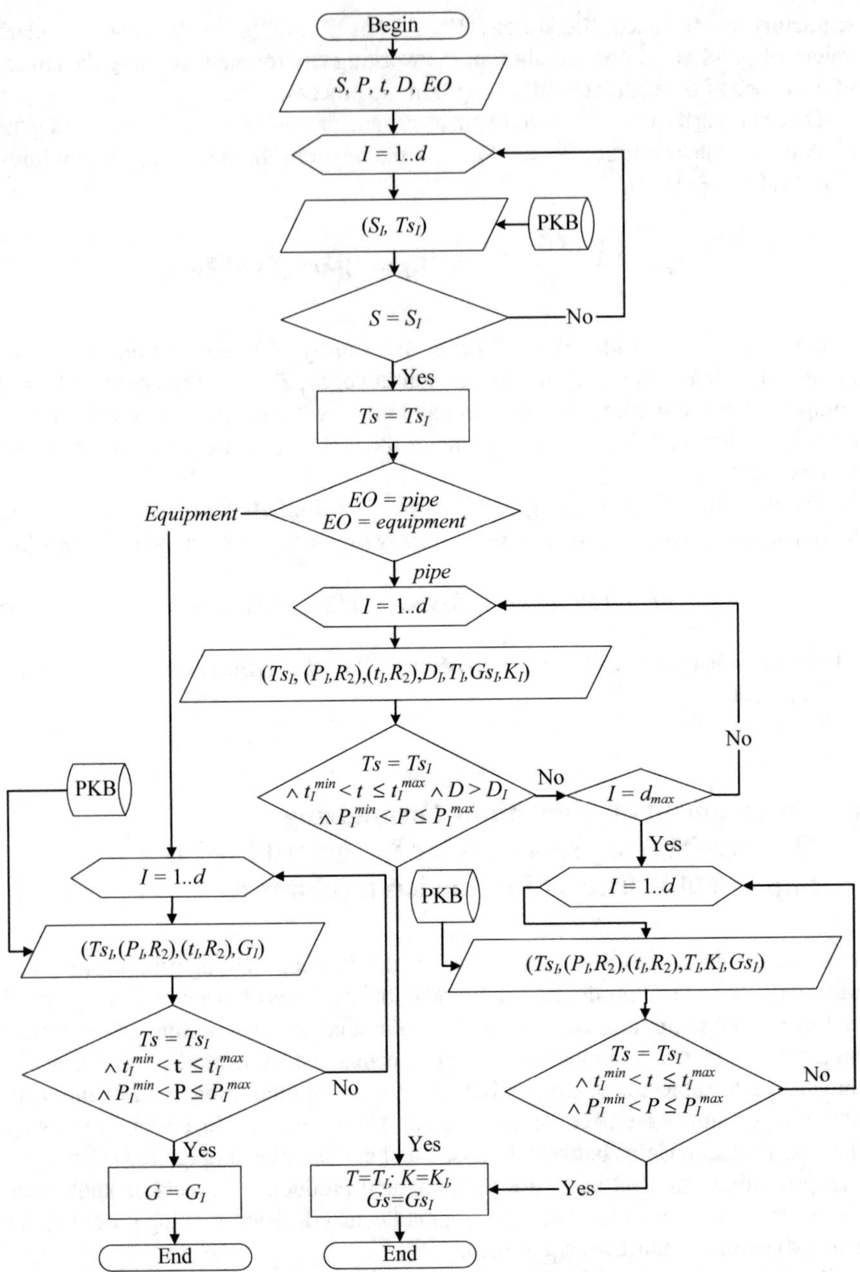

Fig. 3 Block diagram of heuristic-computational algorithm for determining the characteristics of the classification of chemical production equipment according to the degree of danger of the working environment

knowledge base developed with the help of production models of knowledge representation; D—a diameter determining a type of a pipeline; T and K—a type and a category of a pipeline; Gs—working environment group by hazard class. An algorithm differs in that the given values of characteristics of working environment (S), pressure (P) and temperature (t), as well as the diameter (D) of the pipeline with the help of frame and production models of knowledge about the characteristics of chemical production equipment, permits to automate the definition of the working environment group; type, group and category of pipelines, as well as groups of vessels in accordance with the requirements of regulatory and technical documentation.

7 Conclusion

The main scientific contribution of the work is representation of the life cycle and ILP of chemical industries equipment as a single system of interrelated organizational and technological processes. An applied approach allows not only to improve the quality of ILP, as an organizational and technological process, but also eliminate duplication of most processes and the operations performed at all the stages of the life cycle of chemical production equipment.

As a result of the work with the help of the methods of system analysis there were developed:

- SADT-life cycle models of chemical production equipment as a system of interrelated organizational and technological procedures of ILP. The models differ by taking into account the complex relationships between the procedures for making intellectual decisions at different stages of life cycle of chemical production equipment and with the help of special models and algorithms they allow to fulfill the complex automation of ILP, as well as to ensure the interaction of all the subjects of the life cycle of chemical production equipment in a single information space.
- Frame models of knowledge representation about chemical production equipment. The models are distinguished by taking into account the general technical, technological and structural characteristics of chemical production equipment, which allows automating: an image of a process facility as a single structure; an image of equipment units in 3D and 2D modes; making intellectual decisions and formation of technical documentation for ILP of chemical production equipment.
- Production models of knowledge representation on decision-making processes for integrated logistics support of chemical production equipment, which differ by display of system interconnections between regulatory and technical requirements, characteristics of chemical production equipment and working environment that allows to automate selection and calculation of the required characteristics of chemical production equipment, in accordance with the requirements of regulatory documents.
- Heuristic-computational algorithms which enable to automate decision-making processes for integrated logistics support of chemical production equipment. The

algorithms differ by displaying the system relationships between heuristic and computational operations within the ILP procedures, as well as the use of frame and production models of knowledge representation, which allows automating intellectual decision-making processes for ILP of chemical industries equipment.

Developed and described in the article models of knowledge representation, and heuristic-computational algorithms will allow to carry out complex automation of many procedures for making intelligent decisions on ILP of chemical industries equipment. Their inclusion in the system of management of the cyber-physical system of the life cycle of chemical production equipment will ensure high quality of ILP of this equipment, and, consequently, high profitability and industrial safety of the chemical industry enterprises.

References

1. Elena Nenni, M.: A cost model for integrated logistic support activities. Adv. Oper. Res. **2013**, 1–6 (2013)
2. Martin, P., Kolesár, J.: Logistic support and computer aided acquisition. J. Logistics Manag. **1**, 1–5 (2012)
3. Meshalkin, V.P., Moshev, E.R.: Modes of functioning of the automated system "pipeline" with integrated logistical support of pipelines and vessels of industrial enterprises. J. Mach. Manuf. Reliab. **44**(7), 580–592 (2015)
4. Moshev, E.R., Meshalkin, V.P.: Computer based logistics support system for the maintenance of chemical plant equipment. Theor. Found. Chem. Eng. **48**(6), 855–863 (2014)
5. Wu, D., Olson, D.L., Dolgui, A.: Artificial intelligence in engineering risk analytics. Eng. Appl. Artif. Intell. **65**, 433–435 (2017)
6. Russell, S.J., Norvig, P.: Artificial Intelligence: A Modern Approach, 3rd edn. Prentice Hall, New Jersey (2010)
7. Guo, F., Zou, F., Liu, J., Wang, Z.: Working mode in aircraft manufacturing based on digital coordination model. Int. J. Adv. Manuf. Technol. **98**, 1547–1571. http://dx.doi.org/10.1007/s00170-018-2048-0 (2018) (Springer, Cham)
8. Kim, H., Han, S.: Interactive 3D building modeling method using panoramic image sequences and digital map. Multimedia Tools Appl. **77**(20), 27387–27404. http://dx.doi.org/10.1007/s11042-018-5926-4 (2018) (Springer, Cham)
9. Cheng, J., Liu, Z., Yu, X., Feng, Q., Zeng, X.: Research on dynamic modeling and electromagnetic force centering of piston/piston rod system for labyrinth piston compressor. Proc. Inst. Mech. Eng. Part I: J. Syst. Control Eng. **230**(8), 786–798 (2016)
10. Pretorius, P.J.: How integrated is integrated logistics. S. Afr. J. Ind. Eng. **8**(2), 11–16 (1997)
11. Comelli, M., Gourgand, M., Lemoine, D.: A review of tactical planning models. J. Syst. Sci. Syst. Eng. **17**(2), 204–229 (2008)
12. Yevstratov, S.N., Vozhakov, A.V., Stolbov, V.Y.: Automation of production planning within an integrated information system of a multi-field enterprise. Autom. Remote Control **75**(7), 1323–1329 (2014)
13. Khouri, S., Bellatreche, L.: Design life-cycle-driven approach for data warehouse systems configurability. J. Data Semant. **6**(2), 83–111. http://dx.doi.org/ https://doi.org/10.1007/s13740-017-0077-8 (2017) (Springer, Cham)
14. Bertoni, M., Bertoni, A., Isaksson, O.: A value-driven concept selection method for early system design. J. Syst. Sci. Syst. Eng. **27**(1), 46–77 (2018)
15. SAP Software & Solutions. http://go.sap.com/index.html (2015). Accessed 14 Dec 2015

16. Isogen®. Automatic piping isometrics from 3D plant design systems. http://www.alias.ltd.uk/ISOGEN_main.asp (2015). Accessed 21 Jan 2015
17. AVEVA. Software Solutions for the Plant Industries. http://www.aveva.com/en/Products_and_Services/AVEVA_for_Plant.aspx (2015). Accessed 25 Jan 2015
18. CEA Systems. Plant-4D-Plant Engineering Solution. http://www.ceasystems.com/plant-4d-plant-engineering-solution (2014). Accessed 1 Jul 2014
19. iMaint.—CMMS Software—Service Management Software. http://www.imaint.com/en (2014). Accessed 5 Apr 2014
20. Lu, J., Zhu, Q., Wu, Q.: A novel data clustering algorithm using heuristic rules based on k-nearest neighbors' chain. Eng. Appl. Artif. Intell. **72**, 213–227 (2018)
21. Moshev, E.R., Romashkin, M.A.: Development of a conceptual model of a piston compressor for automating the information support of dynamic equipment. Chem. Petrol. Eng. **49**(9–10), 679–685. http://dx.doi.org/10.1007/s10556-014-9818-9 (2014). Springer, Cham
22. Menshikov, V., Meshalkin, V., Obraztsov, A.: Heuristic algorithms for 3D optimal chemical plant layout design. In: Proceedings of 19th International Congress of Chemical and Process Engineering (CHISA-2010), vol. 4, pp. 1425. Prague (2010)

Simulation of the Multialternativity Attribute in the Processes of Adaptive Evolution

Semen Podvalny and Eugeny Vasiljev

Abstract This article is devoted to the expansion of the ideas of evolutionary cybernetics to the problems of cyber-physical systems design. The main objective of such design is to reproduce the ability of an adaptive evolution that is proper to biological systems in the cyber-physical systems. This ability specific to biologic systems provides their sustainable development in a wide range of criteria of their functioning. More and more, the principle of variety of the processes running simultaneously in a complex system becomes the principal mechanism of realization of adaptive evolution. Mathematical representation and the analysis are made of this mechanism of variety in biological structures of various level of complexity. For pre-biological structures, the evolutionary value of multialternativity is explained in the processes of their streamlining and self-copying. Evolutionary models of the elementary macromolecules–quasitypes and the model of a syser with linked matrixes are investigated. It is shown below that the emergence and stable existence of pre-biological structures are possible as a result of a variety of the results of copying providing the cross mutational streams as well as the general evolutionary progress of population in general. As a model of the population evolution, its formal representation is offered as the discrete uniform Markov's process altering its state under the influence of complementary streams of events in the external environment and accumulation of a gene pool. For a vector of probabilities of these states the differential equation of Kolmogorov was composed hence, its solution gave the chance to obtain a quantitative assessment of a genetic variety's role as an emergency condition of either the evolution of biological population or its degeneration. The conclusion is made about the significance of the property of multialternativity as the mechanism of realization of the general cybernetic principles of creation the cyber-physical systems.

Keywords Cyber-physical systems · Models of adaptive evolution · Mutational streams · Genetic variety · Principles of multialternativity

S. Podvalny (✉) · E. Vasiljev
Voronezh State Technical University, Moscow av., 14, 394026 Voronezh, Russia
e-mail: vgtu-aits@yandex.ru

A. G. Kravets et al. (eds.), *Cyber-Physical Systems: Advances in Design & Modelling*, Studies in Systems, Decision and Control 259,
https://doi.org/10.1007/978-3-030-32579-4_13

1 Introduction

Generally, the cyber-physical system can be considered as a large-scale distributed control system in which the physical and computing components function in significantly various temporary and spatial dimensions. The distinctive feature of such systems is the variety of behavioral strategy of each component and their close interaction for the purpose of performance of system-wide function [1, 2].

The definition of a cyber-physical system given above indicates the direct analogy of the principles of its construction and functioning with biological systems whereas their unique property is the ability to adaptive evolution. This ability provides biological systems gives them a possibility of steady functioning and development in the changing environmental conditions.

As soon as this analogy exists, it allows using the laws of biological evolution in the design of highly reliable and effective cyber-physical systems, including the prevention critical situations in them which may lead to undesirable consequences.

Such design assumes the process of simulation, the development of models of evolutionary processes in biosystems [3–6]. Creation of these models will undoubtedly allow first to study and then to reproduce the mechanisms and properties of these processes in the cyber-physical systems [7].

The subject of this work is the property is multialternativity of evolutionary process of biosystems [8–12]. The particular mechanisms of realization of the attribute of a variety are offered to comprise the principles of modularity, variety, and division of functions, multileveled of structure and functioning of systems [13].

Recognition of an evolutionary role of the multialternativity principles is not yet supported adequately with the mathematical analysis of an influence of this mechanism on the adaptive properties of biological objects [14–16]. The results of this analysis are given below for both the processes of the evolution of the pre-biological structure preceding the emergence of life and for the higher, genetic level of evolution which determines the processes and the development of species in populations.

2 Models of Evolution of Pre-biological Structures

The most recent idea that the process of emergence of life is actually the process of streamlining and self-copying of the considerably stable molecular elements—the replicators [17–19]. Let's consider the role of multialternativity in this process.

Here is the investigation of the model of pre-biological evolution of macromolecules, or the model of quasitypes [18]:

$$\begin{cases} \dot{x}_i = \left(w_i - \sum\limits_{j=1}^{n} \left(w_j x_j \right) \right) x_i; \\ \sum\limits_{j=1}^{n} x_j = 1; \quad i = \overline{1, n} \end{cases}, \tag{1}$$

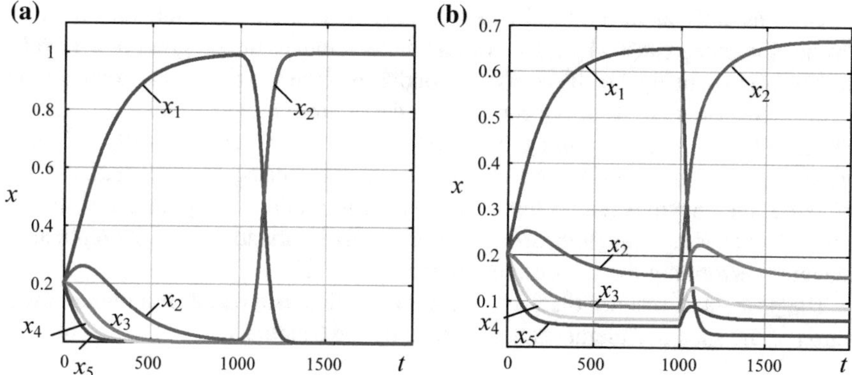

Fig. 1 **a** Process of unstable evolution (1) for a population consisting of five elements with the respective frequencies of occurrence x_1, …, x_5; **b** Formation of a population with a stable variety (2) of the competing elements with the frequencies of occurrence x_1, …, x_5

where w_i—is the fitness degree (selective value) of a macromolecule of i type; x_i—the frequency of occurrence of the i type molecules in a population; $\sum_{j=1}^{n}(w_j x_j)$—an average selective value of all types of molecules; n—a total number of various types of molecules in a population.

The evolution process obtained using this model is shown in Fig. 1a, it demonstrates that the type of molecules possessing the fitness degree above average value forces out all other types up to their full extinction. In case of adverse changes of the environment the death of the dominating species means the completion of an evolution and loss of the initial set of macromolecules considered above.

Now we enter are components into the Eq. (1), these components $\sum_{j}^{n}(\psi_{ij}x_j)$ describing by means of ψ_{ij} coefficients the mutational streams in each i type applied from the molecules of other types j ≠ i [18–20]:

$$\dot{x}_i = \left[w_i - \left(\sum_{j=1}^{n}(w_j x_j) + \sum_{\substack{k=1 \\ k \neq j}}^{n} \sum_{\substack{j=1 \\ j \neq k}}^{n} (\psi_{kj} x_j) \right) \right] x_i + \sum_{\substack{j=1 \\ j \neq i}}^{n} (\psi_{ij} x_j). \tag{2}$$

At the same time even at small values of mutational streams the process of self-copying undergoes the qualitative changes (Fig. 1b).

In Fig. 1b that has been obtained for an extremely small value of the mutations parameter $\psi_{ij} = 0.001$ it is shown that the process of evolution proceeds with a certain distribution of the concentrations of all initial species and the formation of their stable variety of composition. This distributed result received the name of a quasispecies containing the evolutionarily stable variety of macromolecules with a relatively close organization.

Thus, as early as at the level of elementary macromolecules the variety in the results of copying caused by its minor mutations provides the progress of the evolutionary process, being, i.e. the necessary condition of the pre-biological organization.

Let's now consider the more complex model of pre-biological evolution, namely, the model of sysers containing in its own structure the elements of biological organisms: a polynucleotide macromolecule (matrix) storing the information of the structure, the enzymes (catalysts) of translation providing the synthesis of macromolecule's construction elements according to this information, and the replication enzymes performing the self-copying functions [20–22].

The process of macromolecules' self-reproduction as a result of a mutual interaction of the specified elements is described by the equations:

$$\begin{cases} \dot{x}_i = x_i \sum\limits_{j=1}^{n} (a_j y_j) - x_i \sum\limits_{k=1}^{n} x_k \sum\limits_{j=1}^{n} (a_j y_j); \\ \dot{y}_i = x_i \sum\limits_{j=1}^{n} (b_j z_j) - y_i \cdot 2 \sum\limits_{k=1}^{n} x_k \sum\limits_{j=1}^{n} (b_j z_j); \\ \dot{z}_i = x_i \sum\limits_{j=1}^{n} (b_j z_j) - z_i \cdot 2 \sum\limits_{k=1}^{n} x_k \sum\limits_{j=1}^{n} (b_j z_j), \end{cases} \tag{3}$$

where: x_i, z_i, y_i—are the concentrations of nucleotide matrixes, enzymes of broadcasting and enzymes of replication, respectively, in the i type syser, said concentrations are subject to the conditions of constancy of their total amount.

$\sum_{i=1}^{n} x_i = 1$, $\sum_{i=1}^{n} (y_i + z_i) = 1$; a_i—the coefficients reflecting the speed of synthesis of matrixes, and b_j—is the speed of synthesis of the y_j and z_j enzymes.

The numerical solution of a system of the Eq. (3) confirms the lack of evolutionary changes, i.e., the initial concentrations of matrixes do not change.

The reason is the homogeneity of a molecular population. Homogeneity of a population results in its evolutionary stagnation.

Complication of the model (3) with the spatial division of sysers, which means their placements in the phase-isolated clusters called the coacervates, where each coacervate is the concentrate of sizers of only one particular type, thus making it possible with a coacervate viewed separately to rewrite the Eq. (3) as:

$$\begin{cases} \dot{x} = ayx - x(ayx + bzx + dzx); \\ \dot{y} = bzx - y(ayx + bzx + dzx); \\ \dot{z} = dzx - z(ayx + bzx + dzx), \end{cases} \tag{4}$$

where the coefficients a, b, d characterizing the synthesis speeds are individual for each coacervate. The competition between coacervates is described by the equations:

$$\dot{v}_i = \left(w_i - \sum_{j=1}^{n} (w_j v_j) \right) v_i, \tag{5}$$

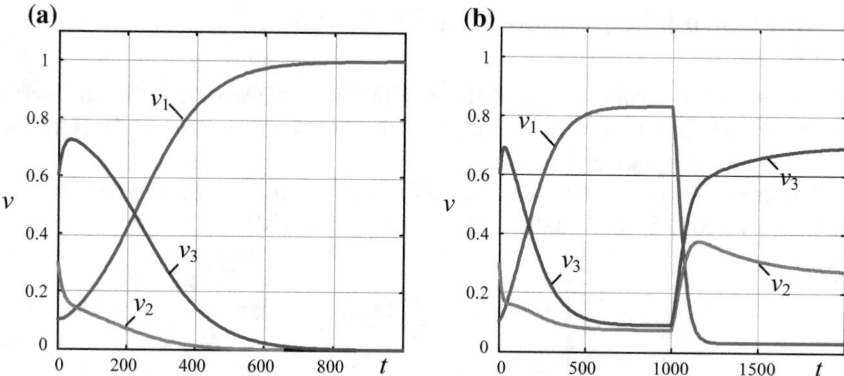

Fig. 2 **a** Emergence of the competition and natural selection as a result of the transition from homogeneous fermental environment to the selective synthesis (5); **b** Formation of a population with a stable variety of the competing elements with v frequencies by way of mutational streams introduction into (5)

in which $\mathrm{v_i}$—is a share of the i type coacervates in their total $\sum_{i=1}^{n} v_i = 1$; $w_i = a_i y_i x_i + b_i z_i x_i + d_i z_i x_i$—the coefficient characterizing the average speed of synthesis of the i type coacervates, and $\sum_{j=1}^{n} (w_j v_j)$—the average speed of synthesis of all n types of coacervates.

The result of the solution of (4), (5) is shown in Fig. 2a and it indicates the emergence of a natural selection of the most adapted types of matrixes, whereas at the same time, owing to the dominating synthesis of macromolecules of one type, a stable evolutionary process of all set of coacervates is not launched. However, the introduction in (5) of minor cross mutational streams $\psi_{ij} = 0.001$ gives a chance to receive the evolution process which is stable against the changes of external conditions (timepoint t = 1000 in Fig. 2b), i.e., here the result received in the analysis of quasitypes is confirmed: the evolutionary process is stable due to the mutations of the selective value of macromolecules.

Thus, the analysis of models of a pre-biological stage of life evolution shows that an emergence and a stable existence of the macromolecules capable to self-replication requires the existence of a number of interconnected conditions:

- the existence of a certain variety of the competing types of macromolecules, including those with a comparatively similar structure;
- intermolecular selectivity (differentiation) of adaptive and replicative mechanisms of molecules of different types;
- compatibility of existence of macromolecules' modifications, providing the cross mutational streams and the general evolutionary progress of a population in general when certain advantages of one element of a population are used by all other elements.

3 Models of a Population-Type Evolution

As Let's pass to the analysis of an attribute of multialternativity on higher, molecular and genetic levels of evolution defining the processes of emergence and development of species in a populations [23, 24].

An empirically established necessity of the genetic variety of populations for their stable existence is fixed formally in M. Eigen's model [18]:

$$\begin{cases} \dot{x}_i = \left(w_i - \sum\limits_{j=1}^{n} \left(w_j x_j \right) \right) x_i; \\ \sum\limits_{j=1}^{n} x_j = 1; \quad i = \overline{1, n} \end{cases}, \tag{6}$$

where xi—is a relative share of the i genotype in a population; w_i—a measure of fitness of the i genotype to the existence in the environment.

If the enter average value of a measure of fitness $\bar{w}$ is taken in consideration, then:

$$\sum_{i=1}^{n} w_i x_i = \bar{w}, \tag{7}$$

we can have the expression for the speed of its change:

$$\dot{\bar{w}} = \sum_{i=1}^{n} w_i \dot{x}_i = \sum_{i=1}^{n} w_i (w_i - \bar{w}) x_i. \tag{8}$$

Taking in account ratio (7) we rewrite (8) as:

$$\dot{\bar{w}} = \left(\sum_{i=1}^{n} x_i w_i^2 \right) - \bar{w}^2, \tag{9}$$

from which it follows that for the whole population in general the speed of change of fitness is proportional to its dispersion, i.e. homogeneous populations in the genetic relation concede in their competitive fight to those populations which comprising significantly the genotypes differing by their fitness. This conclusion is known as Fischer's theorem [17].

However it should be noted that the proof of this theorem stated in (6)–(9) results from the assumption of a static condition of a population at which the variety n of its genotypes is constant. This condition of biological system leads to the fact that, as a result of an unlimited variety of changes of the external environment, the adaptation of genetic resources of a population which «compensates» these changes shall be gradually reduced. The described situation corresponds to the evolutionary stagnation of a population and its subsequent inevitable degeneration, therefore it does not describe real attributes of biosystems. Therefore, model (6) does not correspond to the objective of the research of an evolutionary process [25–27].

For the purpose of elimination of the specified shortcomings of model (6) it is offered to present the process of a populational evolution in the form of the discrete Markov process containing the S_{ij}, i = 0, 1, 2…; j = 0, 1, 2…, N − 1, N conditions, including:

- $S_{0,0}$—the absorbing state designating a degenerate population with zero quantity of individuals $n(t)$ and zero probability of an exit of the population from this state;
- $S_{i,1}, \ldots, S_{i,N}$—conditions of a population that are characterized by the boundlessly growing variety of the i genotypes and the simultaneous growth of the volume of a population which, in turn, is limited by the available food resources to some final number of individuals $n = 1, \ldots, N$.

Let's enter the following parameters to the description of the process: parameter η of the intensity of population's growth, parameter λ specifying the intensity of the stream of changes of the external environment and the intensity n·μ of the accumulation of genetic changes in a population which resists to the stream mentioned here.

The graph corresponding to Markov's model of a population evolution process described above is shown in Fig. 3.

The qualitative analysis of Fig. 3 shows that the population size decreases when the intensity of accumulation of genetic changes n·μ concedes to the intensity of a stream of changes λ of the external environment.

To make a quantitative assessment of this conclusion we use the circumstance stating that the intensity of genetic changes μ is much more than the intensity of population growth η (assuming that each individual in any population contains at least several thousands of genes participating in the mutational changes). This conclusion gives an opportunity to pass to the one-dimensional model of the considered process (Fig. 4) specifying:

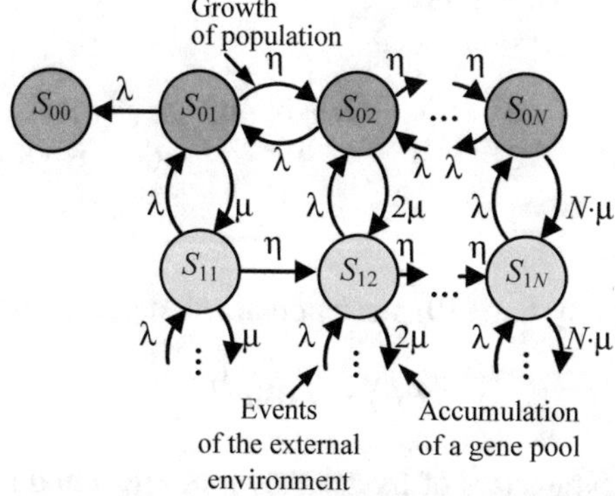

Fig. 3 Evolutionary process graph in the population with a final number of individuals N

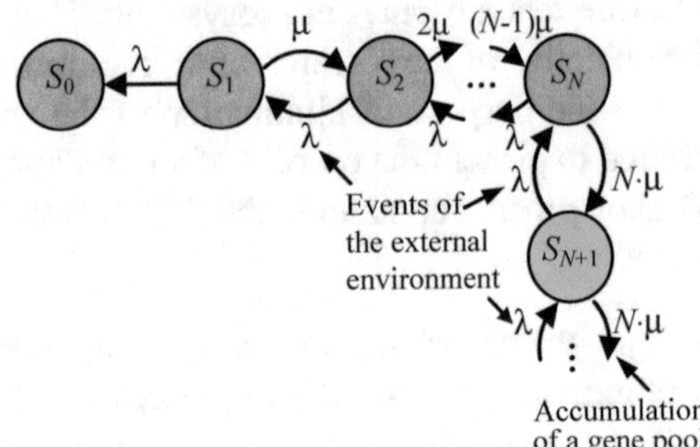

Fig. 4 The one-dimensional graph of the evolutionary process in a population

- S_0—the absorbing state designating a degenerate population with zero quantity of individuals $n(t) = 0$ and zero probability of an exit of the population from this state;
- $S_1,\ldots, S_N$—the states which characterize the growing number n characterizing the size of a population and the corresponding growth of the variety of genotypes i;
- $S_{N+1},\ldots$—the states characterizing a population with maximum value of its size $n = N$. At the same time the accumulation of changes in a gene pool the intensity $N\cdot\mu$, which is necessary for the «compensation» of changes of the external environment shall be continued without any limits.

For the model shown in Fig. 4 we have an opportunity to work out the differential equation of Kolmogorov for the probabilities $P(t) = [p_0(t)\, p_1(t)\, p_2(t)\ldots]$ of the emergence of the corresponding conditions $S_0, S_1, S_2, \ldots$:

$$\frac{dP(t)}{dt} = P(t)\cdot\Lambda, \tag{10}$$

where Λ—is an infinitesimal matrix of a species:

$$\Lambda = \begin{bmatrix} 0 & 0 & 0 & 0 & 0 & \cdots \\ \lambda & -(\lambda+\mu) & \mu & 0 & 0 & \cdots \\ \vdots & \vdots & \vdots & \vdots & \vdots & \cdots \\ 0 & 0 & \lambda & -(\lambda+N\cdot\mu) & N\cdot\mu & \cdots \\ 0 & 0 & 0 & \lambda & -(\lambda+N\cdot\mu) & \cdots \\ \vdots & \vdots & \vdots & \vdots & \vdots & \ddots \end{bmatrix}. \tag{11}$$

The solution of the system (10) can be obtained in general as:

$$P(t) = P(0)e^{\Lambda t}, \tag{12}$$

including all initial parameters of model shown in Fig. 4 and being convenient for the numerical analysis.

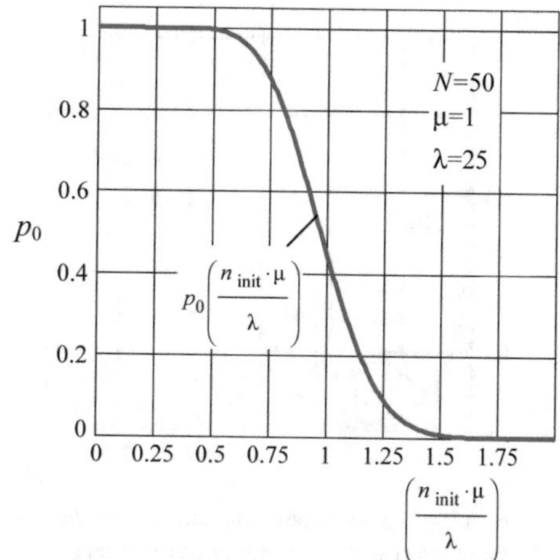

Fig. 5 Dependence of a probability p_0 of the population degeneration on the intensity of a stream of genetic changes accumulation and on the stream of external events

A research of evolutionary process was conducted for the following numerical values of its parameters: population volume N = 50; intensity of a gene pool stream of changes $\mu = 1$; intensity of a stream of changes of the external environment $\lambda = 25$; matrix size $[\Lambda] = 500 \times 500$ corresponds to the volume of the population genotype it is deliberately chosen bigger than the population volume N = 50 which corresponds to actual ratios in real biosystems.

Figure 5 shows the change of probability p_0 of the emergence of an absorbing condition S_0, i.e. the process of degeneration of a population depending on its initial size n_{init} and limited in this case from 0 to 50.

Analysing Fig. 5 we can establish that the initial population size viewed separately is not the defining factor for its successful development: the decisive impact is exerted by the ratio of a total intensity of a stream of genetic changes $n_{init} \cdot \mu$ in a population and the intensity λ of the events in the external environment. It is obvious that as soon as the volume of population of the boundary value N is achieved, the existence of a biosystem can continue as endlessly when the relation $N{\cdot}\mu/\lambda$ is considerably large.

Figure 6 illustrate the changes in a populations with various initial size:

- at $n_{init} = 40$ the size of a population grows up to the maximum value $N = 50$, which is followed by the unlimited growth of its genetic conditions S—from smaller numbers to big ones 60…70….80…100…;
- at $n_{init} = 10$ the changes of a population have an opposite focus, viz., the transition from the condition S_{10} to the states with lower number $S_8 \ldots S_6 \ldots S_1$ and, at last, there is a genetic conditions S_0 with the probability $p_0 = 1$, corresponding to the degeneration of a population.

These obtained results mean that the reliable (in probabilistic sense) maintenance of an evolutionary process in a system is possible at the considerable acceleration of growth of its gene pool in comparison with the speed of change of environmental conditions (in the reviewed example approximately doubled excess is required).

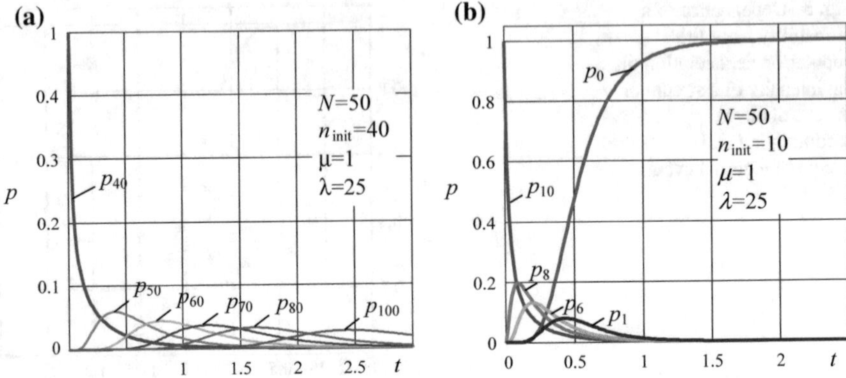

Fig. 6 **a** Growth of population and its gene pool with an initial number $n_{init} = 40$; **b** Degeneration of a population with an initial number of $n_{init} = 10$

Possible growth of population size accompanying this process makes a positive impact on the growth rate of a gene pool. This fact is usually interpreted as an existence of a critical population size, i.e., the lower bound of the number of its individuals required for its evolution.

It should be noted that in the J. von Neumann's theory of artificial self-replicating systems similar idea of critical complexity is used, the «non-return point» of the automatic machine designating the boundary level of system's complexity below which it is doomed to degradation.

The analogies noted above indicate a cybernetic community of the principles of multialternativity that are carried out both in live and in lifeless structures.

Special attention should be paid to the important and often missed attribute of evolutionary process which is essentially defining the possibility of genetic alternatives' accumulation and, as a result, a possibility of natural selection of the best adapted genetic alternatives. This attribute consists in the discretization of the inherited signs following, in turn, from the discretization of genetic structures. Dawkins [28] comes to the conclusion that discretization of gene forms determines its high stability, i.e., its evolutionary «immortality».

An attribute of discretization of genetic structures is generalized now to the principle of the block-and-modular organization and the evolution of molecular and genetic systems according to which the evolution of biological formations of any degree of complexity results from the combination of already existing, stably functioning blocks-modules of the lower level of complexity [29].

4 Conclusion

The comparison of structure and functioning of cyber-physical and biological systems indicates the direct analogy of these processes of adaptive evolution in them.

The analysis of mathematical models of adaptive evolution of pre-biological and population biosystems given above makes it possible to establish that one of the fundamental mechanisms of this evolution is the multialternativity.

At a pre-biological stage of emergence of life this mechanism is expressed in the cross mutational variations of selective value of the macromolecules excluding a complete dominance of only one type of them i.e. providing a mutual but also the competing existence of various types of macromolecules. In the homogeneous environment a stable replication of these molecules becomes impossible.

For biosystems at the level of populations the determining condition of the evolution is a growth of the genetic variety of a population with the growth speed exceeding the speed of a stream of critical events in the external environment. The violation of this condition caused, for example, by the decreasing population size below some critical value, leads to its degeneration.

Thus, the evolutionary principles of multialternativity are the component of the general cybernetic principles of the creation of cyber-physical systems providing the unique adaptive opportunities to these systems similar to live biosystems.

References

1. Lee EA, Seshia SA (2011) Introduction to Embedded Systems—A Cyber-Physical Systems Approach. LeeSeshia.org
2. Fitzgerald, J., Larsen, P.G., Verhoef, M. (eds.): Collaborative Design for Embedded Systems: Co-modelling and Cosimulation. Springer, Berlin, Heidelberg (2014)
3. Krassilov, V.A.: Epistemological approaches to the systemic evolution theory (SET). Bot. Pac. J. Plant Sci. Conserv. **4**(1), 3–6 (2015)
4. Galimov, E.M.: Vernadsky institute of geochemistry and analytical chemistry: scientific results in 2011–2015. Geochem. Int. **54**(13), 1096–1135 (2016)
5. Ponomarenko, A.G., Prokin, A.A.: Review of paleontological data on the evolution of aquatic beetles (coleoptera). Paleontol. J. **45**(13), 1383–1412 (2015)
6. Nikisianis, N., Stamou, G.P.: Harmony as ideology: questioning the diversity–stability hypothesis. Acta. Biotheor. **64**(1), 33–64 (2016)
7. Dmitriev, V.Y.: Evolution of biodiversity: hyperbola or exponent? Paleontol. J. **45**(6), 705–708 (2011)
8. Podvalny, S.L., Vasiljev, E.M.: A multi-alternative approach to control in open systems: origins, current state, and future prospects. Autom. Remote Control **76**(8), 1471–1499 (2015)
9. Podvalny, S.L., Vasiljev, E.M.: Evolutionary principles for construction of intellectual systems of multi-alternative control. Autom. Remote Control **76**(2), 311–317 (2015)
10. Convention on Biological Diversity. United Nations. Treaty Series **1760**(30619), (1992)
11. Visconti P, Elias VV, Sousa Pinto I et al.: Status, trends and future dynamics of biodiversity and ecosystems underpinning nature's contributions to people. The IPBES regional assessment report on biodiversity and ecosystem services for Europe and Central Asia, pp 187–384. IPBES, Bonn (2018)

12. Dyson, F.J.: A model for the origin of life. J. Mol. Evol. **18**(5), 344–350 (1982)
13. Podvalny, S.L., Vasiljev, E.M., Barabanov, V.F.: Models of multi-alternative control and decision-making in complex systems. Autom. Remote Control **75**(10), 1886–1890 (2014)
14. Kato, M. (ed.): The Biology of Biodiversity. Springer, Tokyo (2000)
15. Dreyer, G.D.: Saving biological diversity: an overview. In: Askins, R.A., Dreyer, G.D., Visgilio, G.R., et al. (eds.) Saving Biological Diversity: Balancing Protection of Endangered Species and Ecosystems, pp. 1–11. Springer, Boston (2008)
16. Mammides, C.: European Union's conservation efforts are taxonomically biased. Biodivers. Conserv. **5**, 1291–1296 (2019)
17. Fisher RA (2011) The Genetical Theory of Natural Selection. Oxford University Press
18. Eigen M, Gardiner W, Schuster P at al (1981) The origin of genetic information. Sci Am 244(4):88–118
19. Anderson, P.W.: Suggested model for prebiotic evolution: the use of chaos. Proc. Nat. Acad. Sci. U. S. A. **80**(11), 3386–3390 (1983)
20. Crick, F.H.: The origin of the genetic code. J. Mol. Biol. **38**(3), 367–379 (1968)
21. White DH (1980) A theory for the origin of a self-replicating chemical system. Natural selection of the autogen from short random oligomers. J. Mol. Evol.16(2):121–147
22. Red'ko, V.G.: Mechanisms of interaction between learning and evolution. Biologically Inspired Cogn. Architect. **22**, 95–103 (2017)
23. Plenk K, Bardy K, Höhn M at al.: Long-term survival and successful conservation? Low genetic diversity but no evidence for reduced reproductive success at the north-westernmost range edge of Poa badensis (Poaceae) in Central Europe. Biodivers. Conserv. 28(5):1245–1265 (2019)
24. Broeck, A.V., Cox, K., Melosik, I., et al.: Genetic diversity loss and homogenization in urban trees: the case of Tilia × europaea in Belgium and the Netherlands. Biodivers. Conserv. **27**(14), 3777–3792 (2018)
25. Tilman, D.: Causes, consequences and ethics of biodiversity. Nature **405**, 208–211 (2000)
26. Olvera, J.A.C.: Multi-alternative sequential analysis as a realistic model of biological decision-making. Ph.D. Thesis, University of Sheffield (2012)
27. Zamdborg, L., Spirov, A.V., Holloway, D.M., at al.: Forced evolution in silico by artificial transposons and their genetic operators: the ant navigation problem. Inf. Sci. 306:88–110 (2015)
28. Dawkins, R.: The Selfish Gene. Oxford University Press, USA (2016)
29. Iordansky, N.N.: Functional relationships in the jaw apparatus of the chameleons and the evolution of adaptive complexes. Biol. Bull. **43**(9), 1195–1202 (2016)

Regularization Methods for the Stable Identification of Probabilistic Characteristics of Stochastic Structures

Vladimir Kulikov and Alexander Kulikov

Abstract This chapter covers the method of experimental data analysis and processing in cyber-physical systems for medical monitoring, control of manufacturing processes and management of industrial facilities. The suggested methods are used to develop mathematical models of dynamic systems with stochastic properties for managing complex structural subsystems of cyber-physical systems. The most important computational stage of the simulation is thereat the identification of multimodal (in general) densities of the random variable distribution. A matrix conditioning analysis is herein suggested with minimizing relevant functionals of the identification problem. For the method of identifying multimodal densities of random variable distribution a matrix condition analysis is suggested with minimizing the relevant functionals of the problem. It is shown that under ill-conditioning of the equivalent system of equations an algorithm for regularization of solutions is needed. The regularization of the basic method for identifying distribution densities based on the ridge regression-algorithm (RRA) is proposed and substantiated. The classical RRA is improved and modified for local regularization showing the advantage of the high-order unstable SLAEs over the classical Tikhonov method. The suggested regularization algorithms and programs are universal, applicable to the study of random structures in natural science, biomedicine, and computational mathematics.

Keywords Cyber-physical systems (CPS) · Parameter monitoring · Identification · Multimodal densities · Ill-conditioning of matrices · Ridge regression-algorithm · Minimization of functionals

---

V. Kulikov (✉) · A. Kulikov
Nizhny Novgorod State Technical University n. a. R. E. Alekseev, 24 Minin St., 603950 Nizhny Novgorod, Russia
e-mail: vb.kulikov@yandex.ru

A. Kulikov
e-mail: akulikov@nntu.ru

A. G. Kravets et al. (eds.), *Cyber-Physical Systems: Advances in Design & Modelling*, Studies in Systems, Decision and Control 259,
https://doi.org/10.1007/978-3-030-32579-4_14

1 Statement of the Problem

This chapter is concerned with the study of a universal method for identifying distribution densities of stochastic characteristics and its modification under regularization conditions. Regularization algorithms and methods are universal and applicable to the study of complex structures with random properties (technology, medicine, biology).

Cyber-physical systems (CPS) are in this regard of particular interest including subsystems described by stochastic characteristics. Nonlinear dynamic properties of similar structures, their evolution in terms of random processes and stochastic variables manifest themselves as complex (non-Gaussian, multimodal) distributions to be subject to correct identification [1, 2]. Experimental information for this procedure is usually provided by various sensors including those of the so-called "intellectual nature".

Such information on the law of characteristics distribution may further be used to simulate a mathematical stochastic model for a local control object of a CPS subsystem. The CPS global characteristics may also be identified during the control process by the proposed method just in near real-time for solving any system-wide problems. The considered approach at the level of algorithms and programs has a modular, universal character. It is easily integrated into existing cyber-physical systems. For example, it may be used to create an experimental-mathematical model of the human gastrointestinal tract functioning when analyzing EGEG signals [1], to develop systems for monitoring of critically important parameters of glass melting furnaces or vibration stability of aircraft engines during their testing and operation.

The considered methods for identification of distributions are being continuously evolved, improved. In particular, it is suggested to complete the method for identifying multimodal densities of random variable distribution [1–3] with a matrix conditioning analysis while minimizing basic functionals with algorithmic integration of a solution regularization module there into (in case of ill-conditioning of a problem).

As shown by computational experiments, some peculiarities in the singular spectrum of matrices of equations equivalent to minimized functionals may cause inadmissible distortions of solutions.

Any probable ill-conditioning of matrices may result in unstable and erroneous decisions making the procedure of restituting the random values distribution law to be incorrect. Such peculiarities include the cases when a rapid decrease in singular values to the level of 10^{-12} and below is noted in the singular matrix spectrum. Furthermore, such instability may be caused by the assignment of an unreasonably large number of steps in minimizing the functional.

When regularizing problems of identifying the random values distribution laws, it is suggested to apply the classical RR-algorithm (epsilon-structuring).

The RR-algorithm is a version of the Tikhonov regularization method. In general, this is an algorithm for solving one-dimensional inverse problems based on the regularization method for Fredholm integral equations of the 1st kind.

The Tikhonov regularization method [4] consists in the reduction of an integral equation to a system of linear algebraic equations the solution thereof is sought under restriction to the solution norm value. The selection of the restriction value depending on the perturbation value is a major problem in the regularization method [5–12]. The desired solution for an integral equation is represented as a piecewise constant function, i.e. vector φ, which coordinates are values of function $\varphi(t)$ at n points on the segment $[a, b]$.

Using the numerical integration formula the integral equation is replaced by the system of equations $R\varphi \approx y$, where R is a relevant matrix of size lxn, y is a vector composed of l independent measurements of the observed function. Many publications are devoted to solving unstable systems of algebraic equations with approximately specified right-hand sides; in particular, there should be noted [13–15].

Since the RR-algorithm (ridge regression-algorithm) is methodologically most developed as a stochastic regularization method (within the framework of a multidimensional linear regression problem), the choice of this regularization method as a more common one (as compared to the classical Tikhonov regularization algorithm) is regarded to be reasonable and correct [16].

In future chapters, there will also be studied some original and modified methods for the regularization of singular decompositions in solving ill-conditioned systems of linear equations taking into account the above features. These approaches are implemented in the form of algorithms and software (MATLAB) for the method of identifying multimodal densities of random variables distribution.

2 Method of Solving the Problem

So, when minimizing the functional, it is necessary to solve the system of equations under constraint conditions [17] $\|\varphi\|^2 = \sum_{i=1}^{n} \varphi_i^2 \le r^2$, that is equivalent to minimizing the smoothing functional

$$M^{\alpha}(\varphi) = \|y - R\varphi\|^2 + \alpha \|\varphi\|^2,$$

where α is the constraint-compliant regularization parameter (Lagrange multiplier). In terms of matrix operators, a regularized normal system is solved

$$(R^T R + \alpha E)\varphi = R^T y \tag{1}$$

The RR-algorithm consists of the following steps:

1. ε—net is introduced in the system solution space, which consists of the following vectors
$$V = \sum_{j=1}^{n} \frac{\varepsilon \mu_j}{\lambda_j} \psi_j$$

where λ_j and ψ_j are matrix eigenvalues and eigenvectors R^TR, μ_j are random integral values, ε is a scalar parameter assigning a step to ε-net.

2. With a fixed value of the parameter α, the normal system (1) is solved, ε-net node is sought closest to the solution found for φ_α.
3. Further, the quality of the φ_α solution is assessed with fixed α, which is composed of the following two constituents: the estimate of the mean risk functional value achieved on the vector V_α (the closest node of ε-net) plus the estimate of a change in the value of the same function in the transition from φ_α to V_α. The final estimate [17] takes the form

$$J(\alpha) = \left[\frac{\frac{1}{l}\,\|y - R\,V_\alpha\|^2}{1 - \sqrt{\left[k\left(\ln\frac{l}{k} + 1\right) - \ln\eta\right]/l}} \right]_\infty + \frac{1}{l}\,\|R(V_\alpha - \varphi_\alpha)\|^2\,,$$

where $1 - \eta$ is the likelihood of this estimate validity, k is the number of linearly independent nodes of ε-net satisfying the condition $\|V\| \le \|\varphi_\alpha\|$.

The regularization parameter value α and the constraint valuer is determined by the minimization of the estimate of $J(\alpha)$ for α and ε, thereunder the resulting solution φ_α is optimal for a specified experimental data volume.

It is important to note that the RR-algorithm plots for regression problems ridge estimates which are shifted but best with respect to the standard mean square deviation (with some constraints by the value of an input data error) than the least-squares method in the classical formulation [17]. These estimates are more stable than least-squares estimates since the reduction in error variance caused thereby is more than it is required for the compensation of offset entered.

The author's modification of this algorithm is suggested in this chapter using the MATLAB package, namely:

- the replacement of the algorithm for finding eigenvalues and eigenfunctions (implemented through the standard EIGEN procedure in the FORTRAN system) by the MATLAB package algorithm for computations by means of orthogonal similar transformations with a matrix in the upper Hessenberg form and using the QR-algorithm of Francis and Kublanovskaya;
- the replacement of the algorithm for solving the regularized SLAE (1) to find φ_α with the transformation to a triangular form by the algorithm for system matrix pseudo-inversion using the SVD decomposition;
- the generalization of the classical RR-algorithm with the scalar parameter of global regularization to a local version (this is a major distinctive feature), then α becomes a random vector of a relevant dimension (n) with various components.

In the first two options, we get more economical and fast-acting algorithms to simplify the software design procedure and also have the possibility to control the stability of solutions.

In the latter case, a new original regularization option is suggested, when the found optimal value α^* serves as a scale factor to specify n samples of a uniformly distributed random variable on the segment [0, 1]. In this case at each step of the cycle, the arrangement of the random vector projections in descending order is performed and local regularization parameters are formed, the maximum component thereof is approximately equal to (but does not exceed) α^*.

Further, this vector shall be substituted into (1) (where matrix "castling move" of α and E has been made) to find a locally regularized solution for φ_α^* and to calculate the empirical risk value on the resulting solution:

$$I_э = \frac{1}{l}\left\| y - R\varphi_\alpha^* \right\|^2. \quad (2)$$

The stochastically optimal solution of system (1) shall be finally determined by minimization (2). An approach associated with the simulation of disturbances and the analysis of the subsequent reaction of a structure or a system is widely used in computational practice. These methods are based on the classical Monte Carlo algorithms and their current modifications [18].

To verify the work of the RR-algorithm and regularization programs, the solutions of test SLAEs were performed with rectangular matrices of higher dimensionality. The maximum matrix size was (315 × 190). As a test, there was studied the integral equation solution using random-noise distorted measurements of the right-hand side [17].

The integral equation algebraization reduces to a system with the number of equations M = 210, number of unknowns K = 126. Matrix elements were calculated by the following formula

$$\begin{aligned} &A(I, J) = [0.164 * (I - 1) - 0 * 328 * (J - 1)]_+; \\ &I = 1, \ldots, 210;\ J = 1, \ldots, 126; \\ &[\ldots]_+ = \{x, x \geq 0; 0, x < 0\}. \end{aligned} \quad (3)$$

All non-zero matrix elements lie below the main "quasi-diagonal", matrix A is extremely ill-conditioned, rank (A) = 105, cond (A) = Inf (an infinitude in MATLAB). Some results of solutions of SLAEs with matrix A (210 × 126) are shown in Fig. 1.

Note that for version (b) the optimal regularization parameter $\alpha = 0.005$, the discrepancy with the optimal regularization parameter is d = 4.2549e−005. In version (c) the perturbation level for each of six projections is ~7%, $\alpha = 0.50$, d = 0.0025; for (d) $\alpha = 0.05$, d = 1.863e−004. The calculation time of a regularized solution for each version is no more than 1.3 s.

As we can see in the Fig. 1 in the test example the "exact" solution has a locally smooth but "impulse" form. The number of sign variations reaches six. Such problems are quite difficult for the correct solution recovery under conditions of right-side projection distortion [19–22]. The suggested algorithms demonstrate sufficiently high efficiency and speed.

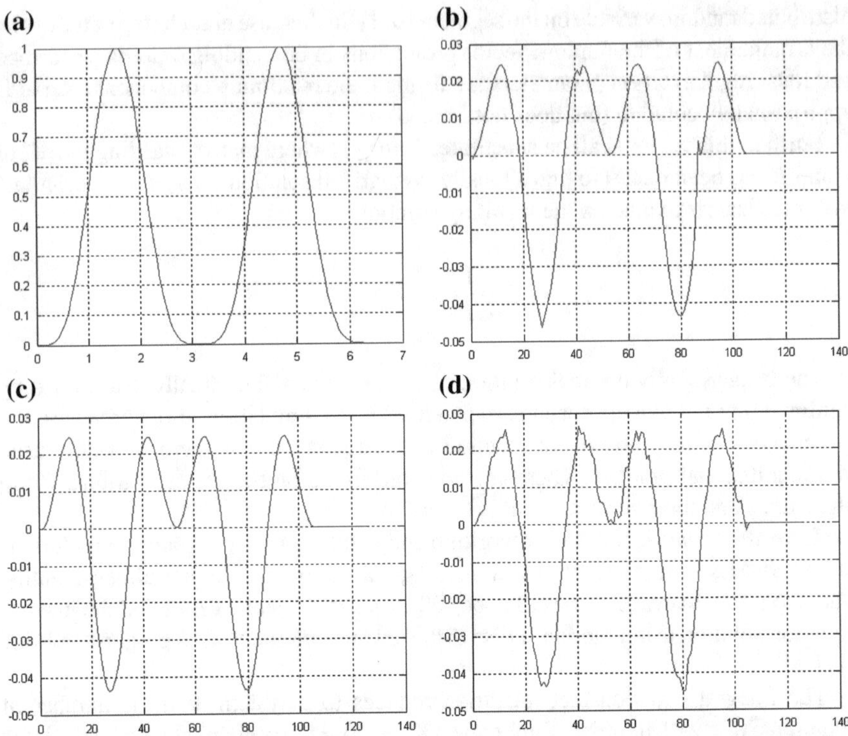

Fig. 1 Verification of the RR-algorithm for regularizing solutions of SLAEs with matrix A (210 × 126): **a** the graphic chart for projections of the right-side unperturbed vector Y on the segment [0, 2π]; **b** the graphic chart for projections of the SLAEs solution in case of no Y perturbation; **c** the solution for deterministic distortion of six projections of *Y*; **d** the solution for a random perturbation of all projections of *Y* (the normal interference distribution law with zero mean and m.s.d. equal to 0.001); on the x-axis for (**b–d**)—solution projections readings

3 Algorithms and Singularities of Problem Solution for RR-Algorithm Modification

The original regularization version, when the found optimal α^* value serves as a large-scale factor in the formation of local regularization parameters, has passed testing on very complex examples.

Computational experiments in solving SLAEs with an approximately specified right sideshow that this algorithm is not always effective. The high algorithm accuracy is limited by the structure of the singular spectrum of SLAE matrices as well as by the conditioning number (not more than 10^{12}). Based on the accumulated experience the author suggests for complex problems a more perfect algorithm (in the general case of vectorial one) of RR-regularization.

The essence thereof is as follows. The stage is preserved when local regularization parameters are formed and their maximum component is approximately equal to

α^*. A situation may herewith occur when the "tail" of the regularization vector contains, just by chance, too small parameters. The procedure of "cutting off" these parameters founds the basis for a new algorithm. The randomness of the α^* value is simultaneously taken into account, which is determined by the random nature of vector y of system R $\varphi \approx$ y (measurement errors or a parametric uncertainty factor in the experiment).

To this end, the found optimal α^* value enters the "stochastic cycle" (of 50–100 iterations) for multiple calculations of:

1. local regularization parameters (uniform distribution);
2. a solution of the relevant regularized SLAE;
3. finding the empirical risk value $I_э$—formula (2);
4. determining a super-optimal vector of regularization parameters.

At each step of the cycle, firstly, a vector of local regularization parameters is calculated; secondly, to the α^* value, a proportionally small random variable is added distributed according to the normal law with small variance and zero means. For example, $\alpha^* = \quad 0.500, CL\ (i) = \alpha^* \ + 0.015 * randn$. Where CL(i) is the i-th threshold value of regularization parameters, $\sigma = 0.015$—m.s.d. of perturbation. Further, all local regularization parameters smaller than CL(i) shall be replaced by this computed constant. The algorithm performs this operation at each step of the stochastic cycle. The selection of the optimal regularization vector occurs as before in minimizing the functional (2).

Thus, a stochastically stable and super-optimal solution is chosen: there are no too small incidental values of local parameters. The level of smallness in replaced parameters depends on the initial singular spectrum of a matrix. In fact, in the final local parameters there is a combination of the following two distributions and two versions:

- the uniform distribution for the "upper" parameters, when *CL* (*i*) is smaller than α^* and the "Gaussian threshold" for the "lower" parameters, i.e. *CL*(*i*) has the normal distribution and determines the constant, which replaces small regularization parameters;
- when the random (Gaussian) value *CL*(*i*) is higher than α^*, we have one global constant of the regularization parameter (the vector shifts to a super-optimal scalar).

The new algorithm of the vector RR-regularization has passed multiple tests on a set of test problems of higher complexity, proved the high efficiency and computational accuracy for extremely ill-conditioned systems of equations. This algorithm is recommended for matrices with conditioning number cond (A) > 10^{12} at significantly higher levels of interference for the right side of the SLAE.

Many numerical results described below illustrate the new algorithm and its superiority as compared to the SLAE solution algorithm by the classical Tikhonov method.

We describe now the "technology" and stages of the proposed super-optimal algorithm on the example of the above problem of solving an integral equation using random-interference distorted measurements on the right-hand side. The algorithm is implemented in MATLAB.

The complicated system A x = Y is solved with the number of equations M = 315 and number of unknowns K = 190. Matrix A elements are calculated by formula (3) with I = 1, …, 315; J = 1, …, 190. The right side, as before, is assigned by the expression Y = Y0 = $\sin^4(x/2)$, $0 \leq x \leq 2\pi$. Matrix A is extremely ill-conditioned, rank (A) = 157, cond (A) = Inf.

Some results of solutions for SLAE with matrix A (315 × 190) are shown in Figs. 2, 3, 4, 5 and 6. The initial perturbation for Y is thereat an interference with a normal distribution law—zero mean and s.m.d. equal to 0.003 (i.e. Y = Y0 + 0.003*randn). The interference level is here three times higher than for the problem with A (210 × 126).

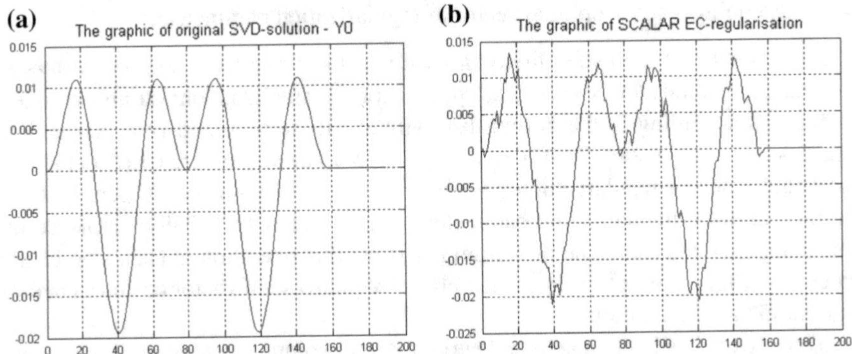

Fig. 2 Verification of algorithms when solving SLAEs with matrix A (315 × 190): **a** the graphic chart of the SLAE solution by the SVD algorithm in case of no *Y* perturbation; **b** "scalar" regularized RR-solution with the initial perturbation of all *Y* projections; on the abscissa axis—readings of solution projections

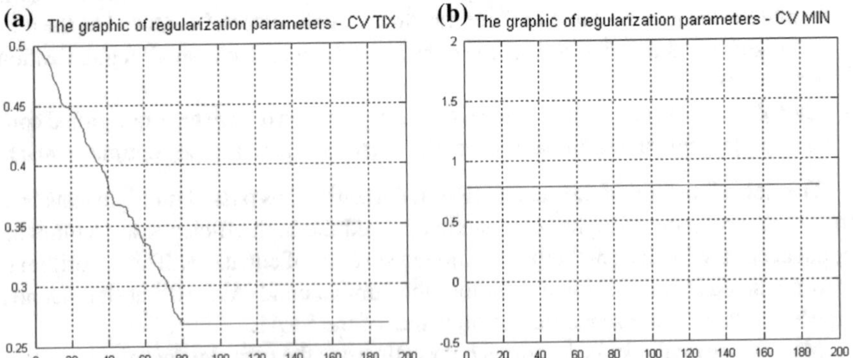

Fig. 3 RR-algorithm for vector regularization of solutions of SLAEs with matrix A (315 × 190): **a** the graphic chart of regularized local parameters with the initial *Y* perturbation—the optimization in the cycle by the classical Tikhonov functional; **b** the graphic chart of the super-optimal regularization vector found by the minimum of the functional $I_Э$; $CL_{OPT} = 0.687$

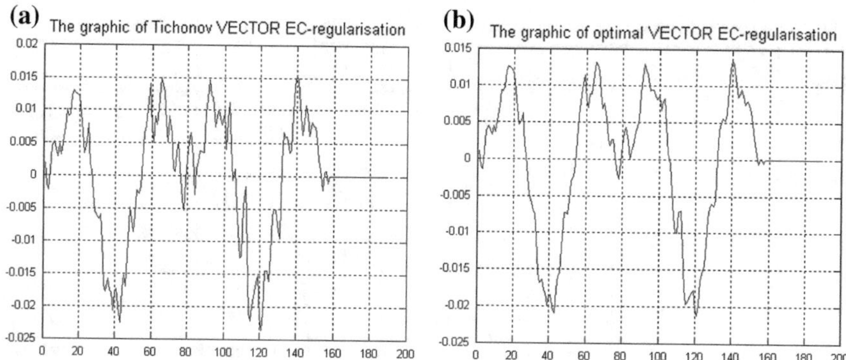

Fig. 4 Regularized solutions of the modified RR-algorithm for SLAEs with matrix A (315 × 190)—(test perturbation $Y = Y0 + \mathbf{0.005} * \text{randn}$): **a** for local regularization parameters (Fig. 3a)—optimization in the stochastic cycle by the Tikhonov smoothing functional; **b** when the supra-optimal vector of regularization parameters (Fig. 3b) found on the minimum of the functional (f.2)

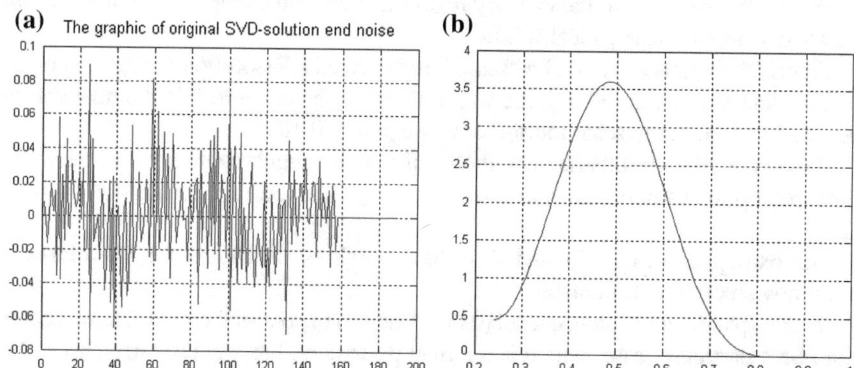

Fig. 5 Verification of solutions of SLAEs with matrix A (315 × 190): **a** the graphic chart of the SLAEs solution by the standard SVD algorithm without regularization—the initial perturbation level of *Y*; **b** the identified distribution density of *CL* values in a cycle of 100 iterations for the supra-optimal RR-regularization algorithm

Vectors of regularized parameters found as a result of optimizations (according to A. N. Tikhonov and the super-optimal RR-algorithm) are used for solutions of SLAEs with other perturbation levels, for example, Y = Y0 + 0.005 * randn. In this case, the quality of the synthesized regularization algorithm is verified on random independent variations. It is assumed that the test interference levels may exceed the initial level.

We comment now on the results of a computational experiment, compare them to the method of obtaining normal pseudo-solutions of SLAE (the method of singular decompositions). Figure 2a represents such a solution for an exactly defined right part

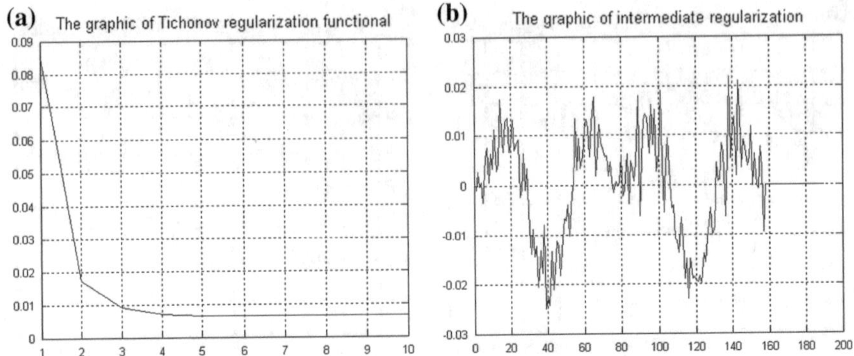

Fig. 6 Verification of the EC algorithm in solving SLAEs with matrix A (315 × 190): **a** the graphic chart for values of the Tikhonov functional when solving the SLAE—the initial perturbation level of Y); **b** the graphic chart for the system solution at the minimum of the classical Tikhonov functional

of the test system. Due to the primary regularization of the singular decomposition, we have a smooth nearly ideal solution.

Figure 2b depicts a graph of a "scalar" regularized RR-solution for optimal $\alpha^* = 0.500$. Note that the initial α value was assigned to be equal to 5.0; the final one to be equal to 1.10^{-4}; the scale multiplier was equal to 0.10.

Figures 3b and 4b illustrate the choice of the best regularization parameters and the super-optimal solution by minimizing the empirical risk functional using formula (2).

For example,—(Figs. 3a and 4a) show versions of choice for the classical Tikhonov smoothing functional.

With a specific interference level (0.003 * randn) the optimal regularization vector shall be a particular case, which is a global parameter. For small interferences, the algorithm provides the limiting properties of the regularization principle.

Numerous tests show that the proposed super-optimal RR-algorithm surpasses the classical method.

Figure 5a shows the SLAE solution using the standard SVD algorithm (MATLAB) with the initial perturbation level of Y. The saw-tooth "spread" of solutions even with weak perturbations makes it impossible to use the SVD algorithm without regularization.

The method for identifying the laws of random variables distribution elaborated by the author of this chapter enables to restore the distribution density for CL values in an iterative cycle of 100 steps (Fig. 5b), thus, confirming the correctness of the suggested algorithm and its modifications.

Figure 6 provides the perception of what solution may be obtained if in the RR-algorithm the φ_α solution quality and the Lagrange multiplier are assessed not by the functional $J(\alpha)$ but by the classical Tikhonov smoothing functional [4]. The minimum of this function is either weakly expressed in the case of small values of the regularization constant or there is an asymptotically slow decrease of the functional $M^{\alpha}(\varphi) = \|y - R\varphi\|^2 + \alpha \|\varphi\|^2$.

It is seen that the suggested versions of the RR-algorithm modifications surpass the classical methods. If necessary, the modified solutions of the RR-algorithm may be supplemented with a smoothing procedure of the arithmetic average of two adjacent values. The total calculation time of the results given is no more than 10 s.

Regularization software modules are built into the basic module for identifying random distribution densities in order to improve the operational reliability under conditions when it is a priori known that a complex structure is identified with many extremes of the multimodal distribution.

The elaborated algorithms are used in higher intricate applied problems. In particular, in restoring the distribution density of pore sizes in polymer membranes when studying the morphology of their surfaces by atomic force microscopy method (AFM) using the SPM-9700 scanning probe microscope (Shimadzu, Japan). Based on the proposed method, algorithms and programs, the results obtained confirm the multimodal density of the distribution of pore sizes and depth profiles.

The maximum scanning field of a microscope is an area of 30 × 30 μm in size. The characteristic geometric dimensions of polymer membranes measured by the AFM method are 3–10 μm, the pore sizes are several orders smaller. The obtained results are stated in detail in the author's dissertation thesis.

In biomedicine, the considered method has shown high-resolution properties by the example of monitoring and stochastic analysis of EGEG signals from the actual clinical practice [1]. The identification of local signal sections has revealed the universal characteristics of fractional Brownian processes—the multimodal distribution of their parameters.

And multimodal stochastic characteristics contain information about the non-equilibrium, non-linear dynamics of the state of the body organs and systems. In this case, based on the density of local sections of the process there are calculated moment functions, entropy characteristics, the studied therapy process is monitored and expert health management systems are formed.

4 Results and Conclusions

The classical RR-regularization method is modified and implemented in the MATLAB package, evaluated in test problems of relatively high dimensionality (matrices of several hundred lines and columns).

It is shown that to obtain stable solutions with minimizing the basic functionals (at the stage of solving the SLAEs), the algorithm regularization is required.

The regularization of the basic method for identifying distribution densities based on the RR-algorithm was proposed and validated.

The classical RR-algorithm is developed within the framework of the original author's methods for vector regularization versions.

Computational experiments in solving SLAEs with matrices having a conditionality number of 10^{12} or more confirmed the effectiveness of regularization algorithms.

The proposed approaches are important for the RR-algorithm development in local regularization under conditions of giving a priori information about the type of the identified distribution law and its structural features [23].

The scope of application for the developed regularization methods and programs covers flying object control systems (spacecraft, aeronautics), the regulation and control of complex technological processes, the monitoring of medical parameters, the analysis of operators of singular decompositions and multidimensional data arrays of computational mathematics [1–3].

Acknowledgements The study was carried out with the financial support from the Russian Fund for Basic Research as part of a research project № 19-07-00926_a.

References

1. Kulikov, V.: The identification of the distribution density in the realization of stochastic processes by the regularization method. Appl. Mathem. Sci. **9**(137), 6827–6834 (2015)
2. Kulikov, V., Kulikov, A.: Applied problems of the identification and regression analysis in stochastic structures with multimodal properties. ITM Web of Conferences, vol. 6, article 03008 (2016)
3. Kulikov, V.B., Kulikov, A.B., Khranilov, V.P.: The analysis of stochastic properties of the SVD decomposition at approximation of the experimental data. Procedia Comput. Sci. **103**, 114–119 (2017)
4. Tikhonov, A.N., Arsenin, V.Y.: Methods of Ill-Conditioned Problems Solution. Nauka, Moscow (1986). (in Russian)
5. Bertero, M.: Regularized and positive-constrained inverse methods in the problem of object restoration. Opt. Act. **28**(12), 1635–1649 (1981)
6. Engl, H.W.: Regularization Methods for Inverse Problems. Kluwer Academic Publishers (1996)
7. Golub, G.H.: Generalized cross-validation as a method for choosing a good ridge parameter. Technometrics **21**, 215–222 (1979)
8. Urmanov, A.M.: Information complexity-based regularization parameter selection for the solution of ill-conditioned inverse problems. Inverse Prob. **18**(2), L1–L9 (2002)
9. Karajiannis, N.B.: Regularization theory in image restoration—the stabilizing functional approach. IEEE Trans. Acoust. Speech Sign. Process. **38**(7), 1155–1179 (1990)
10. Lukas, M.A.: Comparison of parameter choice methods for regularization with discrete noisy data. Inverse Prob. **14**(2), 161–184 (2000)
11. Reginska, T.: A regularization parameter in discrete ill-posed problems. SIAM J. Sci. Comput. **17**(6), 740–749 (1996)
12. Neto, F.D.M.: Introduction to inverse problems with applications. Springer, Cham (2013)
13. Gebali, F.: Solving Systems of Linear Equations. Algorithms and Parallel Computing. Wiley (2011)
14. Hansen, P.C.: Rank-Deficient and Discrete Ill-posed Problems. Numerical Aspects of Linear Inversion. SIAM, Philadelphia (1998)
15. Hoang, N.S.: Solving ill-conditioned linear algebraic systems by the dynamical systems method. Inverse Probl. Sci. Eng. **16**(5), 617–630 (2008)
16. Hoerl, A.E., Kennard, R.W.: Ridge regression: biased estimation for non-orthogonal problems. Technometrics **12**, 55–67 (1970)

17. Vapnik, V.N.: Algorithms and Programs for Recovery of Dependence. Nauka, Moscow (1984). (in Russian)
18. Binder, K.: Monte Carlo Simulation in Statistical Physics. An Introduction. Springer, Berlin (1992)
19. Goos, P.: Optimal Design of Experiments. A Case Study Approach. Wiley (2011)
20. Hansen, P.C.: Regularization tools: a Matlab package for analysis and solution of discrete ill-posed problems. Numer. Algorithms **6**(1), 1–35 (1994)
21. Siebertz, K.: Statistische Versuchsplanung. Design of Experiments. Springer, Heidelberg, Dordrecht, London, New York (2010)
22. Wang, Y.: Optimization and Regularization of Computational Inverse Problems and Applications. Springer, (2011)
23. Voskoboinikov, Y.E., Mukhina, I.N.: Local regularizing algorithm for high-contrast image and signal restoration. Optoelectron. Instrum. Data Process. **3**, 41–48 (2000)

Outlier Detection in Predictive Analytics for Energy Equipment

Alexander Andryushin, Ivan Shcherbatov, Nina Dolbikova, Anna Kuznetsova and Grigory Tsurikov

Abstract The method of data preprocessing used to predict the technical condition of power equipment is described. Preprocessing implemented using neural networks allows us to identify and eliminate outliers in the investigated data. An example illustrating the proposed method of processing big data using bagged trees algorithm, support vector machines and artificial neural networks is shown.

Keywords Predictive analytics system · Energy · Data preprocessing · Neural network · Bagged trees · Support vector machines

1 Background

At present, the organization of repairs of energy equipment in its actual state is an extremely urgent task [1]. In this regard, there is the task of forecasting the technical condition of energy equipment for the timely implementation of measures to maintain it in proper technical condition. The construction of an accurate prediction of the time of failure of equipment (failure) will allow planning appropriate measures and resources (financial, human, materials and machinery) to bring a specific piece of equipment to the proper technical condition [2]. For forecasting, characteristics based on expert knowledge formulated during the operation of equipment by heuristics and the assumption that the dynamics of changes in a number of parameters indicate equipment failure with a certain certainty after a certain time interval is mainly used. Such systems are called predictive systems. Predictive analytics systems are increasingly used in various fields due to the increasing capabilities of computer technology for the software implementation of mathematical methods for processing large data arrays. Big data are often not homogeneous [3]. Therefore, methods for processing big data (Big Data) [4] and predictive analytics systems are inextricably linked.

A. Andryushin (✉) · I. Shcherbatov · N. Dolbikova · A. Kuznetsova · G. Tsurikov
Moscow Power Engineering Institute, 14 Krasnokazarmennaya Street, Moscow 111250, Russia
e-mail: andriushinav@mpei.ru

A. G. Kravets et al. (eds.), *Cyber-Physical Systems: Advances in Design & Modelling*, Studies in Systems, Decision and Control 259,
https://doi.org/10.1007/978-3-030-32579-4_15

In this study, big data will be understood as unstructured telemetry information tied to real-time readings and coming from ASKUE-level and SCADA systems, as well as accounting systems (for example, fixing defects, diagnostic results, etc.), MES (Manufacturing Execution System), ERP (Enterprise Resource Planning,), EAM (Enterprise Asset Management) [5]. Big data and opportunities for their processing should provide the power industry with a qualitative leap forward in the near future [6]. In fact, the ability to process large data (along with the use of data from highly qualified experts) can become one of the competitive advantages of a modern energy enterprise. For most predictive analytics systems, the following structure is characteristic [7]: data import into the system; data conversion and preparation; clustering to identify the data that then enter the predictive (predictive) model; prediction of trends and failures based on a predictive model. Predictive analytics systems and big data processing methods are part of the technological basis of the industrial revolution or Industry 4.0 [8].

In the predictive analytics systems, various methods are used to obtain and predict the technical condition: the k-means algorithm and the auto-regression model for predicting parameter values [8]; machine learning systems based on artificial neural networks (INS) [9]; multi-agent approach [10]; deep learning methods [11] and others.

2 General Scheme of Data Processing

The quality of solving problems with the use of SPA predictive analytic systems in the energy sector and the efficiency of functioning of other enterprise systems using forecast data depends on the processing of data that is implemented in modern predictive analytic systems technological equipment for energy enterprises (Fig. 1).

The collection of large data is fraught with difficulties since they come from various lower-level systems, for example, SCADA and/or AMR systems, as well as higher levels, such as PLM, MES, EAM. At the same time, the data itself is non-uniform, has different types, the structure of transmission and methods of storage. One of the key subtasks for obtaining a forecast of the technical condition and other

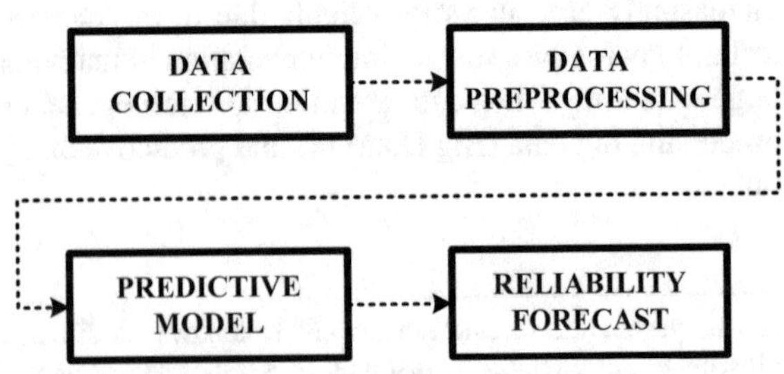

Fig. 1 Processes of processing big data in predictive analytics systems

indicators characterizing the operation of energy equipment in various modes should actually be solved using the Industrial Internet of Things technology (IIoT) [12] and goes beyond the scope of this work.

For the purpose of making predictions based on retrospective data, machine learning systems based on artificial neural networks (INS) can be used [13].

The methodology IDARTS (Intelligent Data Analysis and Real-Time Supervision) uses solutions based on knowledge and rules used for making decisions, identifying faults, potential deviations and other events that are critical for equipment operation [10]. The practical implementation of IDARTS is a multi-agent system where there are a number of agents who are responsible for monitoring equipment, monitoring subsystems combining several pieces of equipment, etc. Agents are used, for example, also in [9].

Methods of deep learning, for example, multilayered neural networks of various types can be used to monitor operating conditions, identify incipient defects, diagnose root causes of failures [14], diagnose and classify equipment faults [15]. The work [3] also discusses the need to form an intelligent maintenance strategy, which will determine the current state of the equipment during its operation to predict the maintenance activities. It is noted that a general recurrent neural network can be used to predict the propagation of defects and estimate the remaining service life of elements and subsystems.

In this case, the key subtask is data preprocessing, the purpose of which is to obtain a general set of data as the basis for the synthesis or correction of a predictive model and a forecast. The incoming data should be filtered out of outliers, noise, and errors, as well as tied to the real time scale.

There are various methods and algorithms of emissions (and anomalies in general) searching that can be used in the predictive analytics systems of energy industry enterprises. These methods include:

- iterative methods or methods of convex hulls constructing [16], which eliminate the need to fit a specific type of distribution (the constructed convex hull is removed from consideration because the points that belong to it are considered emissions). The complexity of these methods exceeds the rest;
- metric methods (for example, the LOC—Local Outlier Factor method [17]) are based on the estimated distance of the sample element from a subset of the elements closest to it. These methods do not work well with samples that simultaneously contain two regions—a dense arrangement of parameter values and a rarefaction region [18];
- clustering—splitting into clusters can be a way to isolate anomalies, in the sense that outliers will belong to a small cluster that contains few values [19];
- spectral decomposition [18] uses the principal component analysis method (PCA) and splits the set of parameter values into subsets of normal, anomalous values and a subset of noise;
- classification considered in this paper allows us to assign measurements to one of two classes (in the simplest case)—normal and anomalous values, and preliminary information (i.e. training set) about the outliers is needed to construct a classifier [20].

Data preprocessing is important for obtaining a good forecast. For example, we will demonstrate the implementation of the data emission search, as one of the stages of data preprocessing necessary for predicting the combined-cycle power.

3 Method of Outlier Detection

As a baseline for predicting the power of the CCGT, the dependence of the total power of the steam and gas turbines on the following indicators is used: ambient temperature, pressure in the steam turbine condenser, relative humidity and atmospheric pressure. The listed parameters are presented in the form of 9586 averaged per hour values measured during the six years of the station operation at full load (from 2006 to 2011).

To build a regression model, it is necessary to exclude outliers from the data set, i.e. items that are significantly different from the rest. Subsequently, data outliers should be classified into data obtained as a result of unreliability of meter readings and anomalies.

The described problem of detecting outliers of data was solved, like a classification problem, by applying the following machine learning algorithms: ANN algorithms; decision trees begging algorithm; support vector algorithm (SVM).

Initially, a sample of the data needed to train the classification model was made. To determine the composition of the sample was carried out correlation analysis (Fig. 2).

It can be seen from the figure that the ambient air temperature and pressure in the turbine condenser have the greatest effect on the power values: the correlation coefficients are 0.95 and 0.87, respectively. Thus, having excluded from the study information on atmospheric pressure and relative humidity of the air, the composition of the training sample was determined. In addition, for learning, after presenting the data in three-dimensional space, 100 elements were selected, among which it was visually easy to determine the values of the parameters, markedly different from the others.

The resulting dataset was labeled using the GOST R ISO 16269-4-2017 methodology as follows: power indicators, defined as outliers, were marked with 1; all other values are marked with 0.

The feasibility of using the GOST methodology is that it is based on international ISO standards used to search for outliers. The essence of this technique is to present the available power versus temperature and pressure as linear regression and further search for outliers by components Y, X1, and X2, where Y is the power value, X1 is the ambient temperature X2 is the pressure in the steam turbine condenser.

Potential outliers for Y are determined by finding the studentized error and comparing it with the Student's t distribution with the following parameters: confidence level 0.995; the number of degrees of freedom $k = n - p - 2$, where n is the number of elements in the sample, p is the number of variables affecting the values of Y.

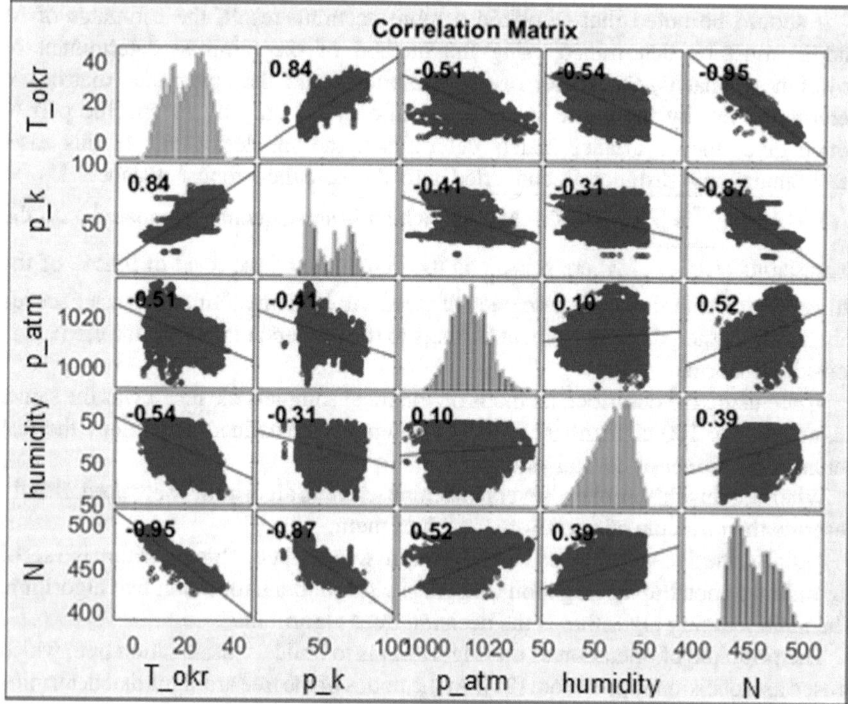

Fig. 2 Correlation analysis of investigated parameters

The potential outliers for components X are found by determining the parameter h, which characterizes the distance between the value taken by the random variable X at the i-th point and the arithmetic average of all n values taken by X. The sample element is considered an outlier if the value of this parameter is less than the result of the expression $h \leq 2 * (p + 1)/n$.

To establish the fact that potential outliers are real, the following values are determined: Cook distance; DFFITS indicator. Cook's distance is subsequently compared with the F-distribution indicator, characterized by parameters: a 50% level percentile, a $p + 1$ value, a $n - p - 1$ value. In turn, the indicator DFFITS is compared with the result of the expression $2 * \sqrt{(p + 1)/n}$.

Another method for determining emissions, approved by the ISO standard, is the method of determining the distance of Mahalanobis. This distance characterizes the location of the studied element from the data set relative to the average value of the set.

To determine the distance of Mahalanobis, initially, it is necessary to make a p-dimensional vector X, consisting of observations: $X = \left(X_1, X_2, X_3 \ldots, X_p\right)^T$. The following actions determine the matrix of mean values for each observation: $M = \left(M_1, M_2, M_3 \ldots, M_p\right)^T$, and the covariance matrix, Σ of size $p \times p$. Thus, the required distance: $M_D = \sqrt{(X - M)^T * \Sigma^{-1} * (X - M)}$.

It should be noted that to obtain a more accurate result, the estimates of M and Σ must be determined using the method of the minimal determinant of covariance. That is, the vector of average values and the covariance matrix are determined not by the entire set of observations, but only by its specific part h, which gives the covariance matrix determinant the smallest value. In this case, the Mahalanobis distance is converted into the so-called robust distance: $D_R = \sqrt{(X - M_{MCD})^T * \sum_{MCD}^{-1} *(X - M_{MCD})}$, which is subsequently compared with the expression: $\sqrt{\aleph^2_{0.995,p}}$, where $\aleph^2_{0.995,p}$ is the percentile of the level of 99.5% of the chi-square distribution with p degrees of freedom. If the resulting distance exceeds this criterion, then the data element belongs to the outliers; otherwise, the element is considered normal.

When using the described methods on identical samples, the result was the same, i.e. among the 100 elements studied, 9 elements were defined as outliers, the rest belonged to the set of normal values.

When solving the problem, several machine learning algorithms were used. Briefly describe the principle of operation of each of them.

Initially, the decision bagged trees algorithm was applied. This algorithm is a set of algorithms: bootstrap aggregation (bagging) algorithm and decision tree algorithm. The main learning algorithm is the decision trees algorithm.

The principle of operation of this algorithm is to build a classification tree, which is used as a decision support tool [21]. At the nodes of the tree are signs that determine the value of the objective function. The objective function in the problem in question provides the answer: is this element of the sample an outlier or not. Its values, 0 or 1, are written on the leaves of the tree. The remaining nodes of the decision tree are attributes that differentiate possible classification paths. For correct classification, it is necessary to sequentially pass the decision tree to the leaves and obtain the corresponding value of the objective function [22]. This will allow you to identify a specific sample element as an outlier, or refer it to the set of normal values used, subsequently, to create a predictive model.

An additional algorithm in the presented set of methods that are needed to improve the accuracy of classification is bootsrap-aggregation (bagging). This algorithm randomly varies the training sample and allows you to build several different decision trees.

After that, certain weights are assigned to each value of the objective function and the so-called voting between trees is performed. The results obtained, as a result, are averaged, and the classification accuracy is improved.

4 Implementation and Results

When working on a training set, the accuracy of the algorithm was 92%, that is, 2 outliers from 9 and 90 normal elements from 90 were correctly identified (Fig. 3). In addition, the resulting classification model was tested on a control sample consisting

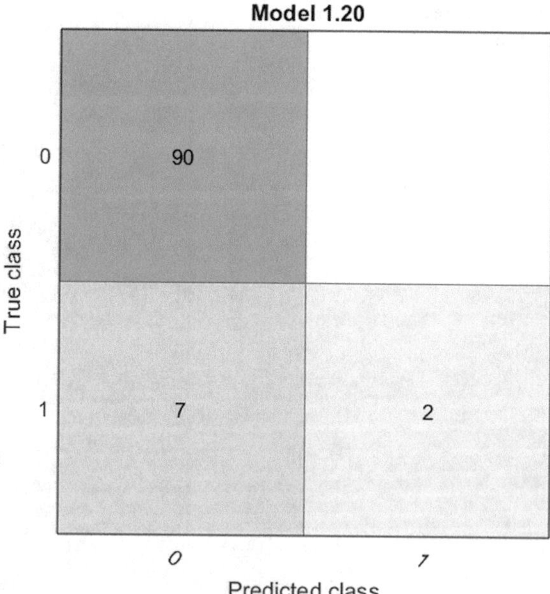

Fig. 3 The result of bagged trees algorithm

of fifty elements. As a result of testing, the algorithm made 6% errors, i.e. correctly classified 47 elements.

Also, to solve the problem, an SVM (support vector machine) algorithm with a polynomial kernel function was used [23]. The essence of this method consists in the separation of two classes, in this task: outliers and normal values, using a hyperplane. Those vectors that most closely lie to the hyperplane and contribute to its construction are support vectors. In general, if there is a training sample $\{(x_1, y_1), \ldots, (x_m, y_m)\}$, $x_i \in R^n$, $y_i \in \{0, 1\}$, then the classifying function $F(x) = \mathrm{sign}(\langle w, x\rangle + b)$ is built, where w is the normal vector to the separating hyperplanes, b—auxiliary vector. Thus, if the value of the function $F(x) = 1$, then the element is classified as an outlier. In the case of $F(x) = 0$, the element refers to the set of normal values. In this case, when solving the classification problem, it is necessary to choose such values of the vectors w and b so that the distance from the hyperplane to each class is maximum.

However, the above classifying function is used in the case when two classes can be divided linearly. In the framework of the task under consideration, the class of outliers and the class of normal elements cannot be linearly separated, they have the so-called linear inseparability. To solve this problem, it is necessary to transfer all elements of the training sample to the space X of a higher dimension, where this sample will be linearly separable. In this case, you must enter the kernel function: $k(x, x') = \langle \varphi(x), \varphi(x')\rangle$, which will display the data in a higher dimension [24]. In this problem, the polynomial form of the kernel function is used: $k(x, x') = (\langle x, x'\rangle + \mathrm{const})^d$. Thus, the classifying function dividing the data into outliers and normal elements take the form: $F(x) = \mathrm{sign}(\langle w, \varphi(x)\rangle + b)$. The accuracy of the support vector algorithm was 88% on the training set. In this case, as shown in Fig. 4,

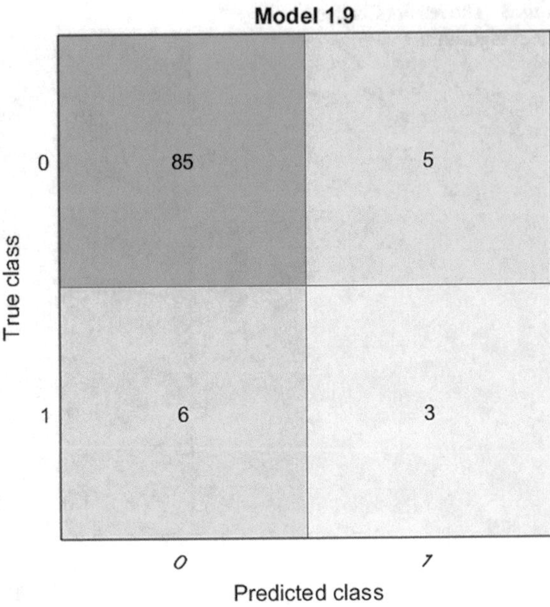

Fig. 4 The results of SVM algorithm

the number of correctly defined normal elements was 85 of 90, and the number of correctly defined outliers was 3 of 9. Similar accuracy was obtained when testing the algorithm on a test sample.

Finally, the problem was solved using the artificial neural network (ANN) algorithm [25]. For outliers classification, a neural network of direct propagation without feedback was configured (a multilayer perceptron with the number of neurons in the hidden layer equal to 800). This architecture of the ANN was obtained in the course of preliminary studies. A training sample was sent to the ANN input, then, in a hidden layer, each element was assigned a certain weight and the incoming information was processed, after which the classification data was sent to the network output. In the event that any error was detected at the output of the network, weights were adjusted and then the final answer was compiled.

As a result of the neural network operation, all outliers were correctly identified, i.e. classification accuracy was 100%. In Table 1 shows a fragment of a table with the results of the neural network on the test sample. In Fig. 5 a diagram with the results of the work of the ANN on a set of training and test samples is shown. As you can see, among the one hundred sampling elements, 9% of the outliers are determined correctly, the remaining 90% of non-outliers data are also correctly identified. The total accuracy, in the end, was 100%.

Table 1 Identification of emissions using a neural network

Ambient temperature (°C)	Pressure in the steam turbine condenser (kPa)	Electric power (MW)	Identification of emissions by ISO	Emission identification with neural networks
8.34	4.077	480.48	0	0.0001
23.64	5.849	445.75	0	0.0005
29.74	5.69	438.76	1	0.9983
19.07	4.969	453.09	0	0.0003
11.8	4.066	464.43	0	0.0044
13.97	3.916	470.96	0	0.0067
22.1	7.129	442.35	0	0.0023
14.47	4.176	464	0	0.0002
31.25	6.951	428.77	0	0.0006
6.77	3.818	484.31	0	0.0045
29.3	7.004	426.25	1	0.9924

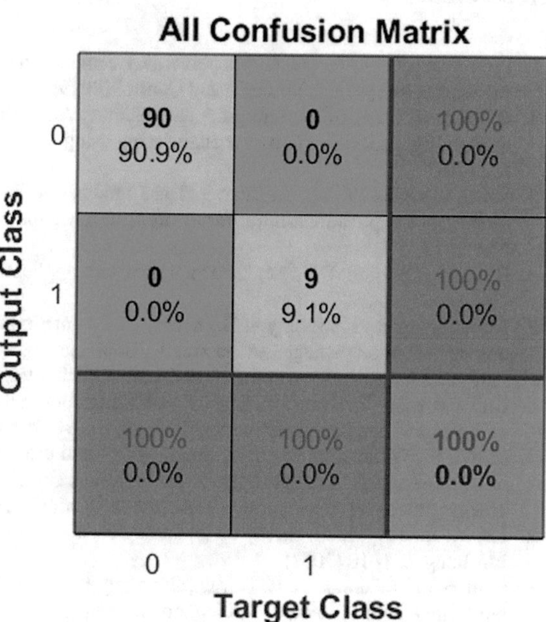

Fig. 5 The result of neural network algorithm

5 Conclusion

Thus, it can be concluded that to improve the efficiency of forecasting the parameters, modes and technical conditions of power engineering equipment in the implementation of predictive analytics systems, it is advisable to use neural network algorithms for searching and eliminating outliers in data. At the same time, it is worth noting that a quick and efficient method for determining emissions is a classification in which a training sample can be quickly organized using algorithms based on international standards. As a further direction of work, it is necessary to indicate the synthesis of algorithms for predicting the values of the current parameters characterizing the modes of operation of power engineering equipment.

Acknowledgements This work is supported by the Russian Science Foundation under grant № 19-19-00601.

References

1. Protalinsky, O., Khanova, A., Shcherbatov, I. Simulation of power assets management process. In: Studies in Systems, Decision and Control (2019)
2. Protalinsky, O.M., Shcherbatov, I.A., Stepanov, P.V.: Identification of the actual state and entity availability forecasting in power engineering using neural-network technologies. J Phys Conf Ser (2017)
3. Wang J., Zhang W., Shi Y., Duan S., Liu J.: Industrial big data analytics: challenges, methodologies, and applications. https://arxiv.org/ftp/arxiv/papers/1807/1807.01016.pdf. Accessed 16 Mar 2019
4. Bi, Z.M., Cochran, D.S.: Big data analytics with applications. J. Manag. Anal. **1**(4), 249–265 (2014)
5. Kumaraguru S., Kulvatunyou B., Morris K.C.: Integrating real-time analytics and continuous performance management in smart manufacturing systems. In: Proceedings of the IFIP Advances in Information and Communication Technology, pp. 175–182 (2014)
6. Chidambaram, V., Evans, H., Etheredge, K.: Big data: is the energy industry starting to see real applications? Supply Chain Manag. Rev. **12**, 62–64 (2015)
7. Silipo, R., Winters, P.: Big data, smart energy, and predictive analytics time series prediction of smart energy data. https://files.knime.com/sites/default/files/inline-images/knime_bigdata_energy_timeseries_whitepaper.pdf. Accessed 18 Mar 2019
8. Lu, Y.: Industry 4.0: a survey on technologies, applications and open research issues. J. Ind. Inf. Integ. **6**, 1–10 (2017)
9. Shin, S.J., Meilanitasari, P.: Developing a big data analytics platform for manufacturing systems: architecture, method, and implementation. International J. Adv. Manuf. Technol., 1–42 (2018)
10. Peres, R.S., Rocha, A.D., Leitao, P., Barata, J.: IDARTS—towards intelligent data analysis and real-time supervision for Industry 4.0. Comput. Ind., 1–12 (2018)
11. Wang, J., Ma, Y., Zhang, L., Gao, R.X., Wu, D.: Deep learning for smart manufacturing: methods and applications. J. Manuf. Syst. **48**, 144–156 (2018)
12. Rose, K., Eldridge, S., Chapin, L.: The internet of things: an overview. Internet Soc., 7 (2015)
13. Shina, S.J., Wooa, J., Rachuri, S.: Predictive analytics model for power consumption in manufacturing. In: Proceedings in 21st CIRP Conference on Life Cycle Engineering, pp. 153–158 (2014)

14. Park, J.K., Kwon, B.K., Park, J.H., Kang, D.J.: Machine learning-based imaging system for surface defect inspection. Int. J. Precis. Eng. Manuf.-Green Technol. **3**(3), 303–310 (2016)
15. Zhao, R., Yan, R., Chen, Z., Chen, Z., Mao, K., Wang, P., et al.: Deep learning and its applications to machine health monitoring: a survey. https://arxiv.org/pdf/1612.07640.pdf. Accessed 2 Apr 2019
16. Johnson, T., Kwok, I., Ng, R.T.: Fast computation of 2-dimensional depth contours. In: Proceedings of the ACM KDD Conference, pp. 224–228, New York, NY, USA, 27–31 Aug 1998
17. Breunig, M.M., Kriegel, H.P., Ng, R.T., Sander, J.: LOF: identifying density-based local outliers. In: Proceedings of the 2000 ACM SIGMOD International Conference on Management of Data, pp. 93–104, Dallas, TX, USA, 16–18 May 2000. ACM, New York, NY, USA (2000)
18. Marti, L., Sanchez-Pi, N., Molina, J.M., Bicharra Garcia, A.C.: Anomaly detection based on sensor data in petroleum industry applications. Sensors **15**, 2774–2797 (2015)
19. Loureiro, A., Torgo, L., Soares, C.: Outlier detection using clustering methods: a data cleaning application. In: Malerba, D., May, M. (eds.) Proceedings of KDNet Symposium on Knowledge-based Systems for the Public Sector (2004)
20. Upadhyaya, S., Singh, K.: Classification based outlier detection techniques. Int. J. Comput. Trends Technol. **3**(2), 294–298 (2012)
21. Ting, K.M.: An instance-weighting method to induce cost-sensitive trees. IEEE Trans. Knowl. Data Eng. **14**, 659–665 (2002)
22. Weiss, G., Provost, F.: Learning when training data are costly: the effect of class distribution on tree induction. J. Artif. Intell. Res. **19**, 315–354 (2003)
23. Tang, Y., Zhang, Y.Q., Chawla, N.V., Krasser, S.: SVMs modeling for highly imbalanced classification. IEEE Trans. Syst. Man Cybern. B Cybern. **39**(1), 281–288 (2009)
24. Wu, G., Chang, E.Y.: Class-boundary alignment for imbalanced dataset learning. In: Proceedings of the ICML Workshop on Learning from Imbalanced Data Sets (2003)
25. Fedorov, E.E.: Artificial neural networks: monograph, 317 p. DVNZ DonNTU, Krasnoarmeysk (2016) (in Russian)

Ontology-Based Model of User Activity Data for Cyber-Physical Systems

Tatiana Shulga, Alexander Sytnik, Nikita Danilov and Denis Palashevskii

Abstract The chapter focuses on the issue of modeling of user interaction with the graphical interface of cyber-physical systems. The primary subject of the presented study is users activity data, that is the user's actions with system graphical interface and their characteristics. This data is collected during the process of software testing or experimental operation. An overview of existing ontologies of domain «Information Systems Graphic Interface» is given. We propose an open model of user activity data in the form of ontology based on the OWL 2 DL language. The main classes, properties, and axioms of this model are covered in the report. This model differs from the other existing ontologies in that it is focused on user activity data rather than the interface and its elements. In addition, it is based on description logic SHOIQ (D), which makes it possible to draw logical conclusions in the process of analysis by experts of the system usability. We also present the structure of software developed for user activity data collection which allows filling the ontology with specific data on the user experience with the interface. As an example, we describe possible tasks where the model and collected data can be used.

Keywords User activity data · User interface · Usability · OWL ontology · Description logics

---

T. Shulga (✉) · A. Sytnik · N. Danilov · D. Palashevskii
Yuri Gagarin Saratov State Technical University of Saratov, 77 Politechnicheskaya Street, Saratov 410054, Russia
e-mail: shulga@sstu.ru

A. Sytnik
e-mail: as@sstu.ru

N. Danilov
e-mail: nikita_danilov@outlook.com

D. Palashevskii
e-mail: palash.denis@gmail.com

A. G. Kravets et al. (eds.), *Cyber-Physical Systems: Advances in Design & Modelling*, Studies in Systems, Decision and Control 259,
https://doi.org/10.1007/978-3-030-32579-4_16

1 Introduction

An essential step in the development of cyber-physical systems is the process of evaluating the system for usability for the end user. Usability, according to ISO 9241-11:2018 [1], is the extent to which a system, product or service can be used by specified users to achieve specified goals with effectiveness, efficiency, and satisfaction in a specified context of use. Evaluation of the usability of the developed system is mostly performed during the testing process or trial operation and consists of two stages. The first stage aims at collecting data in the process of interaction with the graphical interface of the system. This data includes actions performed by users (mouse cursor movements, mouse button clicks, keyboard keys presses, and so forth) and detailed characteristics of these actions (cursor coordinates, frequency of presses, used keys). Such data is indicated by the well-established term "user activity" [2–4]. The second stage is the analysis of this data by an expert in order to identify problems regarding usability.

Ontologies have been proposed as models describing the process of user interaction with the graphical interface of the system in recent years [5]. A brief overview of some of those models is given in this chapter. In this chapter, ontology is defined as a formal model of knowledge representation in a certain domain. The main elements of ontologies are classes (basic concepts, entities), properties (connections between concepts, predicates) and axioms (constraints on relationships between classes and properties) [6]. The trends in knowledge engineering field suggest that web ontologies, as models of knowledge representation, have several advantages [7, 8]. They are published on the web using standard languages, which significantly expands the scope of their application and simplifies the process of appropriation and integration of knowledge. In addition, ontologies are easily extensible. Any researcher may not only freely use the model but also, if necessary, extend and modify it to fit his needs.

The de facto standard for representing ontologies on the web (that is, ensuring their openness) is the languages and formats offered by the W3C. In particular, the most popular language for representing ontologies is OWL 2 language. This language relies on the description logic, which allows for any ontology using the language to utilize existing reasoners and tools.

In this chapter, we propose an open model of user activity data in the form of an ontology-based precisely on the OWL 2 DL language. At the same time, we are developing a model using the SHOIQ (D) description logic. It makes it possible to build logical conclusions when conducting a broad class of studies by experts who are involved in the problems of analysis of the activity of users of cyber-physical systems.

2 User Interface Ontologies Overview

We analyzed several ontologies describing the user interface and user interactions with it [9–13]. We will dwell on the description of the most comprehensive and significant of them.

Timothy John Berners-Lee, an English scientist, one of the creators of the world wide web, the author of the concept of the semantic web proposed the User Interface model (UI) [9]. The ontology is considerably simple (24 classes, 27 properties, and 4 individuals), but it includes important basic concepts from the domain of user interfaces, such as Form, Field, and so on. The ontology representation in the form of a graph is shown in Fig. 1.

Valeria Gribova in her chapter "The ontology model Graphical user interface", published in 2005 [10], proposed a much more complicated model, since the main purpose of its use is the development of a declarative user interface model based on universal standards of ontologies and the following automatic generation of executable interface code. To describe the user interface model, she developed a detailed ontology model "graphical user interface," which describes the interface elements, their properties and the relationship with each other to establish a dialog based on screen forms with the user. The underlying general ontology includes more than 50 classes, where each class contain 10 or more properties. Among the various types of visual tools of a graphical user interface (GUI), there are two main groups—windows and window controls, and three additional groups—control panels, window menus, and auxiliary tools. A fragment of the model is presented in Fig. 2.

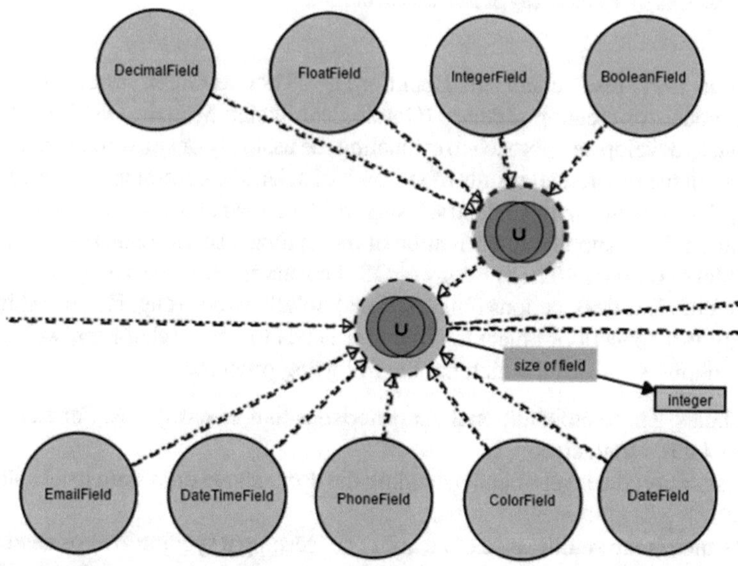

Fig. 1 Fragment of the ontology "User Interface"

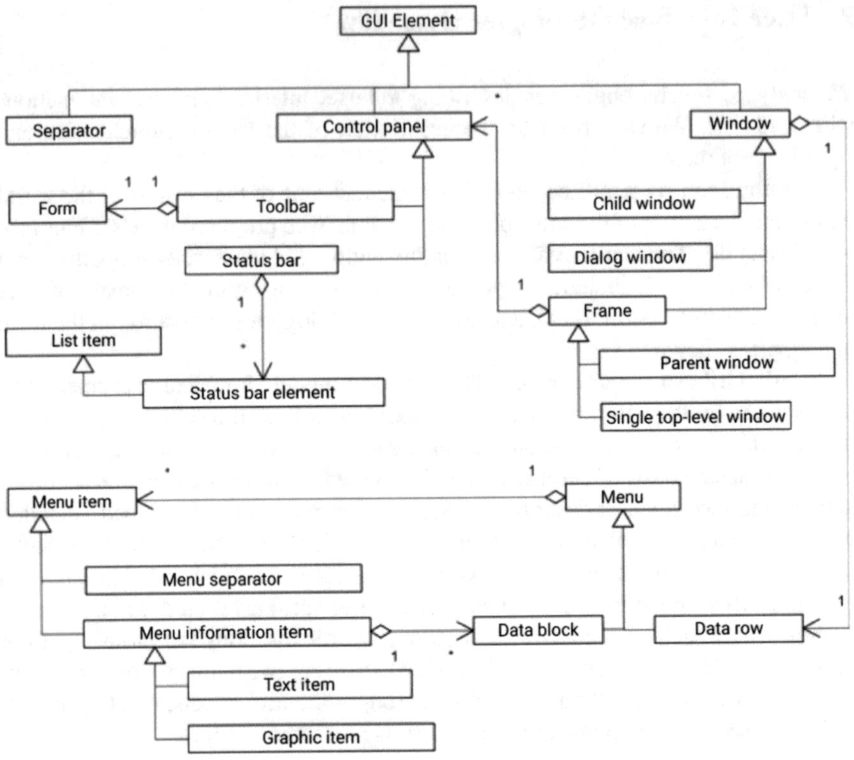

Fig. 2 Fragment of the hierarchy of elements of the GUI

Back in 1997, researchers Ann Blandford and Thomas Green suggested an ontological model for creating sketches (Ontological Sketch Models, OSM) [11]. They set a goal to develop an approach to evaluating the usability of software, which would be based on the theoretical results of research of scientific communities but could be used by design teams in the industry. Using the described approach, the expert forms a structured, but a simple representation of the analyzed software based on a simplified model of ontology. The ontology of OSM covers three aspects of system design: entities (eng. Entities), actions (eng. Actions), relationships (eng. Relationships).

Entity is a concept or object that the user needs to know (character, word, paragraph, column width) and which has the following properties:

- Attributes. These are additional characteristics that an entity has. For example, a symbol has a font, size;
- Accessibility; the level of understanding the user, device or system level and common;
- Relevance is the significance of an entity concerning a specific area or device. For example, the word is relevant to the writing area of the text, and the scrollbar is relevant to the text editor;

- Visibility determines if a user can see an entity, whether it exists conditionally or transparently;
- Disguise defines whether the entity has a clear name or symbol.

Action is what the user can do. An action has the following properties:

- Name of the action;
- The entities which the action is performed on;
- The effect, obtained after committing the action;
- Context which is miscellaneous contextual information.

Relationship describes a connection between entities. It has the following properties:

- Type of relationship. For example: "consists of," "affects," "restrict" or any other;
- Entities that are connected with each other by the relationship.

Therefore, most of the existing ontologies, and, in particular, those that were described above, are models of the user interface or ways of user interaction with the interface. But they are not suitable for accumulating user activity information for the purpose of its following analysis. Besides, well-known ontologies do not use descriptive logic, which does not allow these models to be used to form logical conclusions in the course of conducting usability research.

3 User Activity Ontology

The main object of research in assessing the process of user interaction with the graphical user interface of cyber-physical systems, including ease of use, is the so-called user activity. User activity refers to actions performed by a user in the process of interacting with the graphical interface of the system and characteristics of these actions. Based on the analysis of existing standards in the sphere of usability (for example, [1]), we have formalized the concepts of the subject area, identified the classes required for presenting user activity data and the relationships between them. The following is the description of the main classes of the proposed ontology.

`Session` class describes the period during which the user actively interacts with the system. A session is determined by the time of the beginning and end of the time interval, which is described by the `hasStartDateTime` and `hasEndDateTime` properties, respectively. All events and other necessary information are being saved during an active session.

The `Device` class describes the device on which the session was performed. A session is bound with the device by the `wasPerformedOn` property; in turn, the device is connected with the sessions launched on it by the `wasUsedIn` property. The class is designed to store only necessary information about the physical device, like, for example, the device name, which can be set with the property `hasName`.

The user who performed the session is described by the corresponding User class. Like the Device class, the User class stores the minimal essential information about who executed the session. This information, in particular, includes the user's name, defined by the hasName property. The session is associated with the user who performed it by the wasPerformedBy property. The user's sessions are encompassed under the performed property.

The base Event class is used to store information about the concrete actions of the user in the graphical interface of the program system. The hasDateTime property defines the time of each event.

To determine the kind of the event, the model implements a set of subclasses of the Event class. At the top level of this classes hierarchy, the Event class is subdivided into subclasses Action (ActionEvent) and Command (CommandEvent).

The subclasses of the ActionEvent class represent basic types of user action events that come from mechanical operations of the user in the interface, like for example, cursor movements, clicks, key presses, or interactions with a touch screen. At the same time, these actions are not necessarily associated with the functions of the cyber-physical system.

This way, the KeyboardActionEvent subclass is used to describe the interaction with a keyboard, as well as any other device that relies on the use of keys and buttons. An additional subclass of TypeKeyboardActionEvent is introduced to specify text input events.

For systems, interaction with which is built on the use of a pointer (cursor) in combination with a computer mouse, trackpad, trackball, and other devices, there is a general MouseActionEvent subclass. Its subclasses represent the common types of events associated with the use of this method of interaction. In particular, the ClickMouseActionEvent subclass describes a button press event (click). The SingleClickMouserEvent and DoubleClickMouseEvent subclasses are separated and describe single and double clicks, respectively, as the most common variations of this type of events. The MoveMouseActionEvent class describes pointer movement.

In the dedicated TouchActionEvent subclass there are the events that are related to the usage of a touchscreen. The TapTouchActionEvent subclass describes the user touching a touchscreen surface, which is substantially similar to the ClickMouseActionEvent event. Subclasses of this event are SingleTapTouchActionEvent, DoubleTapTouchActionEvent, LongTapTouchActionEvent, HoldTapTouchActionEvent. They describe different options for tapping which are: quick single tap, double tap, long tap, and touch and holds, respectively.

The SwipeTouchActionEvent subclass describes user interactions with an interface when a user moves one or multiple fingers across the screen.

The CommandEvent subclass describes a command (GOST 34.320-96) which is a user action intended for interacting with the functions of the cyber-physical system.

The complete hierarchy of events types in the model is shown on Fig. 3.

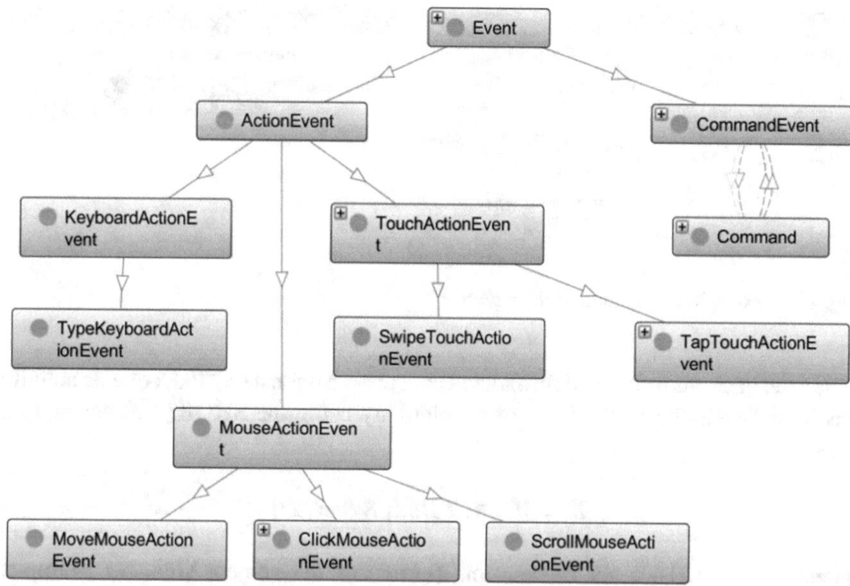

Fig. 3 Hierarchy of events types

We also have introduced new concepts such as `Region` which is a piece of the user interface (for example, the entire window, or its particular part) and region `Variation`. The definition of variations is necessary when analyzing an adaptive user interface when the appearance and functionality of a region are changing depending on its size. For example, large buttons with detailed labels can be replaced by smaller buttons with icons and with no labels. The variation is defined by a combination of the width and height parameters of the image, or their ranges. This is done using the data properties `hasWidth`, `hasHeight`, and also `hasMinWidth`, `hasMaxWidth`, `hasMinHeight`, `hasMaxHeight`. The variation is associated with the region by the `isContainedIn` property. All variations of the region are determined by the contains property.

All events are associated with the region of the user interface in which they were executed and the variation that was active at the time of the event occurrence of the property `isContainedIn`. The collection of Variation events is determined by the contains property. In addition, the `hasInRegionX`, `hasInRegionY` properties are defined, which allows specifying the coordinate of the point in the variation of the region (Fig. 4) where the action was performed. This is used for events related to the pointer and the touch screen.

In total, we have identified 28 concepts, 27 connections between the concepts and 47 restrictions imposed on the common use of concepts and connections between them (for example, `Session wasPerformedBy exactly 1 User`, which means "A session can only be performed by one user").

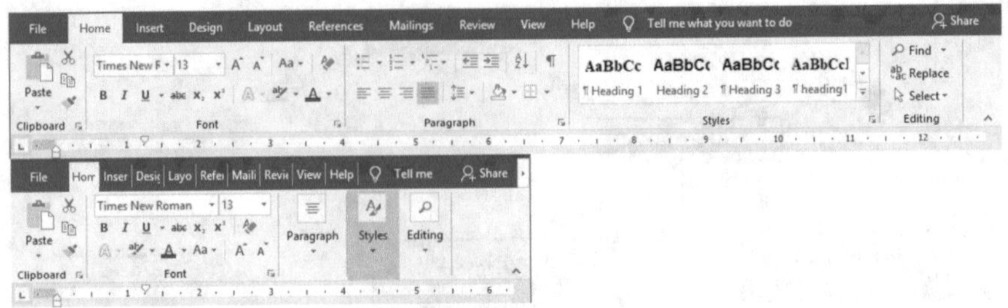

Fig. 4 An example of variation of the region

On the basis of introduced formalized concepts an open data model of user activity has been developed in the form of terminology using the SHOIQ (D) description logic:

$$K = \{C, R, TBox, RBox, D\}$$

where $C = \{C_i\}$—a set of concepts (terms) of the subject area, for example: `Session`, `Device`, `User`, etc.; $R = \{r_i\}$—a set of domain roles (relationships between concepts), for example: `hasName`, `contains`, etc.; *TBox*—a terminology of concepts, a set of terminological axioms of the form $C \sqsubseteq D$ or $C \equiv D$, where C and D are arbitrary concepts; *RBox*—a terminology of roles, i.e. set of terminological axioms of the form $R \sqsubseteq S$ or $Tr(R)$, where R and S are arbitrary roles and D—the set of domains (data types) corresponding to the one defined by the W3C in the XML Schema Definition Language (XSD).

The model is also presented in the form of the ontology using the language OWL 2 DL, which is based on description logic. The significance of this way of representations is that ontologies published on the web in standard formats simplify the process of distributing knowledge and reusing it. In addition, ontologies are clear to read, since they can be visualized in the form of graph (Fig. 5). Classes (concepts) are displayed as graph vertices, and the links between them (roles) are shown as arcs. This ontology is explained in more details in [14].

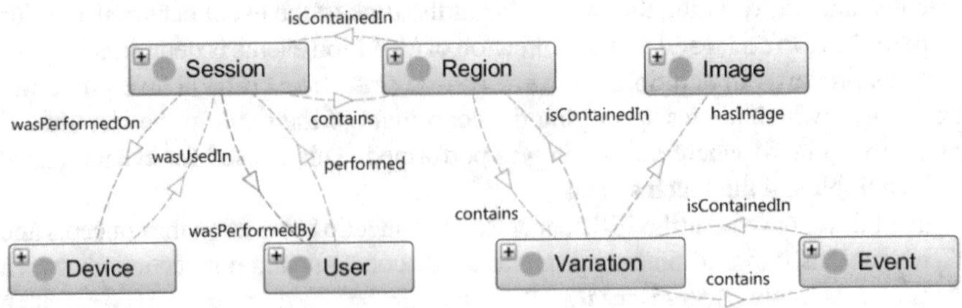

Fig. 5 The main classes of the ontology

Description logic allows us to apply the ontology (knowledge base) for a logical analysis during the usability research conducted by experts. Let's assume that we have the logical concept "Did user X used the command Y?". In order to verify if the concept is valid, it is required to run the following expression in the given knowledge base.

$$\exists X \exists Y\ \mathrm{User}(X) \wedge \mathrm{Command}(Y) \wedge \exists s.\,\mathrm{performed}(X, s) \wedge \exists e.\,\mathrm{contains}(s, e) \\ \wedge \exists e.\,\mathrm{isCommand}(e, Y)$$

The developed model is open and easy to extend or modify for representing specific user activity data required by the experts during the process of solving a wide range of tasks of analyzing user interaction with the system through a graphical user interface. The collection of this data can be performed by the developers of any system automatically by embedding special modules in the software for recording user activity.

4 An Example of Software for Populating the Ontology

An example of such a module is the specialized application developed by the authors [15]. It consists of two main parts: the module for data collection and the module for data visualization and analysis

The main functions of the complex:

- data collection of user activity with the module that can be embedded in any software based on the .NET platform;
- visualization of activity data in the form of heat maps, with the ability to customize parameters (color scheme (palette); the radius of the circle; intersection distance);
- search for consecutive patterns of user activity and the calculation of the values of their support, with the ability to customize parameters (classification function, maximum pattern length).

Thus, the expert receives a toolkit that allows him to identify problems of the user interface quickly and more thoroughly. The program complex consists of three modules (Fig. 6).

- User Activity Client Library (UACL) is a software module for collecting user activity data that is being embedded in the software in which data is to be collected.
- User Activity Data (UAD) are XML files generated by the UACL module during application operation and activity data collection.
- User Activity Viewer (UAV) is a program for downloading, visualizing and analyzing user activity data.

This software solution was developed for use in graphical applications on the .NET Framework. WPF (Windows Presentation Foundation) applications are supported. Data collection module is a .NET Framework library named UserActivity.CL.WPF.

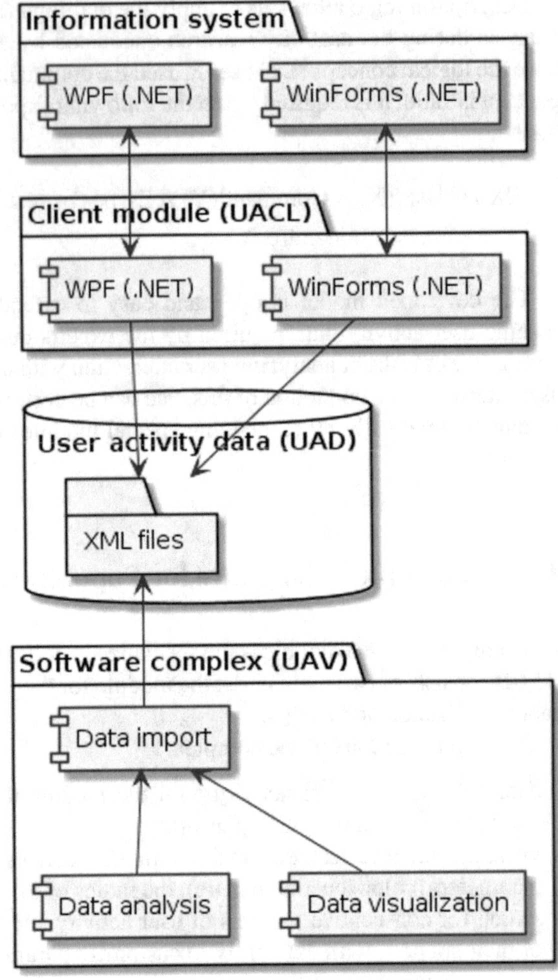

Fig. 6 Component diagram of the development software complex

Data visualization and analysis module is a graphical WPF application which is intended for reading data from UAD XML files with it further visualization and analysis. It supports two ways of representing events which are events list and heatmap. It is also capable of search for repetitive event patterns in users' actions.

5 Approbation of the Model

The proposed model was tested in solving two types of problems: the production of heat maps of user activity [16], and search for repetitive patterns of user activity [17]. In solving the latter task, in addition to the proposed model, the methods of intellectual analysis of associative rules and sequential patterns [18] were used, which allowed experts to detect previously unknown patterns of user interaction with information systems at lower time costs.

The developed model can be used for cyber-physical systems of various classes, for example, for conducting experiments related to the graphics interface of modern aircraft control systems. The actions of the pilot in the process of controlling the aircraft or its onboard equipment are divided into certain phases. In general, it is the detection and recognition of information, the evaluation of the situation, decision-making, the implementation of actions. At all stages of this process, errors are possible due to various factors of an objective and subjective nature, which require subsequent correction. An example is a situation when the pilot has the intention to request a map of the terrain at a specific scale to assess the situation. The pilot executes the corresponding command, but makes an error and gets the wrong scale of the map.

To adjust the scale, the pilot needs to spend additional time resources to execute another command with increased attention. Since we are talking about the interaction of the user with the graphical interface and the execution of commands, the algorithm for finding repetitive activity patterns allows experts to detect such patterns of frequent and fast map switching. It can be argued that a change in the information management system can reduce the time costs for the pilot to obtain the necessary information, which may affect the outcome of a critical situation, as the crew has more time to analyze the situation and make a right decision.

6 Conclusion

A data model in the domain of software user activity is proposed. Formalization of the concepts of the subject area has been done. Classes that are necessary for the presentation of user activity data and properties representing relationships between them have been defined. The proposed model can be used to collect user activity data and further analysis of this data by experts, as well as for automated analysis using description logic. Approbation was performed on two types of tasks related to the analysis of user activity, involving the use of software built on top of the model. The advantage of the proposed model is its extensibility; any expert can add the classes or properties he needs to work more efficiently. Authors express willingness to cooperate in order to expand and refine the existing model.

The results of the work were used when performing research in the project part of the state assignment of the Ministry of Education and Science of the Russian Federation in the field of scientific activity—task No. 9.2108.2017/PC "Development and experimental development of theoretical bases for the use of complexes with unmanned aerial vehicles up to 500 kg when performing search and rescue operations on the water".

References

1. ISO 9241-11:2018. Ergonomics of human-system interaction—Part 11: usability: definitions and concepts. https://www.iso.org/obp/ui/#iso:std:iso:9241:-11:ed-2:v1:en. Accessed 05 Apr. 2019
2. Cuomo, D.L., Bowen C.D.: Stages of user activity model as a basis for user-system interface evaluations. Proc. Hum. Factors Soc. **2**, 1254–1258 (1992)
3. Yang, D., Zhang, D., Zheng, V.W., Yu, Z.: Modeling user activity preference by leveraging user spatial temporal characteristics in LBSNs. IEEE Trans. Syst. Man Cybern. Syst. **45**(1), 129–142 (2015)
4. Woerndl, W., Manhardt, A., Schulze, F., Prinz, V.: Logging user activities and sensor data on mobile devices. In: Atzmueller, M., Hotho, A., Strohmaier, M., Chin, A. (eds.) Analysis of Social Media and Ubiquitous Data. MUSE 2010, MSM 2010. Lecture Notes in Computer Science, vol. 6904. Springer, Berlin, Heidelberg (2011)
5. Paulheim, Heiko, Probst, Florian: Ontology-enhanced user interfaces: a survey. Int. J. Semantic Web Inf. Syst. **6**, 36–59 (2010). https://doi.org/10.4018/jswis.2010040103
6. Vocabularies. W3C. https://www.w3.org/standards/semanticweb/ontology. Accessed 05 Apr. 2019
7. Muñoz, I., Zambrana, M.R.: Applying ontologies to terminology: advantages and disadvantages. Hermes J. Lang. Commun. Bus. **51**, 65–77 (2013). https://doi.org/10.7146/hjlcb.v26i51.97438
8. Munir, K., Sheraz Anjum, M.: The use of ontologies for effective knowledge modelling and information retrieval. Appl. Comput. Inf. **14**(2), 116–126 (2018). ISSN 2210-8327. https://doi.org/10.1016/j.aci.2017.07.003
9. Berners-Lee, T.: A user interface ontology (ui). Linked Open Vocabularies (2014). http://lov.okfn.org/dataset/lov/vocabs/ui. Accessed 05 Apr. 2019
10. Gribova, V., Tarasov, A.: The ontology model of the "Graphical User Interface" domain. Informatika i sistemy upravlenija **1**(9), 80–90 (2005)
11. Blandford, A., Green, T.: OSM: an ontology-based approach to usability evaluation. In: Proceedings of Workshop on Representations. Queen Mary & Westfield College (1998)
12. Chen, X., Kim, T.W., Chen, J., Xue, B., Jeong, W.: Ontology-Based representations of user activity and flexible space information: towards an automated space-use analysis in buildings. Adv. Civil Eng. **2019**, 1–15 (2019). https://doi.org/10.1155/2019/3690419
13. Shahzad, S.K., Granitzer, M., Helic, D.: Ontological model driven GUI development: User Interface Ontology approach. In: 2011 6th International Conference on Computer Sciences and Convergence Information Technology (ICCIT), Seogwipo, pp. 214–218 (2011)
14. Sytnik, A.A., Shulga, T.E., Danilov, N.A.: Ontology of the "Software Usability" domain. Proc. Inst. Syst. Program. RAS **30**(2), 195–214 (2018)
15. Shulga, T., Danilov, N.: The complex of problem-oriented programs for analyzing data activity of software users. Certificate of legal registration of computer program No. 2018662773 of 15 Oct 2018
16. Danilov, N., Shulga, T., Frolova, N., Melnikova, N., Vagarina, N., Pchelintseva, E.: Software usability evaluation based on the user pinpoint activity heat map. Adv. Intell. Syst. Comput. **465**, 217–225 (2016)
17. Sytnik, A., Shulga, T., Danilov, N., Gvozdjuk, I.: Mathematical model of software user activity. Programmnye produkty i sistemy **31**(1), 79–84 (2018)
18. Danilov, N.A., Shulga, T.E., Sytnik, A.A.: Repetitive event patterns search in user activity data. In: Proceedings of the 2018 IEEE Northwest Russia Conference on Mathematical Methods in Engineering and Technology (MMET NW), 10–14 Sept 2018, pp. 92–94. Saint Petersburg Electrotechnical University "LETI", St. Petersburg, Russia (2018)

Selection of Components of a Composite Material Under Fuzzy Information Conditions

I. V. Germashev, M. A. Kharitonov, E. V. Derbisher and V. E. Derbisher

Abstract The task of selecting components of a composite material from an indefinite set in the system "polymer matrix + filler + ingredients" is considered. The initial information is fuzzy, therefore the formalization of the initial data and further analysis were carried out using fuzzy numbers. At the first stage, a parametric space is formed, describing the composite material as a multicomponent physical system. Then, the index of compliance of each system parameter with the specified physicomechanical requirements was calculated, which made it possible to go over to the relative dimensionless real values characterizing each component of the composite material. The standard weighted voting procedure was used as an aggregate function. The obtained numerical values reflect an integral measure of the conformity of the ingredient as a component of the designed composite material. The results of theoretical analysis expand and complement the mathematic support of profiled intelligent decision support systems in the scientific and technical activities of the constructor and technologist when modeling technical objects in conditions of uncertainties, including linguistic ones. The chapter provides a general and particular solution.

Keywords Problem of choice · Intellectual system · Mathematical programming · Software · Modeling · Fuzzy numbers · Optimization · Composite material · Decision making

I. V. Germashev (✉) · M. A. Kharitonov
Volgograd State University, Volgograd, Russia
e-mail: germashev@volsu.ru

M. A. Kharitonov
e-mail: kharitonov@volsu.ru

E. V. Derbisher · V. E. Derbisher
Volgograd State Technical University, Volgograd, Russia
e-mail: derbisher1@yandex.ru

V. E. Derbisher
e-mail: derbisher2@vstu.ru

A. G. Kravets et al. (eds.), *Cyber-Physical Systems: Advances in Design & Modelling*, Studies in Systems, Decision and Control 259,
https://doi.org/10.1007/978-3-030-32579-4_17

1 Introduction

Let us give a rationale for the problem. At present, fillers and modifying additives are widely used for the ingredient control of the properties of a composite material (CM) in the presence of a basic high-molecular matrix. Essentially high-molecular compounds as individual substances are not recycled today—they are used to prepare process mixtures. This particularly applies to multi-component CM of complex composition. At the same time, the goals of directional, operational, economical regulation and optimization as the composition of the "polymer matrix (up to 100%) + filler (0–95%) + additives (0–45%)" and processing technology are pursued, and a wide range of operational characteristics [1, 2].

The solution of the multifactor task of optimizing the composition of CM, which is fuzzy in its essence, today should be based on the "composition–property–quality–application" dependence and provide for the choice of a certain amount and composition of ingredients with the necessary technical functions. Modern sets of ingredients today are represented very widely, so the design result is ambiguous engineering solutions.

From the standpoint of the theory, raw CM, in other words, mixtures, should be considered as complex multifactorial, multi-component physical systems (PS), in which the composition, physico-chemical structure of PC and the interaction between ingredients and the polymer matrix, depend not only on their nature but also from the technological processes of their processing. They are difficult to give to theoretical multifactorial modeling and identification and forecasting. Here the research tool is the methods of system analysis.

It is clear that materials produced on the basis of polymers also fall into this circle and are complex heterogeneous systems, which, according to processing technology and structural features, can be divided according to established views [3, 4] about the following groups:

- fiber-filled;
- dispersion-filled;
- with interpenetrating structures of continuous phases;
- mixed;
- volumetric ("3D");
- layered ("2D");

Taking into account the special modern scientific and technical and consumer interest in nanotechnology, we separately distinguish CM of composition "polymer matrix–additive (filler)", which use 2D and 3D nano-substances—graphene, carbon nanotubes and similar fillers-ingredients. They make it possible to obtain CM with a number of unique properties [5, 6].

In this work, the fillers are not considered in detail—their range is relatively limited and in this part there is a detailed set of theoretical and practical developments [7–10]. Another thing is supplements. Today, their diversity, offered by the chemical industry and under development, is very large and constantly growing [5, 11, 12].

Superconcentrates, for example, are widely represented (2–5% of the mass of the polymer matrix) [13, 14]. However, the number of problems associated with their properties and the properties of a PC as a system object operating under conditions of uncertainty also grows.

In general, PC additives on the production market can be divided into two classes: technological (processing) and modifying (functional) related to the operation of the polymer product. Let us note the main most common goals of their use in PC [15–17] (the most relevant for modern conditions in bold type):

- generation and satisfaction of new needs of society;
- technology optimization;
- giving specific effects;
- durability extension (weather resistance, hydro resistance, frost resistance, crack resistance, fatigue, chemical resistance, etc.);
- increase fire resistance;
- the expansion of the intervals of physico-mechanical characteristics;
- expansion of technical and ergonomic functions;
- regulation of shrinkage;
- improvement of surface properties;
- improvement of environmental friendliness;
- improvement of aesthetic properties;
- acceleration of drying and a set of normalized properties;
- saving resources.

The most important technical functions of additives with an indication of the approximate content (without detail) are presented in Table 1.

The formation of a complete and relevant recipe with regard to the above goals and properties has the feature that, for example, the virtual (pre-project) composition of the CM, like the real one, is also a multi-component and multi-factor complex system with an uncertain number and quality of connections which is incomplete in nature

Table 1 Types of additives in the PC

S. No.	Technical function of the additive	Content, mass %
1	Flame retardants	0–3.0
2	Antiseptics	0–2.0
3	Antistatic	0–3.0
4	Inhibitors	0–2.0
5	Dyes (pigments)	0–2.0
6	Hardeners	0–10.0
7	Plasticizers	0–45.0
8	Rheological	0–3.0
9	Lubrication	0–2.0
10	Stabilizers	0.1–5.0
11	Other	0–5.0

and requires engineering support in technology for the implementation of the full algorithm "CM design—practice" by means of consistent design and experimental work. This, in addition to the above, may be the definition of additional conditions, the nature of the co-ingredients, their concentration, technological and operational characteristics, etc.

In connection with the above, let us point out that the management of the properties of CM for creating new competitive technologies is possible only in a combination of theoretical and empirical methods, while taking into account the multi-componentity of CM and the variety of technologies for its processing, the following vectors of design work are possible:

- pre-experimental heuristic identification of CM as a system and analysis of material properties;
- design of the polymer matrix and its production technology (polymerization, polycondensation, modification, crystallization, structuring, etc.).
- the choice of composition and structure by varying the ingredients;
- regulation of technical functions and concentration of ingredients, taking into account the interests of society.

The proposed route for developing a current CM in these conditions is shown in Fig. 1.

Here, the ever-increasing complexity of CM with the intention of eliminating the existing and emerging disadvantages, inaccuracies in the formulation of the tasks themselves and the goals of their solution, some unstable properties of the polymer matrices themselves and especially the ingredients, is of great importance. Today, only the number of types of additives in the composition of the CM can be more than thirty [18]. In this connection, in this area there are many questions and complex decision-making problems under uncertainty [19].

Taking into account the basic concept of this work, let us point out that the current stage in world practice in solving theoretical problems in the field of designing CM to support decision making increasingly uses mathematical models and methods aimed at structuring and summarizing the initial data to identify patterns of behavior for systems under certain constraints and optimization of control actions. These models, in turn, serve as mathematical support for information systems used by specialists in planning, designing, and optimizing technological systems in the broad sense. One of the options for identifying the properties of a CM involves solving a mathematical programming problem, in which the consumer quality of a CM is the target criterion.

In this context, the management of QM properties can be differentiated according to the following directions:

- variation of goals;
- variation of consumer functions;
- choice of polymer matrix;
- the choice of the physicochemical nature and quantity of ingredients;
- variation of the composition;

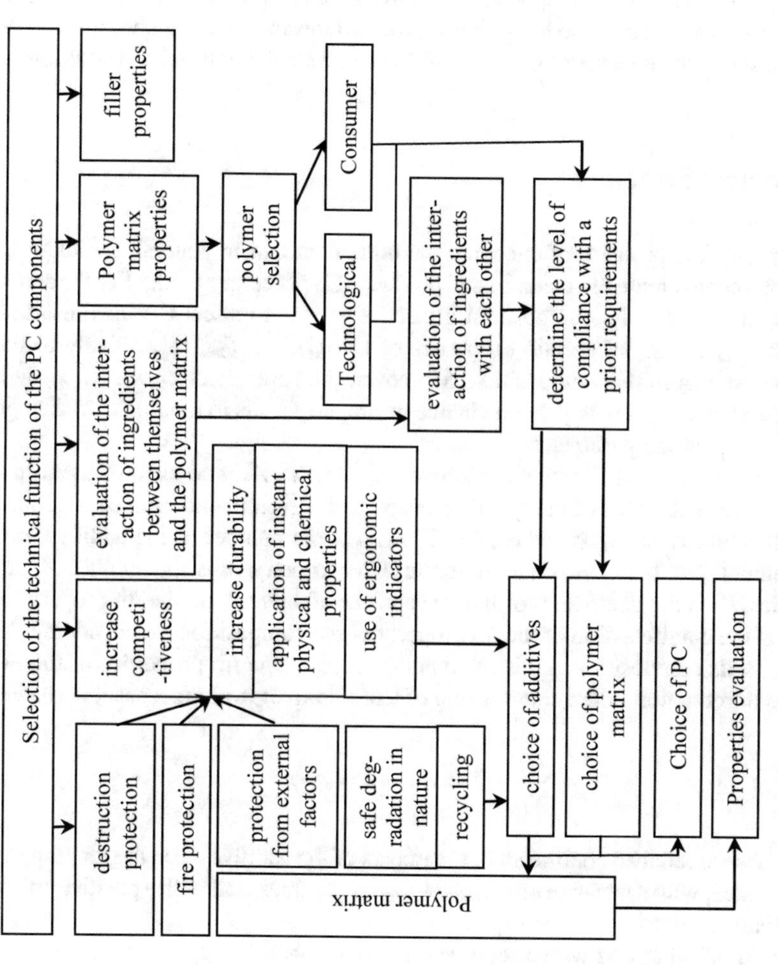

Fig. 1 The route of designing CM with new consumer functions based on the analysis of their technical functions

– variation of technology;
– variation of operating conditions.

Certain achievements are on each segment.

One of the actively developing in this and other similar areas where multicomponent systems are used is the use of fuzzy mathematics and its methods to formalize the problem and data in terms of the mathematical model being developed. Here, relatively recently [20], new opportunities have opened up for solving the problems of PS management. Thus, the foregoing shows the relevance of the theoretical choice of both a separate and a set of CM ingredients to create the optimal composition.

2 Problem Statement

We formulate the problem of choosing the ingredient CM in general. Let there be some polymer to create the material and product. It is necessary to identify the active additives included in its composition in such a way as to affect the manifestation of the $x_1, x_2, \ldots, x_m$ of certain properties of $Q_1, Q_2, \ldots, Q_m$, respectively to the properties shown in this composition. Moreover, the parameters $x_1, x_2, \ldots, x_m$ will be normalized in such a way that a change in property leads to a purposeful change in the corresponding parameter.

Let here be a set of functional additives $A_1, A_2, \ldots, A_n$, and each of them in a certain way affects some of the specified properties. This influence will be described by the formula $x'_{ij} = a_{ij}x_j$, where $i = 1, \ldots, n$ is the number of the additive, n is their amount, $j = 1, \ldots, m$ is the number of CM property, m is the number of CM properties, a_{ij} is the coefficient of influence of the additive A_i on the jth property of CM, x_j is the manifestation of the jth property of the composition before adding the additive to its composition, x'_{ij} is the manifestation of the jth properties of the PC after the introduction in the composition of CM additives A_i. In this case, we have

$$\frac{x'_{ij}}{x_j} - 1 = a_{ij} - 1 = q_{ij},$$

where q_{ij} is the reduced coefficient of the impact of the additive A_i on the jth property of CM, which, with a negative effect, takes negative values, and with a positive effect, takes positive values.

When modifying CM with a set of additives, we obtain

$$q_j = \frac{x'_j}{x_j} - 1, \tag{1}$$

where q_j is a certain reduced coefficient describing the cumulative effect of all additives on the jth property, x'_j is the manifestation of the property Q_j after modifying with the complex of additives $A_1, A_2, \ldots, A_n, j = 1, \ldots, m$.

In the context of solving this problem, we note that modern theoretical and experimental achievements in the physicochemical chemistry of polymer composite materials (PCM) do not yet allow determining q_j for an arbitrary set of additives. This process is carried out by empirical methods so far and only for a specific composition of CM with a specific polymer matrix. In the present case, the task is to identify the q_j estimate, which allows at least an approximate assessment of the cumulative effect on the polymer and PCM of the additive set in the selected composition, that is, it is conceptually necessary to give a theoretical assessment of the effect of the complex (set) of additives $A_1, A_2, \ldots, A_n$ on the totality of $q_1, q_2, \ldots, q_m$.

If, for the evaluation of (1), we assume that the maximum possible cumulative effect is the sum of the possible technical effects of the actions of additives,

$$q_j \le \sum_{i=1}^{n} q_{ij} = \bar{q}_j, \tag{2}$$

where $\bar{q}_j = \sum_{i=1}^{n} q_{ij}$ is the upper limit of the possible cumulative effect, $j = 1, \ldots, m$.

On the other hand, assuming that each additive does not affect the action of the others, we obtain

$$q_j \ge \frac{1}{n}\sum_{i=1}^{n} q_{ij} = \underline{q}_j. \tag{3}$$

Estimates (2) and (3) require an additional agreement that the complex of additives is formed in such a way as not to impair the cumulative effect, i.e.

$$0 \le \underline{q}_j \le \bar{q}_j.$$

Thus, the cumulative effect of the additive complex can be described by a fuzzy number [20]:

$$\hat{q}_j(x_j) = \exp\left(-\frac{(x_j - \tilde{q}_j)^2}{\delta_j^2}\ln 2\right), \quad \tilde{q}_j = \frac{\bar{q}_j + \underline{q}_j}{2}, \quad \delta_j = \frac{\bar{q}_j - \underline{q}_j}{2}. \tag{4}$$

The task of optimizing the influence of the concentration of additives on the properties of the polymer composition is reduced to the problem of fuzzy quadratic programming

$$r(\hat{c}) = \frac{1}{2}\hat{c}^T D\hat{c} + l^T\hat{c} \to \max_{\hat{c}_i}, \quad \sum_{i=1}^{n} \hat{c}_i \le \hat{p}, \hat{c}_i \le \hat{p}_i^{\max}, \quad \hat{c}_i \ge \hat{p}_i^{\min}, \quad i = \overline{1, n}, \tag{5}$$

where $r(\hat{c})$ is the function of the level of influence of additives on the polymer composition, $\hat{c}^T D\hat{c}$ is the negative definite quadratic form, l is the parameter determined experimentally for the polymer matrix, $\hat{c}_i$ is the concentration of additive A_i, $\hat{p}$ is the limit concentration of all additives in the polymer composition, $\hat{p}_i^{\max}$ is the maximum concentration of additive A_i, $\hat{p}_i^{\min}$ is the minimum concentration of additive A_i required to obtain the physical reaction of the material.

3 An Example of Solving the Problem

Let us give an example of calculating the level of influence of additives on the polymer composition. Let the composition of the polymer composition, consisting of a polymer matrix and four additives, be given. Moreover, the influence of each of the additives on the property of the polymer composition, which is expressed by dependencies

$$\hat{q}_{ij} = -h_i\hat{c}_i^2 + l_i\hat{c}_i,$$

where $h = (-1;-0.5;-1.5;-1.05)$, $l = (0.25; 1.25; 0.375; 0.5)$. In the context of the proposed model, we obtain that $d_{ii} = h_i$, $d_{ij} = 0$, with $i \neq j$, $i = 1, \ldots, 4$, $j = 1, \ldots, 4$.Then we obtain

$$D = \begin{pmatrix} -1 & 0 & 0 & 0 \\ 0 & -0.5 & 0 & 0 \\ 0 & 0 & -1.5 & 0 \\ 0 & 0 & 0 & -1.05 \end{pmatrix},$$

$$\hat{p}^{\min}(x) = \exp\left(-\frac{\left(x-p_i^{\min}\right)^2}{0.41}\ln 2\right), \quad \hat{p}^{\max}(x) = \exp\left(-\frac{\left(x-p_i^{\max}\right)^2}{1.31}\ln 2\right), \tag{6}$$
$$p^{\min} = (0.01; 0.001; 0.03; 0.007)^T,\ p^{\max} = (0.14; 0.15; 0.1; 0.2)^T.$$

In this case, it is assumed that the total share of additives cannot exceed 50%, i.e. $p = 0.5$.

Conducting defuzzification of the components of the problem (5), we obtain the problem of quadratic programming

$$r(c) = \frac{1}{2}c^T Dc + l^T c \to \max_{c_i}, \quad \sum_{i=1}^{n} c_i \leq p,\ c_i \leq p_i^{\max}, \quad c_i \geq p_i^{\min}, \quad i = \overline{1, n}. \tag{7}$$

To find a solution to the quadratic programming problem, we use the Kuhn-Tucker theorem [21]. We create the Lagrange function for the problem of quadratic programming (7)

$$L(c, \lambda) = \sum_{k=1}^{n}\sum_{j=1}^{n} d_{kj}c_k c_j + \sum_{j=1}^{n} l_j c_j + \lambda_i(p_i^{\min} - c_i) + \lambda_{n+1}(p - \sum_{i=1}^{n} c_i). \quad (8)$$

In our case, the objective function and the constraints of problem (7) are continuously differentiable functions with respect to the variable c_i ($i = 1, \ldots, n + 1$), therefore the Kuhn-Tucker theorem can be supplemented with analytic expressions determining the necessary and sufficient conditions for the existence the saddle point of the Lagrange function for problem (7), that is, if the Lagrange function (8) has a saddle point $(c_0; \lambda_0) = (c_{01}, \ldots, c_{0n}; \lambda_{01}, \ldots, \lambda_{0n})$, then at this point there are the following ratios

$$\frac{\partial L(c_0, \lambda_0)}{\partial c_j} \le 0, \quad c_j^0 \frac{\partial L(c_0, \lambda_0)}{\partial c_j} = 0, \quad c_j^0 \ge 0, \quad j = \overline{1, n}, \quad (9)$$

$$\frac{\partial L(c_0, \lambda_0)}{\partial \lambda_i} \ge 0, \quad \lambda_i^0 \frac{\partial L(c_0, \lambda_0)}{\partial \lambda_i} = 0, \quad \lambda_i^0 \ge 0, \quad i = \overline{1, n + 1}. \quad (10)$$

Introducing additional nonnegative variables y_j ($j = 1,\ldots, n$) and z_i ($i = 1, \ldots, n + 1$), converting inequalities (9)–(10) into equalities, as well as artificial variables x_i (i = 1, …, n + 1) in (10), necessary for the existence of a nonnegative basis, we have

$$\frac{\partial L(c_0, \lambda_0)}{\partial c_j} + y_j = 0, \quad c_j^0 \frac{\partial L(c_0, \lambda_0)}{\partial c_j} = 0, \quad c_j^0 \ge 0, y_j \ge 0, \quad j = \overline{1, n}, \quad (11)$$

$$\frac{\partial L(c_0, \lambda_0)}{\partial \lambda_i} - z_i + x_i = 0, \quad \lambda_i^0 \frac{\partial L(c_0, \lambda_0)}{\partial \lambda_i} = 0, \quad \lambda_i^0 \ge 0, \quad z_i \ge 0, \quad x_i \ge 0, \quad i = \overline{1, n + 1}. \quad (12)$$

Thus, in order to find a solution to problem (7), it is necessary to determine a non-negative solution to the system of linear Eqs. (11)–(12), which is equivalent to solving a linear programming problem

$$G(x) = -\sum_{i=1}^{n+1} M_i x_i \to \max_{x_i},$$

$$\frac{\partial L(c_0, \lambda_0)}{\partial c_j} + y_j = 0, c_j^0 y_j = 0, \frac{\partial L(c_0, \lambda_0)}{\partial \lambda_i} - z_i + x_i = 0, \lambda_i^0 z_i = 0,$$

$$c_j^0 \ge 0, y_j \ge 0, j = \overline{1, n}, \lambda_i^0 \ge 0, z_i \ge 0, x_i \ge 0, i = \overline{1, n + 1}. \quad (13)$$

where $M_i >> 1$.

Solving the linear programming problem (13) by the simplex method with an artificial basis, we obtain the optimal composition of the composition

$$c^* = (0.05; 0.15; 0.1; 0.2)^T, \quad r(c) = 0.301.$$

This result suggests that, by varying the concentration of additives, while leaving their composition unchanged, it turns out that the greatest effect will be achieved when the concentration of the first additive is 5%, the second additive is 15%, the third additive is 10%, the fourth additive is 20%.

4 Conclusion

The proposed strategy for choosing an additive in fuzzy conditions can be algorithmized as a mathematical model and used to substantiate decision making when choosing CM compositions based on varying ingredients from the available sets (resources) in order to achieve the maximum possible PCM consumer quality level under specific conditions.

Acknowledgements The reported research was funded by Russian Foundation for Basic Research and the government of Volgograd region, grant no. 18-48-340011 “Quality management of polymer products based on the optimization of formulation of composite materials”.

References

1. Tager, A.A.: Physical Chemistry of Polymers. Scientific World, Moscow (2007)
2. Askadskii, A.A., Matveyev, Y.I.: Chemical Structure and Physical Properties of Polymers. Chemistry, Moscow (1983)
3. Kryzhanovsky, V.K.: Engineering Choice and Identification of Plastics: Study Guide. Scientific Fundamentals and Technologies, Saint-Petersburg (2009)
4. Askadskii, A.A., Khokhlov, A.R.: Introduction to the Physics and Chemistry of Polymers. Scientific World, Moscow (2009)
5. Gudkov, M.A., Bokadarov, S.A.: Identification of the interaction of a polymer and a filler by the method of discriminant analysis. Inter. Sci. Herald Bull. Assoc. Orthodox Sci. **3**(7), 44–46 (2015)
6. Vakulov, N.V., et al.: Prediction of the service life of rubber and rubber products with the help of calculation programs in the MatLab system. In: Rubber Industry: Raw Materials, Materials, Technology, pp. 169–172 (2017)
7. Bobryshev, A.N., Erofeev, V.T., Kozomazov, V.N.: Polymer Composite Materials: A Tutorial. ASV, Moscow (2013)
8. Sokolov, D.V., Nilov, D.Y., Smolyakov, V.M.: Adaptive schemes and topological indices in predicting the properties of polymers. Phys. Chem. Polym. Synth. Prop. Appl. **11**, 141–144 (2005)
9. Askadskii, A.A.: Computational Materials Science of Polymers. Cambridge International Science Publishing, Cambridge (2001)
10. Goldman, A.Y.: Prediction of Deformation and Strength Properties of Polymeric and Composite Materials. Chemistry: Leningrad Branch, Leningrad (1988)
11. Kerber, M.L.: Polymeric Composite Materials: Structure, Properties, Technology. TSOP Professiya, Saint-Petersburg (2014)
12. Germashev, I.V., Derbisher, V.E., Zotov, Y.L., Tsapleva, M.N., Konnova, E.V., Vasiliev, P.M.: Computer-aided designing of active additives for polyvinyl chloride. Plasticheskie Massy Sintez Svojstva Pererabotka Primenenie **7**, 36–38 (2001)

13. Petrov, O.O., Mun'kin, N.I., Ganiev, E.Sh., Simonov-Emelyanov, I.D., Prokopov, N.I., Gervald, A.Y.: Comprehensive evaluation of the characteristics of superconcentrates for dyeing rigid PVC compositions in the process of high-speed extrusion. In: Lomonosov, M. V. (ed.). Herald of MITHT **7**(6), 78–82 (2012)
14. Masterbatches for Plastics. http://www.globalcolors.ru/superkoncentraty-masterbatch/. Accessed 10 Aug 2019
15. Zweifel, H., Maer, R.D., Schiller, M.: Polymer Additives. Handbook Professiya, Saint-Petersburg (2010)
16. Glukhikh, V.V., Mukhin, N.M., Shkuro, A.E., Buryndin, V.G.: Production and Use of Products from Wood-Polymer Composites with Thermoplastic Polymer Matrices: Study Guide. UGLTU, Yekaterinburg (2014)
17. Bazhenov, S.L., Berlin, A.A., Kulkov, A.A., Oshmyan, V.G.: Polymer composite. In: Materials Durability and Technology. Publisher Intellect, Moscow (2009)
18. Polymer Processing Technology: Physical and Chemical Processes: Study Guide for Universities. Publisher Yurayt, Moscow (2018)
19. Germashev, I.V., Derbisher, V.E., Losev, A.G.: Analysis and Identification of the Properties of Complex Systems in the Natural Sciences. Publisher Volgograd State University, Volgograd (2018)
20. Germashev, I.V., Derbisher, V.E.: Properties of unimodal membership functions in operations with fuzzy sets. Russ. Math. **51**(3), 72–75 (2007)
21. Rothlauf, F.: Optimization methods. In: Design of Modern Heuristics. Natural Computing Series. Springer, Berlin, Heidelberg (2011)

Big Data Analysis in Film Production

T. B. Chistyakova, F. Kleinert and M. A. Teterin

Abstract The article analyzes the current trends of digitalization for large innovative industrial production, which are international, large-capacity, distributed in different geographical locations and having several production lines at each plant. Such trends of digitalization as predictive analytics and 6 sigma methodology, which includes Ishikawa diagram and DMAIC (definition, measure, analysis, improvement, control) cycle, are considered. The novelty of the work lies in the application of methods and technologies of intellectual analysis of large industrial data for production of polymeric films and in the application of mathematical models that allow online calculation of uncontrolled consumer characteristics of products (thickness, color of polymeric films) and integrate them into one single system of data mining. Developed software solution includes visualization unit, forecast unit, statistical data analysis unit. Software solution allows us: determine the types of films with the best yield; check the production data for normalcy; calculate process capability index; calculate key performance indicators. Application and testing of the big data analysis system on the example of large industrial Corporation Kloeckner Pentaplast proved its efficiency.

Keywords Polymer films · Predictive analytics · Big data · Machine learning · Statistic analysis

---

T. B. Chistyakova (✉) · M. A. Teterin
Saint Petersburg State Institute of Technology, Moskovsky Prospect 26, 190013 Saint Peterburg, Russia
e-mail: nov@technolog.edu.ru

F. Kleinert
Klöckner Pentaplast Europe GmbH & Co. KG, Montabaur, Germany

A. G. Kravets et al. (eds.), *Cyber-Physical Systems: Advances in Design & Modelling*, Studies in Systems, Decision and Control 259,
https://doi.org/10.1007/978-3-030-32579-4_18

1 Introduction

Productions of polymer films are modern, international, the innovation and are characterized by multiassortment, large capacity; continuity; a large number of data sources (about 35 plants with several production lines on each of them); a large number of different data storages (BDE, Business Warehouse, SAP, OCS) which need to be integrated among themselves; a large number of administrative and production personnel (6300 employees making decisions); in the large volume of the accumulated expert industrial information (one billion records); large number of controlled information (>250 sensors); the total production of equal 1000000 tons of polymeric products a year [1, 2]. The analysis of industrial productions showed that they are characterized: imbalance of classes for the reason that defects on production arise seldom; a complexity and existence of system communications which describe poorly formalized information systems and difficult application entities of management that leads to considerable complication of rules of creation of the formalized information and analytical models describing patterns in data [3–5].

Polymer films found broad application in different areas: in food industry, in medicine and in electronics. The most important consumer characteristics of polymer films are: color; width; film thickness; shrinkage value; existence of black points; existence of destructive (brown) bands; existence of inclusions of unfused polymer; existence of the modifier, cracks (the burst air bubbles) [3, 6]. On process, there are indicators, which can lead to an abnormal emergency situation on an object—reduction of material in a gap between rolls, break of a film [7]. Depending on a scope, different requirements to consumer characteristics of a film are imposed to polymeric film materials (film thickness in the range from 25 to 1200 μm, film width in the range from 100 to 2500 mm).

Digital transformation of the industrial industry consists of 4 stages:

1. Introduction of automated control systems for internal processes of the enterprise (SAP system).
2. Creation of services for cloud computing.
3. Introduction of systems for the analysis of trends and forecasting of data.
4. Introduction of systems for training in precedents (machine learning).

At the current stage digital transformation is at the 4th stage and is planned to complete digitalization by 2020. Introduction of systems for training in precedents consists of 4 under stages: creation of storages of data; to create the module of visualization of data; to increase productivity by means of innovative methods of machine learning and to connect business processes with artificial intelligence.

Task of control for production personnel is search of values of the operating influences and factors of production which provide implementation of requirements to consumer characteristics of a polymeric film.

The purpose of work is development of a program complex which with use of methods of machine learning, methods of the statistical analysis will allow to find set of the operating influences and factors which provide performance of preset values of consumer characteristics of a polymeric film and to visualize set of characteristics of industrial production in a form, ergonomic for administrative and production personnel.

For the solution of objectives the methodology six sigma allowing to improve the existing processes by methods is used: DMAIC (Define, Measure, Analyze, Improve, Control), Isikava's chart, calculation of key performance indicators of process and calculation of indexes of reproducibility of process.

2 The Formalized Description of Polymeric Film Production Process Materials

The formalized description of polymeric film production process materials (see Fig. 1) can be submitted as $P(t) = \{X(t), U(t), Y_1(t), Y_2(t)$, where: $X(t)$—vector of input variables; $U(t)$—vector of the operating influences; $Y_1(t)$—consumer characteristics which are controlled automatically; $Y_2(t)$—consumer characteristics which are calculated by mathematical models$\}$; $X(t) = \{Pt(t)$—type of a polymeric film, $C(t)$—component composition of raw materials, $Cf(t)$—equipment configuration; $Xf(t)$—vector of criteria for comparison of samples of films$\}$; $U(t) = \{UpE(t)$—vector of the operating impacts of work of an extruder; $UpC(t)$—vector of the operating impacts of work of a calender; $UpR(t)$—vector of the operating influences of the pulling device$\}$, where $t = t_1 - t_{adjusted}$.

The vector of input variables includes: $P(t) = \{n(t)$—flow index of the polymer; μ—viscosity of polymer, Pa s; $\alpha(t)$—relative change of the size of a film, calculated

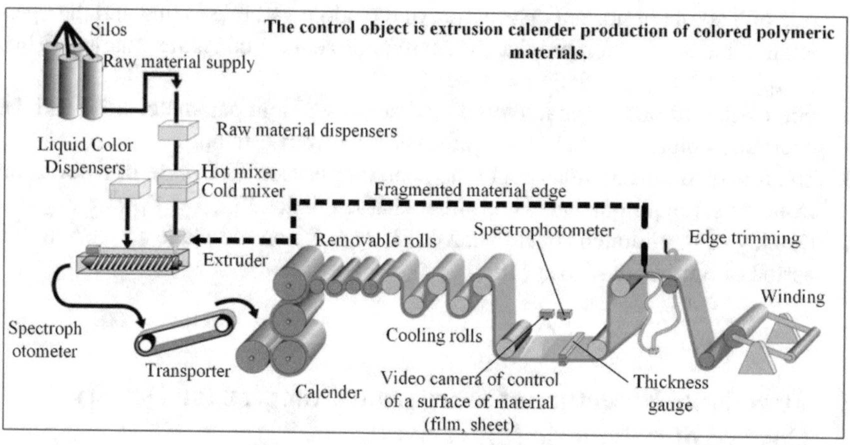

Fig. 1 Description of polymeric film production process

as an equation root; $G_0(t)$—module of elasticity of material} $C(t) = \{M(t)$—composition weight}; $Cf(t)$ {Ln—line number} $Xf(t) = \{Tqe(t)$—requirements for composite uniformity of the extrudate; $Tqk(t)$—requirements to dimensional quality of a film and to quality of a surface of a film (lack of defects: black points, destructive strips, inclusions of unfused polymer etc.)}.

Vector of the operating impacts of work of an extruder consists of $S(t)$—speed of rotation of the screw, turns/min; $V(t)$—Rotating speed of a spiral in a boot funnel, turns/min; $Ts(t)$—screw temperature, °C; $Th(t)$—temperature of heaters, °C, $Tb(t)$—case temperature.

The vector of the operating impacts of work of a four-roll calender includes: $TW(t)$—temperature of rolls, °C; $ToW(t)$—torque of rolls, N m, $Pbend(t)$—bend pressure, Pa; $Poffset(t)$—shift pressure, Pa.

Vector of the operating influences of the pulling device: $Ttor$—temperature of the pulling devices, °C; $Totor(t)$—torque of the pulling devices, N m; $Tpr(t)$—temperature of clamping rollers, °C; $Topr(t)$—torque of clamping rollers, N m; $Tcr(t)$—temperature of the cooling rollers, °C; $Tocr(t)$—torque of the cooling rollers, N m; $Tsr(t)$—temperature of the tempering rolls, °C; $Tosr(t)$—torque of the tempering rolls, N m; $Tt(t)$—temperature of tension rollers, °C.

$Y_1(t) = \{Blp(t)$—quantity of black points on the set surface area; $Hel(t)$—gelik on 10 m^2; $Air(t)$—burst air bubbles on the set surface area; $Dest(t)$—destructive, brown strips on the set surface area; $Inc(t)$—inclusions of unfused polymer, on the set surface area; $Fib(t)$—fibers, on the set surface area; $Th_1(t)$—film thickness, mkm}. $Y_2(t) = \{Shr(t)$—shrinkage size; $Lc(t)$, $ac(t)$, $bc(t)$—rated color coordinates of a finished product; $Th_2(t)$—film thickness, mkm}.

The program complex allows solving the following problems:

1. Using control maps of Shukhart, indexes of a reproducibility of process, and the rule of 3 sigma of normal distribution of a random variable to find and the operating influences, which provide the set requirements to consumer characteristics $Y_{1,adj1}$ $Y_{1i}(t)$ $Y_{1,adj2}$, where $t = t_1 - t_2$ [8, 9].
2. Forecasting of output parameters $Y_{1,i}(t)$ at the set input parameters $X(t)$ and the operating influences $U(t)$ in the period of time from t_2 to t_3.
3. To issue recommendations about the operating influences in case of a deviation from the set requirements to consumer characteristics $Y_{1,adj1} \le Y_{1i}(t) \le Y_{1,adj2}$.
4. Display of multidimensional data $X(t)$, $U(t) \in R^N$ on the plane $z(t) \in R^2$ in the period of time from t_1 to t_2 [10].

3 Functional Structure of a Program Complex for Quality Control of Polymeric Materials

The program complex (see Fig. 2) includes the following components: information support (database of production design characteristics of process, database of parameters of the equipment, database of parameters of material, base of production

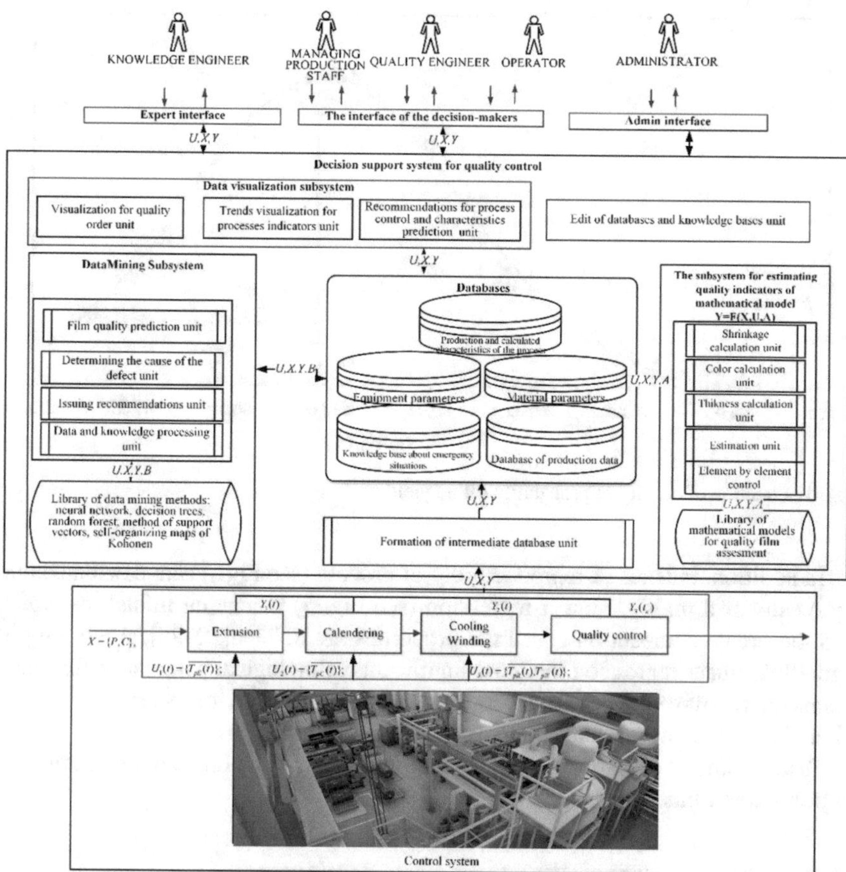

Fig. 2 Functional structure of a program complex for quality control of polymeric materials

data, and knowledge base of emergency situations); subsystem of data visualization [10–15]; module of editing databases and knowledge; subsystem of control and forecasting of quality of polymer films; subsystem of forming of estimated figures of merit of a polymer film.

Testing of operability of the software product happened according to industrial data of the plants of Russia and Germany: for a month of production which contained 500 thousand various data on 250 process parameters on shrinkage (rated size) and black points.

The module of verification of industrial data on normal distribution (see Fig. 3) was tested [16–19]. In case data submit to normal distribution, then for the analysis of data linear algorithms (linear regression, a method of basic vectors, etc.) were used. If the hypothesis of compliance of distribution of probability of result of measurements to the normal law is rejected, then either nonlinear algorithms are used or robust methods of processing of industrial data are applied.

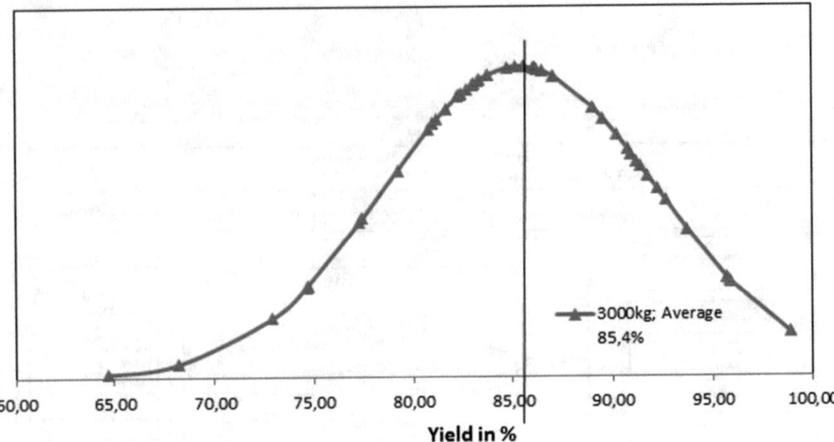

Fig. 3 Function of density of probability for an yield in %

In addition, indexes of reproducibility of process (see Fig. 4) are calculated and by means of a multiple linear regression (see Fig. 5) managing influences which had the greatest impact on consumer properties are defined[20–25]. According to a multiple linear regression the most significant technological parameters for black points are rotating speed of the screw in an extruder, material feed speed in a funnel, high temperature of mix that completely matches expert data.

Testing proved operability of a system of the analysis of big data on production of polymeric films.

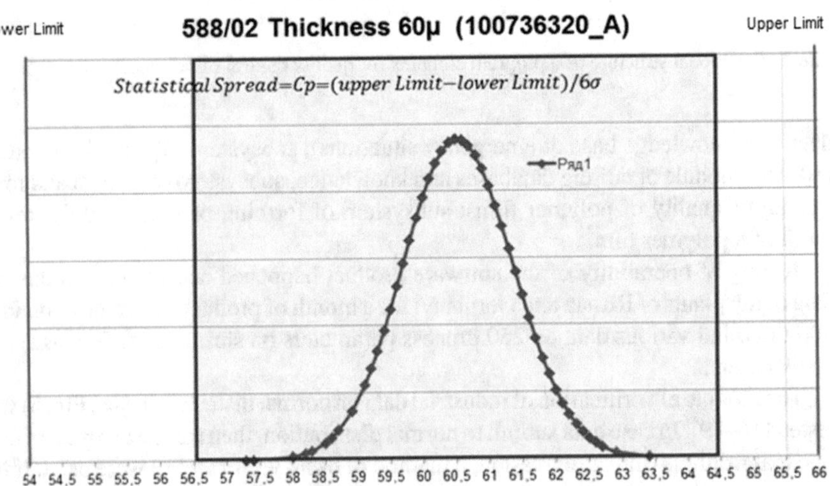

Fig. 4 Interface of work of a subsystem of calculation of indexes of reproducibility of process

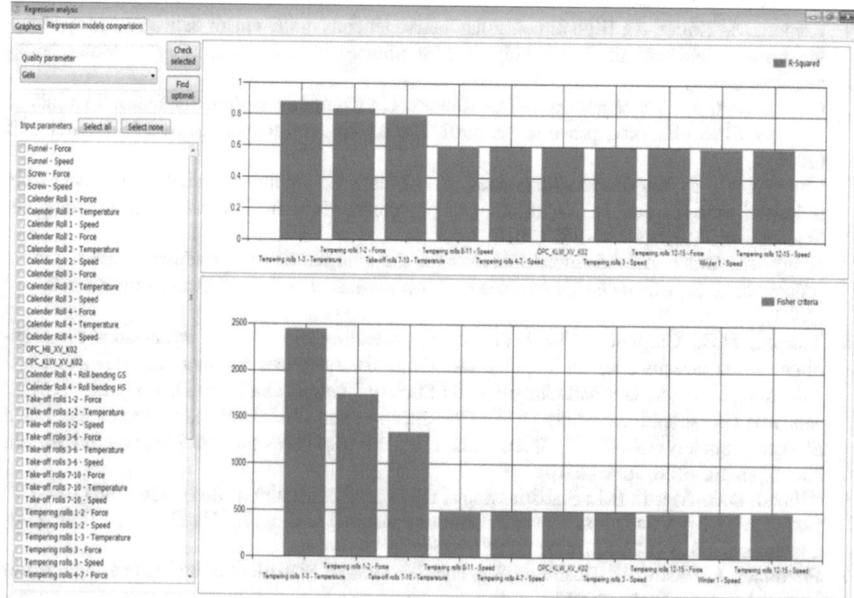

Fig. 5 Search of the most significant process parameters for use of multiple linear regression

4 Conclusion

The program complex, algorithms, which were developed, allow predicting the best supplier of polymeric materials, to visualize data in an ergonomic form, to predict consumer qualities of a polymeric film and to issue recommendations about elimination of emergencies. Use of the developed software product allows increasing the professional level of administrative and production personnel of production.

The software product underwent successful approbation on production of polymeric film materials of the plants of Russia and Germany. The program complex has adaptive flexible architecture thanks to which expansion of its functionality due to development and connection of additional program modules is supported.

References

1. Kohlert, M., König, A.: Advanced polymeric film production data analysis and process optimization by clustering and classification methods. Front. Artif. Intell. Appl. **243**, 1953–1961 (2012)
2. Kohlert, M., Chistyakova, T.: Advanced process data analysis and on-line evaluation for computer-aided monitoring in polymer film industry. J. St. Petersburg State Inst. Technol. Tech. Univ. **29**, 83–88 (2015)
3. Optical control systems GmbH. http://www.ocsgmbh.com

4. Kohlert, M., König, A.: High dimensional, heterogeneous multi-sensor data analysis approach for process yield optimization in polymer film industry. Neural. Comput. Appl. Mater. **26**(3), 581–588 (2015)
5. Chistyakova, T., Araztaganova, A.M., Kohlert, C.: Computer system for thermal shrinkage polymer films obtaining processes control. Herald Kazan Technol. Univ. **19**(17), 101–105 (2016)
6. Chistyakova, T., Teterin, M., Razugraev, A., Kohlert, C.: Intellectual analysis system of big industrial data for quality management of polymeric films. In: Lecture Notes in Computer Science—T.9812, pp. 234–242 (2016)
7. König, A., Gratz, A.: Advanced methods for the analysis of semiconductor manufacturing process data. In: Advanced Techniques in Knowledge Discovery and Data Mining, pp. 27–74 (2005)
8. Tadesse, D.G., Carpenter, M.: A method for selecting the relevant dimensions for high-dimensional classification in singular vector spaces. Advances in Data Analysis and Classification, Springer; German Classification Society - Gesellschaft für Klassifikation (GfKl); Japanese Classification Society (JCS); Classification and Data Analysis Group of the Italian Statistical Society (CLADAG); International Federation of Classification Societies (IFCS), vol. 13(2), pp. 405–426, June (2019)
9. Altman, D.G., Martin B.J.: Statistics notes: the normal distribution. BMJ, **310**, 298 (1995)
10. Cox, Michael AA and Cox, Trevor F. Multidimensional scaling. In: Handbook of Data Visualization, pp. 315–347. Springer (2008)
11. Friedman, J., Hastie, T., Tibshirani, R.: The Elements of Statistical Learning (Springer series in statistics New York) (2001)
12. Korhonen, P.J., Silvennoinen, K., Wallenius, J.: O¨orni A Can a linear value function explain choices? An experimental study. Eur. J. Oper. Res. **219**(2), 360–367 (2012)
13. Belloni, A., Chernozhukov, V., et al.: Least squares after model selection in high-dimensional sparse models. Bernoulli **19**(2), 521–547 (2013)
14. B¨uhlmann, P., et al.: Statistical significance in high-dimensional linear models. Bernoulli **19**(4), 1212–1242 (2013)
15. B¨uhlmann, P., Van De Geer, S.: Statistics for high-dimensional data: methods, theory and applications. Springer Science & Business Media (2011)
16. Candes, E., Tao, T.: The Dantzig selector: statistical estimation when p is much larger than n. Ann. Stat. 2313–2351 (2007)
17. Reid, S., Tibshirani, R., Friedman, J.: A study of error variance estimation in lasso regression. Statistica Sinica **26**, 35–67 (2016)
18. Sala-i-Martin, X.X.: I just ran two million regressions. Am. Econ. Rev. 178–183 (1997)
19. Pavlyshenko, B.: Machine Learning, Linear and Bayesian Models for Logistic Regression in Failure Detection Problems (2016)
20. Tibshirani, R.: Regression shrinkage and selection via the Lasso. J. Roy. Stat. Soc. Ser. B, (1996)
21. Yang, D., Usynin, A., Hines, J.W.: Anomaly-based intrusion detection for SCADA systems. In: 5th International Topical Meeting on Nuclear Plant Instrumentation Controls, and Human Machine Interface Technology. pp. 797–803 (2006)
22. Duncan, A.J.: Quality control and industrial statistics. R.D. Irwin, Homewood, IL (1974)
23. Montgomery, D.C., Runger, G.C.: Introduction to Statistical Quality Control, 5th edn. John Wiley, New York (2008)
24. Hussein, Maher, Allawi Al-Morshedi, Abbas, Shomran, Haideer.: The Comparison Between Shewhart Control Chart, Cusum and EWMA. https://doi.org/10.13140/rg.2.2.34006.34880 (2013)
25. Salah, H., Ahmed, M., Saleh, K., Sherif, A.: Effect of sample size on the performance of Shewhart control charts. Inter. J. Adv. Manuf. Technolo. https://doi.org/10.1007/s00170-016-9412-8 (2016)

Algorithm for Calculating the Reliability of Chemical-Engineering Systems Using the Logical-and-Probabilistic Method in MATLAB

Anastasiya Zakharova, Tatiana Savitskaya and Alexander Egorov

Abstract An algorithm for calculating the reliability using the logical-and-probabilistic method was developed in the MATLAB software package. An algorithm is applicable for inclusion in the module of a cyber-physical system for design and optimization of the chemical-engineering systems reliability. The accuracy of the algorithm was checked for systems of varying complexity with number of elements from 5 to 10. An example of program calculations is also presented. Using the developed algorithm, the reliability of the hydrotreating unit of a catalytic reforming unit under uncertainty was calculated.

Keywords Reliability of technical-engineering systems · Standard software package · Probability of no-failure operation · The logical and probabilistic method · Chemical-engineering system

1 Introduction

Nowadays the actual issue is the introduction of cyber-physical systems in the manufacturing sector [1, 2]. One of the key tasks in the design of such systems is the development of a module for the assessment and optimization of the reliability of complex production systems for the rational distribution of resources, ensuring timely maintenance and safe operation of equipment.

Evaluation of the reliability of complex production systems is an important scientific direction of rational allocation of resources in chemical-engineering systems design, ensuring timely maintenance and safe operation of equipment.

A. Zakharova (✉) · T. Savitskaya · A. Egorov
D. Mendeleev University of Chemical Technology of Russia, Moscow, Russia
e-mail: zakharova.a.y@mail.ru

T. Savitskaya
e-mail: savitsk@muctr.ru

A. Egorov
e-mail: egorov@muctr.ru

A. G. Kravets et al. (eds.), *Cyber-Physical Systems: Advances in Design & Modelling*, Studies in Systems, Decision and Control 259,
https://doi.org/10.1007/978-3-030-32579-4_19

Scientific studies of technical systems reliability are associated with the development of reliability assessment methods, the study of the physics of failures, and the development of design reliability methods. The methods of the theory of random processes, the theory of expert estimates (heuristic prediction), decomposition (equivalent), logical-and-probabilistic, asymptotic, analytical-statistical methods are used to calculate the reliability indices of technical systems. In practice, the methods of simulation and statistical modeling (Monte-Carlo method) are used. Mathematical methods of reliability theory have particular importance since they allow to take into account the specifics of the occurrence and elimination of failures. This specificity is related to the fact that in order to solve reliability problems, understanding of the statistics of failures and those physical, chemical and mechanical processes that lead to changes in the initial indicators of the quality of machines and their elements are necessary. These processes, obeying certain physical laws, have a stochastic nature. Time functions and laws of failure physics are the basis for solving the main problems of reliability [3–16].

Mathematical models for assessing the reliability of chemical-engineering systems differ in structure and implementation complexity. Their comparative analysis allows you to choose the most appropriate method depending on the life cycle and technological structure [17].

2 The Initial Mathematical Formulation of System and Research Methodologies

One of the research tasks was to check the capabilities and usability of standard software for mathematical computation used in solving problems of reliability analysis and chemical-engineering systems [18–20].

In order to assess the reliability of recoverable and non-recoverable chemical-engineering systems with different types of redundancy, several functions for reliability calculation were created in the MATLAB software package [21], which allows an analysis of complex structures without the help of specialized software. In the course of this study, the algorithm for calculating the reliability of the chemical-engineering system (CES) using a logical-probabilistic method was developed for MATLAB.

The logical-probabilistic method for calculating the reliability indicators of a chemical-engineering system is a method based on the description of a mathematical model of a system by functions of logic algebra (FLA), i.e. functions that take only two values, 0 or 1, and are defined by different sets of binary arguments, which can also be only in two incompatible states 0 or 1 [22]. When solving problems of reliability using the general logical-probabilistic method (GLPM), the following steps can be distinguished [23]:

- representation of the system as a finite number of H elements $i = 1,2,\ldots,H$, elements each of which is represented in the reliability model as a simple event with two possible states $\tilde{x}_i = \{x_i, \bar{x}_i\}$ and specific probabilistic parameters $p_i(t)$, the probability of the element's failure-free operation, or $q_i(t) = 1 - p_i(t)$—the probability of element failure;
- determination of logical conditions for the output functions of each element of the system $\tilde{y}_i = \{y_i, \bar{y}_i\}$
- a logically rigorous verbal and graphical (analytical) description of the set X of individual elements of the system and the set of conditions Y of their system functions, which together $G(X, Y)$ form a special functional integrity scheme of the system in question;
- logically rigorous description and specification of the logical criteria of the functioning of the system (1) for the implementation of the main functions and/or the occurrence of hazardous states of the system with the help of individual or group output (integrative) functions;

$$Y_F = Y_F(\{\tilde{y}_i\}, i = 1, 2, \ldots, N) \tag{1}$$

- logical modeling, which consists of constructing a logical function of the system operability (2)

$$Y_F = Y_F(\{\tilde{x}_i\}, i = 1, 2, \ldots, H) \tag{2}$$

Logical modeling allows to determine all combinations of states of elements $\tilde{x}_i, i = 1, 2, \ldots, H$ in which and only in which the system implements its output function F;

- probabilistic modeling, drafting of a polynomial of the calculated probability function (3), which allows analytically strict determining the law of distribution of the system uptime time for the implementation of the output function F, given by the logical criterion of operation.

$$P_F(\{p_i(t), q_i(t)\}, i = 1, 2, \ldots, H; t) \tag{3}$$

The advantages of this method are that it can be used to calculate various complex structures of equipment connection, for example, bridge connection or functional redundancy, and it is applicable to any laws of probability distributions of trouble-free operation.

The main disadvantage of the method, the cumbersome transformations of FLA for complex systems, is leveled by software implementation.

Required initial data:

Table 1 Source matrix describing the structure of a system with bridge link (*CTS*)

i	j				
	1	2	3	4	5
1	0	0	0	0	0
2	0	0	0	0	0
3	1	0	0	0	1
4	0	1	0	0	1
5	1	1	0	0	0

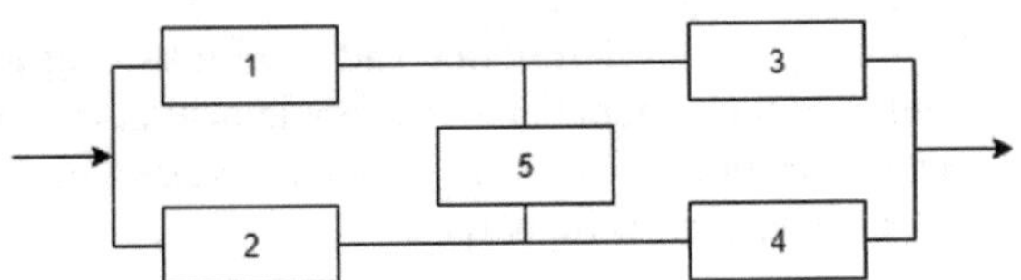

Fig. 1 A system with a bridge link of elements

– the matrix describing the structure of the system (*CTS*);
– vector of probabilities of failure-free operation of system elements (p).

The square matrix of the structure based on the representation of the system in the form of an oriented graph of the "Tree" type. The numbers of the elements from which the material flow exits are located horizontally; located vertically—numbers of the elements into which the material flow enters, i.e. if the flow of matter leaves the jth element and enters the ith element, then $CTS\ (i, j) = 1$, otherwise—0, where j is the number of the matrix column, i is the row number.

Below presented an example of the initial matrix (Table 1) describing the structure for a system with a bridge connection of the elements presented in Fig. 1.

The algorithm is written in the MATLAB language. Consists of the following steps: (Fig. 2)

1. With the help of FLA, matrix A is generated of all possible states of the system. The number of columns of the matrix is determined by the number of elements of the system, the number of rows, l—the number of combinations of the states of the elements, ranging from the case when all elements are in a state of failure (null row of the matrix) to the operability of all the elements (full row of the matrix).
2. The matrix of the *CTS* structure determines the vectors of input (*input*) and output (*output*) elements. The *input* vector consists of the null row numbers of the *CTS* matrix, and the *output* consists of the null column numbers.
3. Next, for each row of the matrix A, the state of the entire system is determined (0 is inoperable, 1 is healthy, functional). The result is written in vector B.
4. Here n is the counter by rows of the matrix A, m is the counter by *input_new* vector, which is formed from the operable input elements in a given set of equipment states, b is the system state for this case (vector B element), r is the counter by rows of the *CTS_new* matrix.

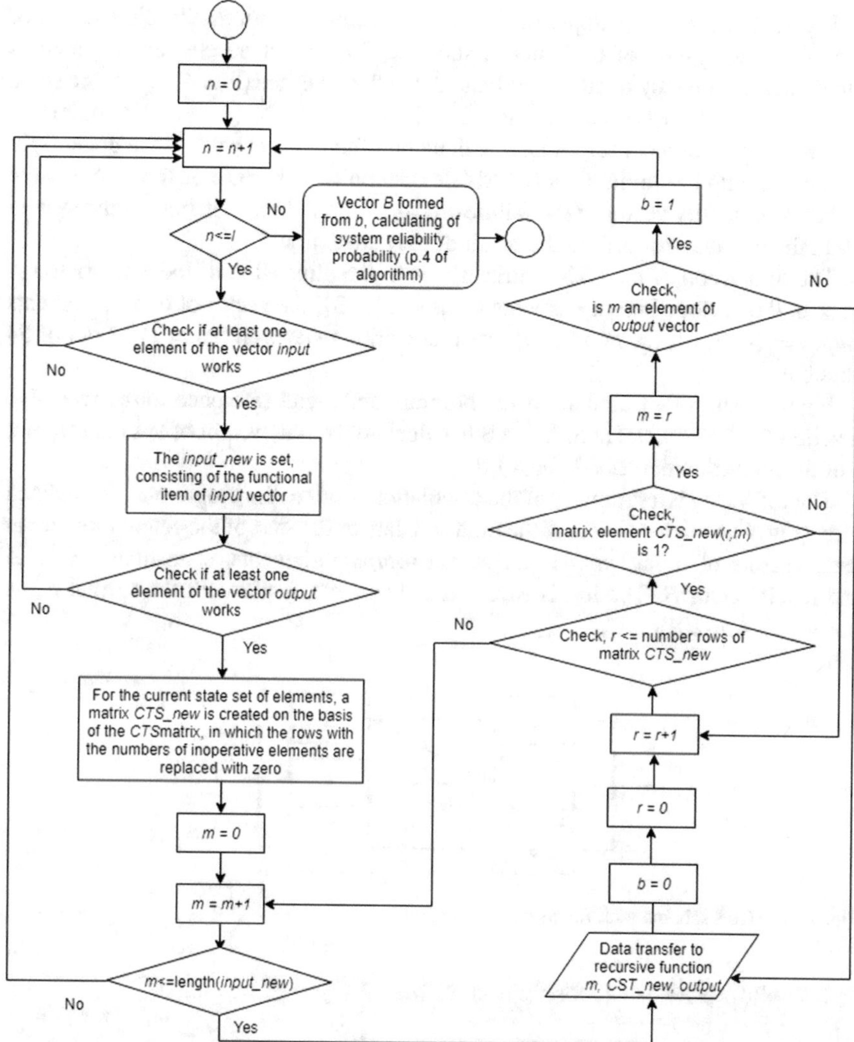

Fig. 2 System availability determining the algorithm for each set of possible states of elements

5. After vector B is generated, all rows for which the corresponding element of vector B is 0 are deleted from matrix A.
6. In the resulting array, the zero values are replaced by the expression $(1 - p_i)$, and the units are replaced by p_i.
7. The multiplication of the elements of the matrix is calculated line by line.
8. Further, all elements of the vector from p. 6 are summarized. The resulting amount is a probability distribution of trouble-free operation for the system.

Figure 4 shows an example of the calculation of the probability distribution of a trouble-free system of 6 elements, shown in Fig. 3, where element 2 performs functional redundancy of elements 1 and 3. On the interface (Fig. 4), the source data on the left shows a list of variables involved in the calculation (upper right corner). In the lower right corner, a call of LPM calculation function showed. In the upper right corner showing the initial data (matrix describing the structure of the CES system and the probability vector of the failure-free operation of the elements of the system *p*). In the variable *ans* passed the result of the calculation.

The dimension of the *CTS* matrix is 6 × 6, the dimension of the state matrix *A* 64 × 6, the vector of input elements *input* = {1,2,3}, the vector of output elements *output* = {4, 5, 6}, the number of possible working states (*B (n)* = *1*) 42 out of 64 possible.

Figure 5 shows a fragment of the program code with reference to the recursive function *SysPerformDet* in MATLAB to calculate the state vector of system *B* (point 3 of the algorithm described above).

The following is a summary of the calculation result of the program in abbreviated form (Fig. 6), which consists of the time and date of the start of the calculation, input data, vectors of input (*input*) and output (*output*) elements, state matrix CTS *A* and result vector *B*. The log is saved in a separate text file with the format name "LPM_27-Apr-2019.txt".

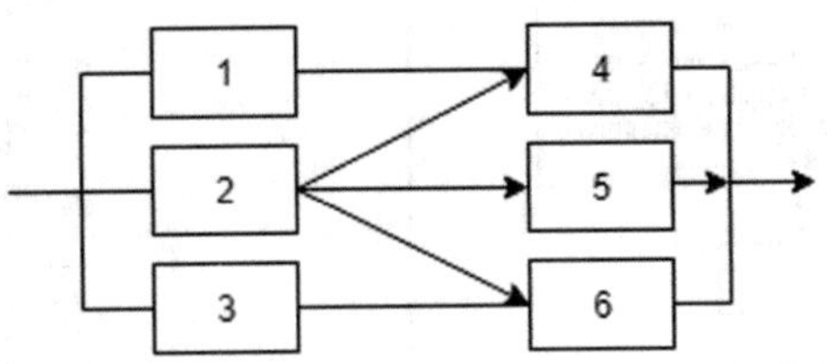

Fig. 3 The CES scheme with functional redundancy

Variables

p

1x6 double

	1	2	3	4	5	6	7
1	0.6000	0.6000	0.5000	0.7000	0.7000	0.8000	
2							

CTS

6x6 double

	1	2	3	4	5	6	7
1	0	0	0	0	0	0	
2	0	0	0	0	0	0	
3	0	0	0	0	0	0	
4	1	1	0	0	0	0	
5	0	1	0	0	0	0	
6	0	1	1	0	0	0	
7							

Workspace

Name	Value	Min	Max
CTS	6x6 double	0	1
ans	0.8500	0.8500	0.8500
p	[0.6000 0.6000 0.5000 ...	0.5000	0.8000

Command Window

```
>> LPM(CTS,p)

ans =

    0.8500

fx >>
```

Fig. 4 An example of the reliability probability calculation of the 6 elements system with functional redundancy using MATLAB

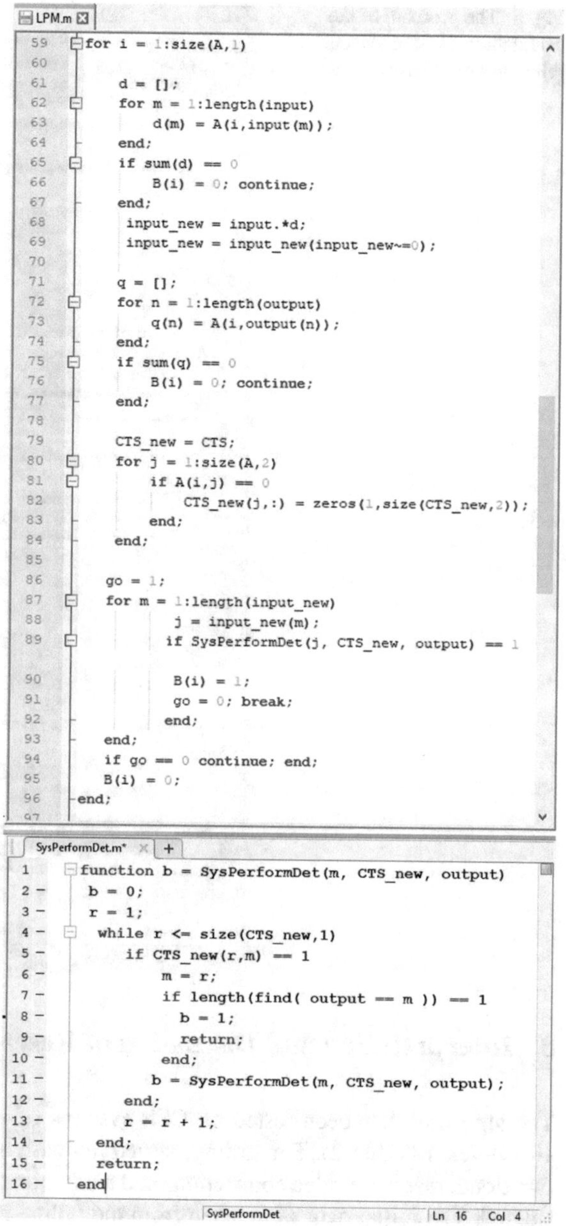

Fig. 5 The fragment of the program code and the recursive function SysPerformDet in MATLAB to calculate the state vector of the system *B*

LPM.m

```
for i = 1:size(A,1)

    d = [];
    for m = 1:length(input)
        d(m) = A(i,input(m));
    end;
    if sum(d) == 0
        B(i) = 0; continue;
    end;
     input_new = input.*d;
     input_new = input_new(input_new~=0);

    q = [];
    for n = 1:length(output)
        q(n) = A(i,output(n));
    end;
    if sum(q) == 0
        B(i) = 0; continue;
    end;

    CTS_new = CTS;
    for j = 1:size(A,2)
        if A(i,j) == 0
            CTS_new(j,:) = zeros(1,size(CTS_new,2));
        end;
    end;

   go = 1;
   for m = 1:length(input_new)
           j = input_new(m);
          if SysPerformDet(j, CTS_new, output) == 1

           B(i) = 1;
           go = 0; break;
          end;
   end;
   if go == 0 continue; end;
   B(i) = 0;
end;
```

SysPerformDet.m*

```
function b = SysPerformDet(m, CTS_new, output)
 b = 0;
 r = 1;
  while r <= size(CTS_new,1)
     if CTS_new(r,m) == 1
         m = r;
         if length(find( output == m )) == 1
           b = 1;
           return;
         end;
       b = SysPerformDet(m, CTS_new, output);
     end;
     r = r + 1;
  end;
  return;
end
```

SysPerformDet Ln 16 Col 4

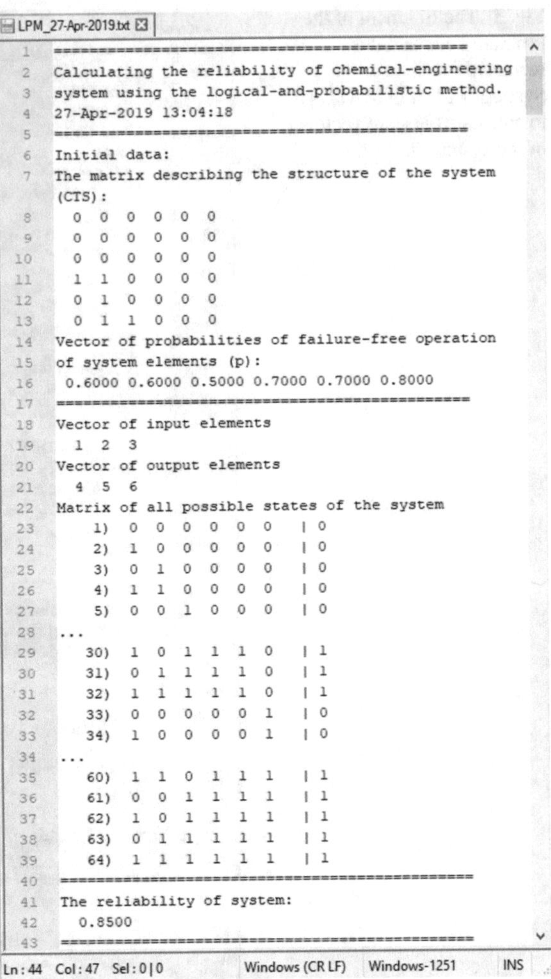

```
LPM_27-Apr-2019.txt
============================================
Calculating the reliability of chemical-engineering
system using the logical-and-probabilistic method.
27-Apr-2019 13:04:18
============================================
Initial data:
The matrix describing the structure of the system
(CTS):
   0  0  0  0  0  0
   0  0  0  0  0  0
   0  0  0  0  0  0
   1  1  0  0  0  0
   0  1  0  0  0  0
   0  1  1  0  0  0
Vector of probabilities of failure-free operation
of system elements (p):
  0.6000 0.6000 0.5000 0.7000 0.7000 0.8000
============================================
Vector of input elements
   1  2  3
Vector of output elements
   4  5  6
Matrix of all possible states of the system
      1)  0  0  0  0  0  0   | 0
      2)  1  0  0  0  0  0   | 0
      3)  0  1  0  0  0  0   | 0
      4)  1  1  0  0  0  0   | 0
      5)  0  0  1  0  0  0   | 0
...
     30)  1  0  1  1  1  0   | 1
     31)  0  1  1  1  1  0   | 1
     32)  1  1  1  1  1  0   | 1
     33)  0  0  0  0  0  1   | 0
     34)  1  0  0  0  0  1   | 0
...
     60)  1  1  0  1  1  1   | 1
     61)  0  0  1  1  1  1   | 1
     62)  1  0  1  1  1  1   | 1
     63)  0  1  1  1  1  1   | 1
     64)  1  1  1  1  1  1   | 1
============================================
 The reliability of system:
   0.8500
============================================
Ln: 44  Col: 47  Sel: 0|0    Windows (CR LF)  Windows-1251  INS
```

Fig. 6 The protocol of the calculation result of system reliability probability

3 Interpretation and Discussion of Research Results

The algorithm has been tested on CES systems of varying complexity from 5 to 10 elements (Table 2). For testing, structures with a series-parallel connection, a functional reserve, bridge connections and a combination of these options were considered. The initial data were taken from the failure rate of the standard equipment [17].

For the calculation of reliability using the developed algorithm, the hydrotreating unit of the catalytic reforming unit LCH-35-11/1000 was considered.

When studying the reliability of CES, there may be uncertainty, for example, in choosing the method of decomposing the structure of CES into various subsystems,

Table 2 Summary table of the results of testing the algorithm on the CES schemes of varying complexity

Structure type	The size of the states matrix A	Vectors input/output	Number of operable states ($B(n) = 1$)
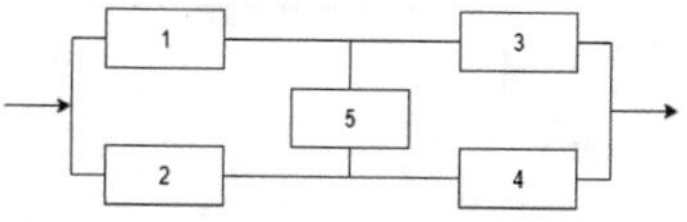	32×5	{1,2}/{3,4}	16
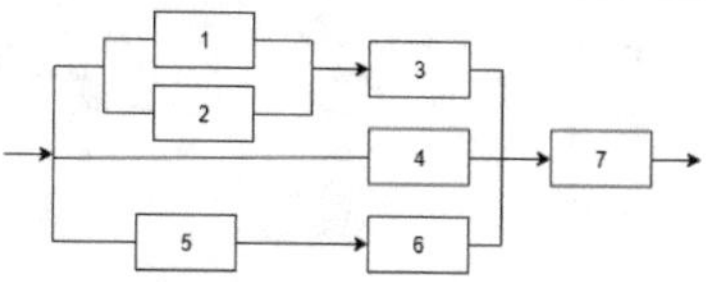	128×7	{1,2,4,5}/{7}	49
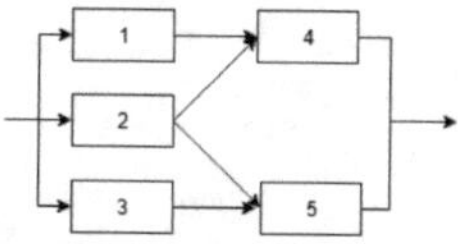	32×5	{1,2,3}/{4,5}	19
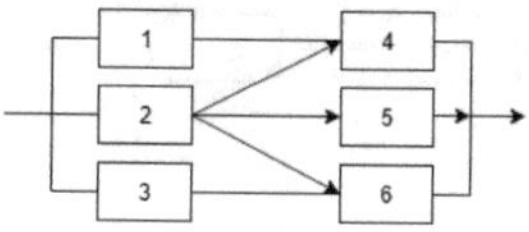	64×6	{1,2,3}/{4,5,6}	42
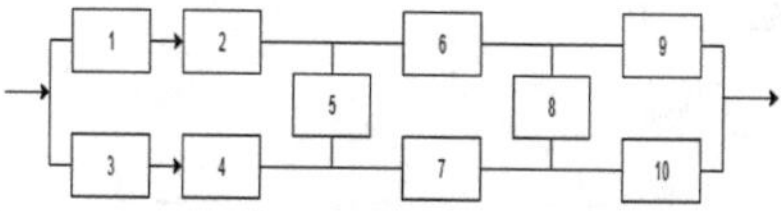	1024×10	{1,2}/{9,10}	181
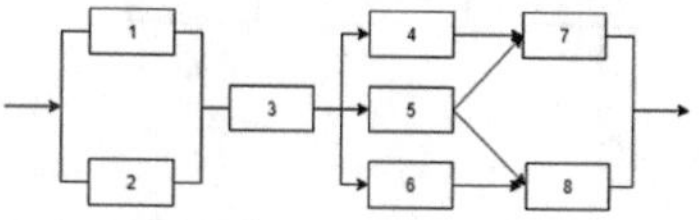	256×8	{1,2}/{7,8}	48

and in the assumptions in the interaction of subsystems, there may be alternative decomposition options.

Figure 7 shows the flowchart. To calculate the unit reliability under conditions of uncertainty, produced a decomposition of the structure in two ways (Figs. 8 and 9).

In the first variant (Fig. 8), the technological chain from the X-101 air cooler to the C-109 separator was presented as a separate unit (Mixing Point) in the technological chain from the H-101/1.2 pumps to the R-101 reactor. The reverse flow from the R-101 reactor to the T-101/1-4 heat exchangers was neglected since it does not affect the reliability indices of the entire system.

In the second variant (Fig. 9), the decompositions presented the entire system sequentially, while the P-101 hydrotreating furnace and the R-101 reactor were

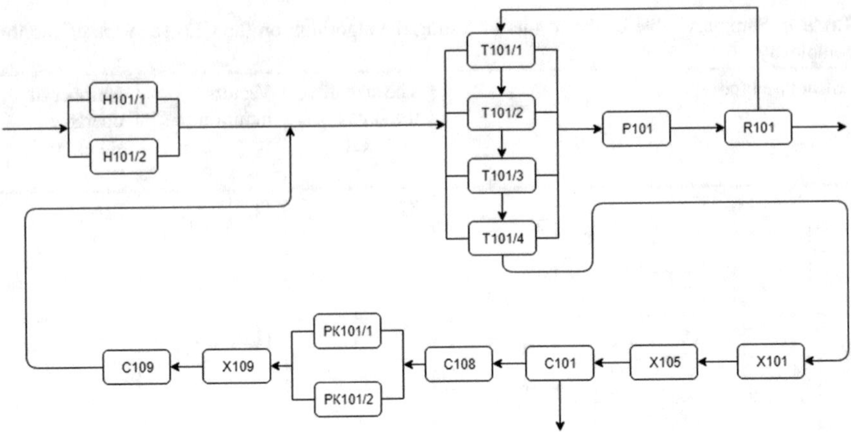

Fig. 7 Flowchart of the hydrotreating unit

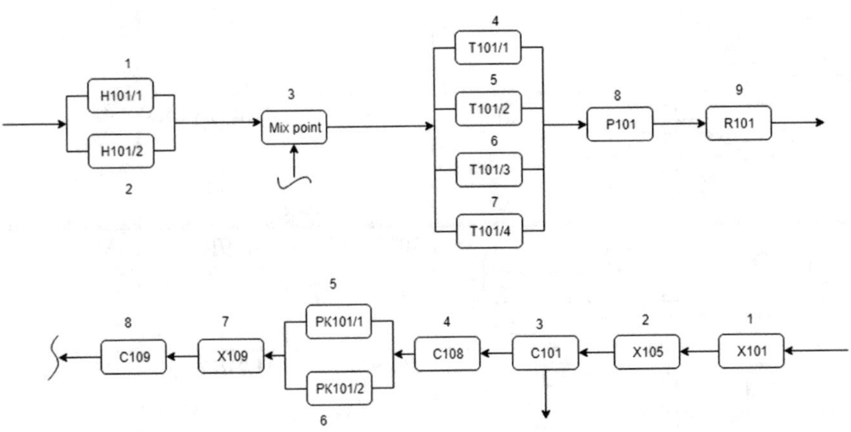

Fig. 8 1st option decomposition

moved to the beginning of the scheme since the order of the devices when calculating the serial connection does not affect the overall system reliability indicator.

Calculations of the system reliability probability were made on a time interval from 0 to 8760 h (1 year) with a step of 365 h. For calculations, the failure rates of the apparatus from Table 3 were used [17].

Figure 10 shows the result of the calculations. The values of reliability with different methods of decomposition were almost the same. However, from a technological point of view, the first version of the decomposition is more correct.

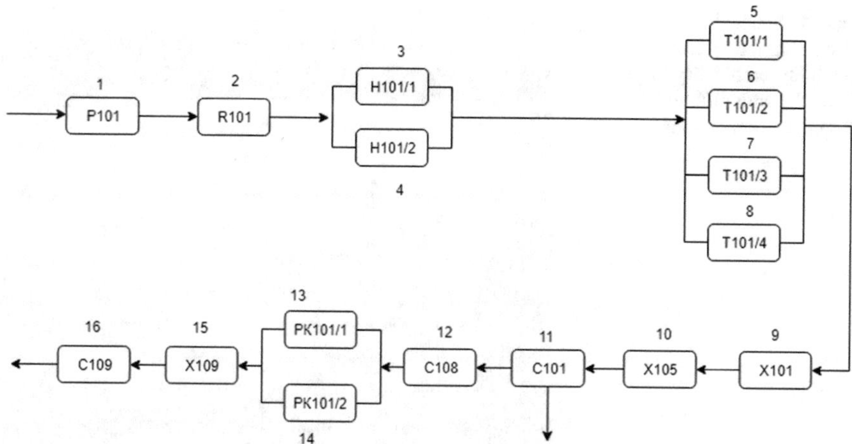

Fig. 9 2nd version of decomposition

Table 3 Values of reliability indicators of hydrotreating unit

Equipment identification	Failure rate ($\lambda \times 10^{-5}$), h^{-1}
Raw material pump H-101/1.2	33
Hydrotreating heat exchanger T-101/1-4	4
Hydrotreating furnace P-101	41
Hydrotreating reactor R-101	11
Hydrogenate air cooler X-101	17.2
Hydrogenate refrigerator X-105	5
Hydrogenated separator C-101	12.5
Reception separator PK-101/1.2 C-108	12.5
Piston compressor PK-101/1.2	13.2
Hydrogen-gas air cooler X-109	17.2
Discharge separator PK-101/1.2 C-109	12.5

4 Conclusion

In the course of the study, an algorithm was developed and implemented for calculating the reliability of CES using a logical-probabilistic method. Confirmed the possibility of using the MATLAB PC to calculate the reliability of production systems. Its only drawback is the need to represent the structure of the CES in a matrix form.

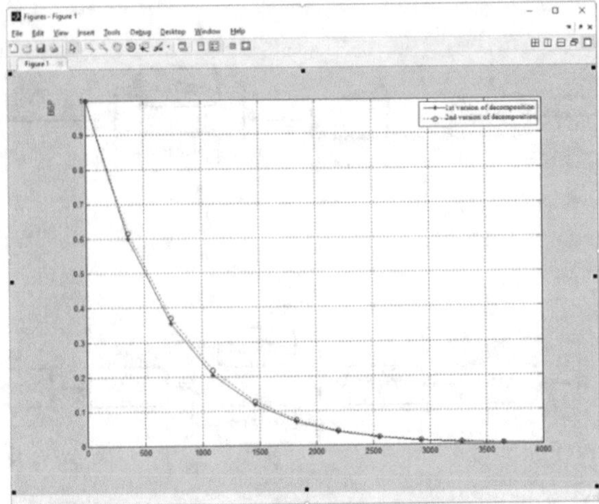

Fig. 10 Schedule of changes in the system of the reliability probability for 5 months with different decomposition options

The developed algorithm is used in conducting scientific research and laboratory work for bachelors and masters in the disciplines of "Fundamentals of the reliability of technical systems" and "Computer systems for designing chemical production".

References

1. Feeney, A.B., Srinivasan, V., Frechette, S.: Cyber-physical systems engineering for manufacturing. In: Industrial Internet of Things: Cyber Manufacturing Systems, pp. 81–110 (2017)
2. Monostori, L., Kádár, B., Bauernhansl, T., Kondoh, S.: Cyber-physical systems in manufacturing. CIRP Annals Manuf. Technol. **65**(2), 621–641 (2016)
3. Okoh, P., Haugen, S.: Improving the robustness and resilience properties of maintenance. Process Saf. Environ. Prot. **94**, 212–226 (2015)
4. Woo, S.W.: The reliability design of mechanical system and its parametric accelerated life testing. In: Handbook of Materials Failure Analysis with Case Studies from the Chemicals, Concrete and Power Industries, pp. 259–276 (2016)
5. Birolini, A.: Reliability Engineering: Theory and Practice, 7th edn, p. 626. Springer, Berlin (2014)
6. Myers, A.: Complex System Reliability: Multichannel Systems with Imperfect Fault Coverage, 2nd edn, p. 238. Springer, London (2010)
7. Gang, W., Wang, S., Augenbroe, G., Xiao, F.: Robust optimal design of district cooling systems and the impacts of uncertainty and reliability. Energy Buildings **122**, 11–22 (2016)
8. Lin, Y., Li, D., Kang, R.: Reliability modelling and simulation of complex systems. In: Chemical Engineering Transactions, pp. 469–474 (2013)
9. Mellal, M.A., Zio, E.: A penalty guided stochastic fractal search approach for system reliability optimization. Reliab. Eng. Syst. Saf. **152**, 213–227 (2016)
10. Abubakar, U., Sriramula, S., Renton, N.C.: Reliability of complex chemical engineering processes. Comput. Chem. Eng. **74**(4), 1–14 (2015)
11. Martowicz, A., Uhl, T.: Reliability- and performance-based robust design optimization of MEMS structures considering technological uncertainties. Mech. Syst. Signal Process. **32**, 44–58 (2012)

12. Guilani, P.P., Sharifi, M., Niaki, S.T.A., Zaretalab, A.: Reliability evaluation of non-reparable three-state systems using Markov model and its comparison with the UGF and the recursive methods. Reliab. Eng. Syst. Saf. **129**, 29–35 (2014)
13. Malygin, E.: Automated Remote Access Laboratory "Design and Exploitation of Technological Systems". In Malygin, E.N., Krasnyansky, M.N., Karpushkin, S.V., Mokrozub, V.G. (eds.) Vestnic TGTU, vol. 6(2). Tambov, pp. 332–335 (2000)
14. Krasnyansky, M.N., Malygin, E.N., Karpushkin, S.V., Chaukin, Y.V., Ostroukh, A.V.: Application of virtual simulators for training students of the chemical technology type and improvement of professional skills of chemical enterprises personnel. Vestnic TGTU, vol. 13(1B), Tambov, pp. 233–238 (2007)
15. An, Y.: Reliable Design and Operations of Infrastructure Systems. https://www.mobt3ath.com/uplode/book/book-17338.pdf (2019). Accessed 20 Feb 2019
16. Shubin, V., Ryumin, Yu.: The Reliability of the Equipment of Chemical and Oil-Refining Industries. Moscow, Himiya, KolosS Publ., p. 359 (2006) (in Russian)
17. Shvetsova-Shilovskaya, T., Kondrat'ev, V., Gorskii, V., Egorov, A., Polekhina, O., Gromova, T., Gamzina, T., Afanas'eva, A., Savitskaya, T., Nazarenko, D., Ivanov, D., Vikent'eva, M.: The Methodology and the Software for Assessment of Reliability and Operational Safety of Chemical Process Equipment: Monograph. Moscow, RHTU im. D. I. Mendeleeva Publ., p. 372 (2016) (in Russian)
18. Strogonov, A., Zhadnov, V., Polesskii, S.: Overview of program complexes by calculation of reliability of complex technical systems. Compon. Technol. **5**, 183 (2007) (in Russian)
19. Egorov, A., Savitskaya, T., Nikitin, S.: The information system of reliability analysis of equipment and chemical technological systems using web technologies. Prikladnaya Informatika—J. Appl. Inf. **11**(4, 64), 30–41 (2016) (in Russian)
20. Meshalkin, V., Moshev, E.: Program complex for life cycle support of pipeline systems of petrochemical companies. Applied Informatics—J. Appl. Inf. **11**(4, 64), 57–75 (2016) (in Russian)
21. Zakharova, A., Savitskaya, T.: Computational Functions Development for Research of Mathematical Models of Chemical-Technological Systems Redundancy: Advances in Chemistry and Chemical Technology, vol XXXII(11, 207). Moscow, RHTU D. I. Mendeleeva Publ., pp. 66–68 (2018) (in Russian)
22. Ryabinin, I.: Reliability and Security of Complex Systems, p. 248. Polytehnika, SPb (2000) (in Russian)
23. Mozhaev, A., Gromov, V.: Theoretical Foundations of the General Logical-Probabilistic Method of Automated System Modeling, p. 143. VITU, SPb (2000) (in Russian)

Cyber-Physical Systems and Digital Twins

Assessment of the State of Production System Components for Digital Twins Technology

T. I. Buldakova and S. I. Suyatinov

Abstract The problem of assessment of the state of production systems is considered. The chapter is suggested applying the technology of digital twins to solve the problem of diagnosing and predicting the state of the components of the production system. The hierarchical structure of modern production is described, as well as the interaction of the production system and its digital twin. The correspondence of the system components and models of their state assessment is indicated. Methods and tools for assessing the state of the components of different hierarchical levels of the production system representation are proposed. As an example, the assessment of the state of stamp-tool production is considered and the models for assessing the state of its components for the digital twin are given. Also, a criterion and method for assessing the state of the upper organizational and technical level of this system are proposed.

Keywords Cyber-physical system · Production element · State assessment · Model · Digital twin

1 Introduction

An assessment of the state of the production system for the purpose of diagnosing it is an important and crucial task. In the process of its solution, they reveal, analyze and evaluate the level of efficiency and development of its various components (equipment, technology, personnel, resources, etc.). The problem is that production is a hierarchical structure, each level of which has its own specifics. At the same time, components of each level contribute to ensuring trouble-free and uninterrupted operation of the entire production. For the diagnosis of different types of components

T. I. Buldakova (✉) · S. I. Suyatinov
Bauman Moscow State Technical University, 2-ya Baumanskaya, 5, Moscow 105005, Russia
e-mail: buldakova@bmstu.ru

S. I. Suyatinov
e-mail: ssi@bmstu.ru

A. G. Kravets et al. (eds.), *Cyber-Physical Systems: Advances in Design & Modelling*, Studies in Systems, Decision and Control 259,
https://doi.org/10.1007/978-3-030-32579-4_20

apply their methods and approaches. The methods of technical diagnostics applied to the lower hierarchical level have the most development [1–3].

There are three main types of technical diagnostics: (1) planned diagnostics—periodic testing of equipment performance according to a predetermined schedule; (2) unscheduled, emergency diagnostics—identifying the causes and conditions that caused the faults, and making informed decisions to eliminate them; (3) monitoring—recognition of the current technical condition of the equipment to predict possible failures. Regardless of what type of diagnosis is implemented, methods and means of its implementation are necessary. Moreover, at present, they are one of the most important factors for increasing the efficiency of using equipment, mechanisms, and machines in the industry [4–6].

At the same time, the problem of assessing the state of upper levels of the production system, in the process of their continuous monitoring, remains relevant.

2 Purpose of Research

The tasks of monitoring and assessing the state of production are particularly relevant for cyber-physical systems. Being fundamentally distributed, such systems are characterized by a high saturation of sensors and actuators, providing automatic operation of production facilities, minimizing maintenance personnel and visual control of equipment operation, especially at lower hierarchical levels of the production process [7–10].

At the same time, despite the high level of automation of information processing and management decision-making, an important element in cyber-physical systems remains a person who makes important decisions at various hierarchical levels of production [11–13]. At the dispatch level, the state of a human operator largely determines the quality indicators of the production process. The staff of the upper organizational and technical level ensures uninterrupted logistical support. For assessing the state of the entire enterprise, it is necessary to monitor the functional state of the upper organizational and technical level.

Currently, the problems of monitoring and predicting the state of the components of the production system can be successfully solved within the concept of digital twins. The concept is based on the technology of mathematical modeling of business processes of a production system and, in particular, the model of the dynamics of executive bodies, including the human operator. The concept of digital twins is a logical continuation of the development of CALS technologies, mathematical modeling and diagnostic models [14–16]. The technical realization of the digital twin became possible due to the development of computer technology, the Internet and wireless sensors.

This study aims to analyze the hierarchical levels of the production system, based on the requirements of assessing its state in the process of monitoring, as well as the development of criteria and methods for assessing the state of the upper organizational and technical level.

3 Representation Levels of the Production System

In the modern production system, it is possible to distinguish several hierarchical levels corresponding to various types of system components. The lower level includes production equipment (machine tools, mechanisms, machine equipment, and other production components), that process raw materials into the finished product following the technological route. At the middle level, we will include workers who support production processes, including the human operator, who manages complex equipment. The upper level of the production system is usually the level of management personnel who make the most responsible decisions on the material, technical and personnel support of production.

The structure of the system "real production—digital twin" is shown in Fig. 1. For simplicity, the division into levels is made on information types and not on the

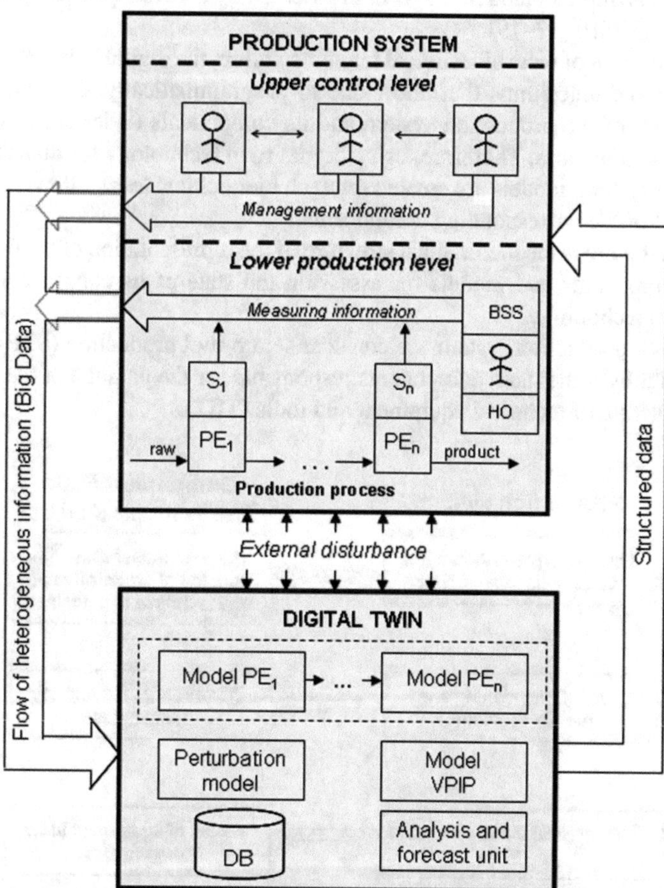

Fig. 1 Interaction of production system and digital twin

types of system components. Therefore, production is represented by a two-level hierarchical system, where the human operator is included in the lower production level.

The upper level of the production system is the organizational and technical level, the level of managerial personnel. Information on this level in the form of planning tasks, specifications, technological maps, standards, and other similar documentation is transmitted to the database (DB) of the digital twin. The lower level represents the sequence of production element (PE), processing raw materials into the finished product in accordance with the technological route. In this case, a human operator (HO) is included in the control loop at the lower level.

The state of production elements is estimated on the basis of signals recorded by various sensors $S_1, \ldots, S_{1n}$. To assess the state of a human operator, biosignal sensors (BSS) of various physiological nature are used. Measuring information recorded by sensors is transmitted via communication channels to a computer center, where a digital twin is implemented in the form of assessed by a virtual physiological image of a person (VPIP) [17–19].

From the point of view of technical implementation, the digital twin is a structured set of data and algorithms that allow one to programmatically simulate the state and behavior of the production system and its components under various external and internal influences. Therefore, using digital twin technology for diagnostics the production system, models are created for each hierarchical level, allowing the state assessment of the corresponding components.

On Fig. 2 there is the hierarchical structure of the representation of the production system components and models for assessing the state of its components for the digital twin technology.

Next, as a production system, we consider stamp-tool production (STP) (Fig. 3). We will highlight the main components responsible for the passage of the order for the manufacture of technical equipment and tools (TET).

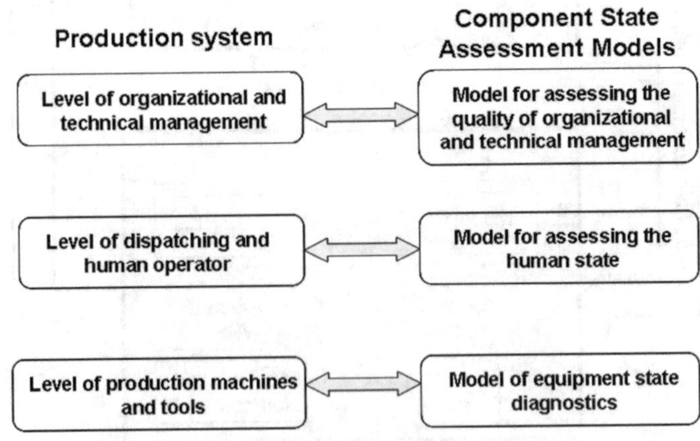

Fig. 2 Compliance of system components and its models

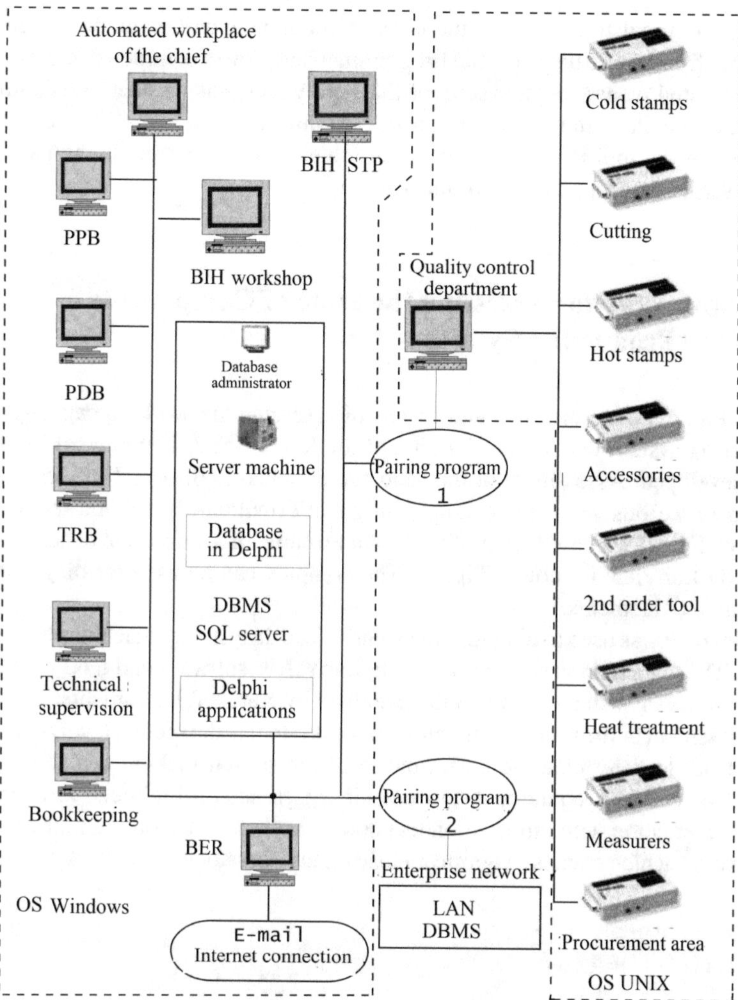

Fig. 3 The interaction of components STP

These include: planning and production bureau (PPB); planning and dispatch bureau (PDB); technology regulation bureau (TRB); bureau of technical supervision; bookkeeping; bureau of external relations (BER); bureau of instrumental house (BIH)—warehouses; quality control department and masters of production sites; technologists and designers engaged in coding cutting tools, stamps and their components.

Since it is impossible to operate personal computers under industrial production conditions, then modernized terminals BDT K 8901 (recorders) are used for operating data collection. The terminals, that control by a server on the Unix platform, are located directly in the workshop near the machines and production equipment.

Data from registrars arrive at the Unix server and are written to the intermediate database, from where they are read programmatically into a common database. Thus, the automated management system of the supply and production of TET allows in real time to collect and process the necessary information, and also to control the presence of the tool in all departments of the enterprise, as well as the passage of orders for its purchase and/or production.

4 Approaches for Assessing the State of Components of the Production System

At present, effective methods and means of assessing the working capacity of the production system under consideration at the lower (third) hierarchical level have been developed. Assessment of the state and diagnostics of operability can be performed in various ways, for example, using information-control and test benches based on SCADA technologies [20, 21]. An example of a diagnostic bench complex based on LabView is shown (Fig. 4). The complex can be used for diagnostics of machine tool elements.

This stand was used to diagnose the spindle feed-in grinding machines of the AGL series. The algorithm is worked out using LabVIEW software and a NI cRIO-9014 microcontroller with a real-time OS developed by National Instruments.

Assessment of the state and diagnosis of the system at the second level is primarily carried out to determine the performance of the person making decisions on the control of complex equipment and machinery. In such a high-tech environment, human interaction with information technology systems is becoming more complex and diverse, which creates a significant load on the operator and can lead to erroneous

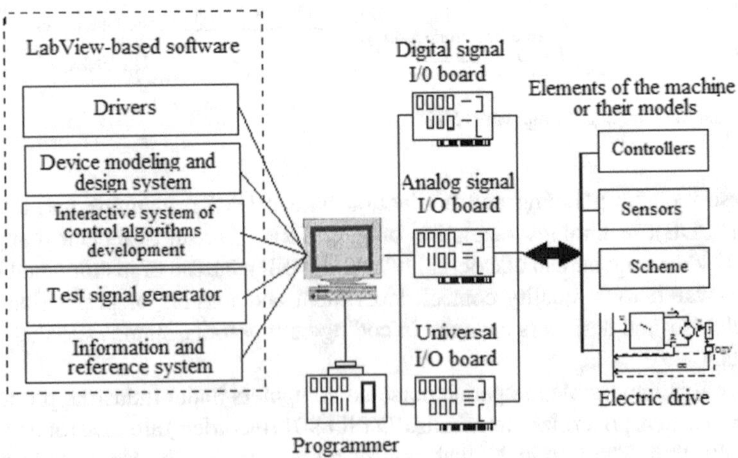

Fig. 4 Diagnostic bench complex

reactions in the control loop. Moreover, the activity of a human operator operating a complex device or equipment is characterized by high psycho-emotional stress, which can also have a negative impact on its performance. Therefore, in cyber-physical systems, it is necessary to constantly monitor the state of not only the technical means and equipment but also the human state.

Issues related to the assessment of the state of a human operator were studied in detail by the authors, for example, in [18, 19, 22, 23].

The first, upper level, associated with organizational and technical management, has a significant impact on the functioning of the entire production system. At this level, the most responsible decisions are made to ensure a clear and uninterrupted operation of the system. Therefore, criteria and methods are needed to assess the state of this level of management.

The production structure includes units of different types that are directly or indirectly related to each other. These various divisions and their relationships determine the complexity of the system. Various methods for estimating the complexity of production are known [24, 25]. Among them, the most generalized complexity estimate is entropy.

Existing approaches to the entropy estimation of production complexity are mainly used to assess the complexity of the equipment used. This assessment makes it possible to simplify production chains and reduce costs accordingly.

In [26], the entropy indicator is used to assess problems with the execution of orders caused by changes in customer requirements and the corresponding increase in the complexity of production. At the same time, the complexity of order execution is determined by the deviation of the planned deliveries from the deliveries modified by the customer. Note that the entropy complexity indicator can also be interpreted as an indicator of system organization, which can be used to assess its state. Therefore, it is proposed to accept the deviation from the planned indicators as initial information for assessing the state of the production system.

Then, to assess the functional state of the upper organizational and technical level of the cyber-physical system, an entropy indicator can be used, which is calculated based on the possible states of the production system. If the production operates in accordance with the planned tasks, then this state of the system corresponds to a certain indicator of organization. Various production disruptions lead to changes in interconnections and intensification of material and information flows. As a result, the entropy indicator of complexity increases, which indicates the occurrence of problems in the organization of production.

Consider the example of the production of cutting tools. Suppose that according to the plan it is necessary to manufacture n_j cutters in a day, where j denotes the number of the day of the planning period, and in fact, m_j cutters are made in a day. Then the daily deviations from the norm will be $d_j = n_j - m_j$. In this case, the value of d_j is an indicator of the state of the organizational and technical system in the manufacture of cutting tools. Then N deviation ranges of d_j are set and a histogram showing the probability of falling into each range is plotted. The entropy indicator is calculated by the known formula:

$$H(D) = -\sum_{i=1}^{N} p(d_i)\log_2 p(d_i). \tag{1}$$

Here H(D) is an indicator of the organization of the system in the production of cutting tools; $p(d_i)$ is the probability of hitting the deviations in the range i.

For a comprehensive assessment of the state of the organizational and technical system, it is necessary to take into account all material and information flows [27]. Then we get

$$H_\Sigma = -\sum_{i=1}^{M}\sum_{j=1}^{N_i} p_{ij}\log_2 p_{ij}. \tag{2}$$

Here H_Σ is an indicator of system organization; p_{ij} is probability that the resource i, (i = 1, …, M), is in the state j, (j = 1, …, N_i); M is number of resources (flows); N_i is the number of possible states for the resource i.

The advantage of the proposed criterion and method for assessing the state of the upper organizational and technical level is the availability of initial information, the ability to identify the most responsible and/or problematic production areas in the monitoring process.

5 Conclusion

Thus, a characteristic feature of cyber-physical systems is information monitoring and assessment of the state of production components of all hierarchical levels. The use of digital twin technology allows you to more effectively assess the state of the production system components.

Methods of diagnostics of the machine park, based on mathematical models of physical processes implemented in digital twins, have received the most development. Although methods of assessing the state of a human operator have a long history, they are still at the research stage. Methods and criteria for assessing the state of the organizational and technical system are at the stage of the problem statement and the choice of mathematical support. The proposed method of state estimation based on the entropy criterion is very promising, given its advantages.

Further research should be directed to the study of alternative options for assessing the state of the production system components. The final stage should be the development of a methodology for integrated assessment of the state of production systems.

References

1. Nikolova, N., Hirota, K., Kolev, K., Tenekedjiev, K.: Technical diagnostic system in the maintenance of turbomachinery for ammonia synthesis in the process Industries. J. Loss Prev. Process Ind. **58**, 102–115 (2019). https://doi.org/10.1016/j.jlp.2019.02.002
2. Efthymiou, K., Papakostas, N., Mourtzis, D., Chryssolouris, G.: On a predictive maintenance platform for production systems. Procedia CIRP **3**, 221–226 (2012). https://doi.org/10.1016/j.procir.2012.07.039
3. Kumenko, A.I.: The improvement modification of rotor unbalance verification technique in monitoring systems and automatic diagnostics. Procedia Eng. **113**, 324–331 (2015). https://doi.org/10.1016/j.proeng.2015.07.273
4. Protalinsky, O.M., Shcherbatov, I.A., Stepanov, P.V.: Identification of the actual state and entity availability forecasting in power engineering using neural-network technologies. J. Phys.: Conf. Ser. **891**(1), 10. Nov 2017, Article 012289 (2017). https://doi.org/10.1088/1742-6596/891/1/012289
5. Protalinsky, O., Khanova, A., Shcherbatov, I.: Simulation of power assets management process. In: Dolinina, O. et al. (eds.) Recent Research in Control Engineering and Decision Making, ICIT-2019. Studies in Systems, Decision and Control, vol. 199, pp. 488–501 Springer, Cham (2019). https://doi.org/10.1007/978-3-030-12072-6_40
6. Lu, Y.: Industry 4.0: a survey on technologies, applications and open research issues. J. Ind. Inf. Integr. **6**, 1–10 (2017). https://doi.org/10.1016/j.jii.2017.04.005
7. Lee, J., Bagheri, B., Kao, H.A.: A cyber-physical systems architecture for industry 4.0-based manufacturing systems. Manuf. Lett. **3**, 18–23 (2015)
8. Hermann, M., Pentek, T., Otto, B.: Design principles for industrie 4.0 scenarios. In: Proceedings of the Annual Hawaii International Conference on System Sciences, Article 7427673, pp. 3928–3937 (2016). http://dx.doi.org/10.1109/HICSS.2016.488
9. Herwan, J., Kano, S., Ryabov, O., Sawada, H., Kasashima, N.: Cyber-physical system architecture for machining production line. In: 2018 IEEE Industrial Cyber-Physical Systems (ICPS), pp. 387–391 (2018). https://doi.org/10.1109/ICPHYS.2018.8387689
10. Koval', V.A., Osenin, V.N., Suyatinov, S.I., Torgashova, O.Y.: Synthesis of discrete controller for construction of a distributed controller of temperature conditions of steam oil heater. J. Comput. Syst. Sci. Int. **50**(4), 638–653 (2011). https://doi.org/10.1134/S1064230711040125
11. Sowe, S.K., Zettsu, K., Simmon, E., de Vaulx, F., Bojanova, I.: Cyber-physical human systems: putting people in the loop. IT Prof. **18**(1), 10–13 (2016). https://doi.org/10.1109/MITP.2016.14
12. Sénéchal, O., Trentesaux, D.: A framework to help decision makers to be environmentally aware during the maintenance of cyber physical systems. Environ. Impact Assess. Rev. **77**, 11–22 (2019). https://doi.org/10.1016/j.eiar.2019.02.007
13. Sharpe, R., Lopik, K.V., Neal, A., Goodall, P., Conway, P.P., West, A.A.: An industrial evaluation of an Industry 4.0 reference architecture demonstrating the need for the inclusion of security and human components. Computers in Industry, vol. 108, pp. 37–44 (2019). https://doi.org/10.1016/j.compind.2019.02.007
14. Skvortsov, V., Proletarsky, A., Arzybaev, A.: Feature recognition module of the CAPP system. In: Proceedings of the 2019 IEEE Conference of Russian Young Researchers in Electrical and Electronic Engineering, ElConRus (2019). http://dx.doi.org/10.1109/EIConRus.2019.8656655
15. Tarassov, V.B.: Enterprise total agentification as a way to industry 4.0: forming artificial societies via Goal-resource networks. In: Abraham, A., Kovalev, S., Tarassov, V., Snasel, V., Sukhanov, A. (eds.) Proceedings of the Third International Scientific Conference "Intelligent Information Technologies for Industry" (IITI'18). Advances in Intelligent Systems and Computing, vol. 874, pp. 26–40. Springer, Cham (2019). http://dx.doi.org/10.1007/978-3-030-01818-4_3
16. Bozhko, A.: Math modeling of sequential coherent and linear assembly plans in CAD systems. In: 2018 Global Smart Industry Conference (GloSIC), pp. 1–5 (2018). http://dx.doi.org/10.1109/GloSIC.2018.8570090

17. Prado, M., Roa, L., Reina-Tosina, J.: Virtual center for renal support: technological approach to patient physiological image. IEEE Trans. Biomed. Eng. **49**(12), 1420–1430 (2002)
18. Suyatinov, S.I.: Criteria and method for assessing the functional state of a human operator in a complex organizational and technical system. In: Global Smart Industry Conference (GloSIC), pp. 1–6. Chelyabinsk, Russia (2018). http://dx.doi.org/10.1109/GloSIC.2018.8570088
19. Buldakova, T., Krivosheeva, D.: Data protection during remote monitoring of person's state. In: Dolinina, O., et al. (eds.) Recent Research in Control Engineering and Decision Making, ICIT-2019. Studies in Systems, Decision and Control, vol. 199, pp. 3–14. Springer, Cham (2019). https://doi.org/10.1007/978-3-030-12072-6_1
20. Qian, P., Zhang, D., Tian, X., Si, Y., Li, L.: A novel wind turbine condition monitoring method based on cloud computing. Renew. Energ. **135**, 390–398 (2019). https://doi.org/10.1016/j.renene.2018.12.045
21. Chattal, M., Bhan, V., Madiha, H., Shaikh, S.A.: Industrial automation control trough PLC and labview. In: 2nd International Conference on Computing, Mathematics and Engineering Technologies, iCoMET (2019). https://doi.org/10.1109/ICOMET.2019.8673448
22. Buldakova, T.I., Suyatinov, S.I.: Registration and identification of pulse signal for medical diagnostics. In: Proceedings of SPIE—The International Society for Optical Engineering, vol. 4707, Article 48, pp. 343–350 (2002)
23. Buldakova, T.I., Suyatinov, S.I.: Reconstruction method for data protection in telemedicine systems. In: Progress in Biomedical Optics and Imaging—Proceedings of SPIE, vol. 9448, Article 94481U (2014). https://doi.org/10.1117/12.2180644
24. Efstathiou, J., Calinescu, A., Blackburn, G.: A web-based expert system to assess the complexity of manufacturing organizations. Robot. Comput. Integr. Manuf. **18**, 305–311 (2002). https://doi.org/10.1016/S0736-5845(02)00022-4
25. Modrak, V., Soltysova, Z.: Novel complexity indicator of manufacturing process chainsand and its relations to indirect complexity indicators. Complexity, Article ID 9102824, pp. 1–15 (2017). https://doi.org/10.1155/2017/9102824
26. Kedadouche, M., Thomas, M., Tahan, A., Guilbault, R.: Nonlinear parameters for monitoring gear: comparison between Lempel-Ziv, approximate entropy, and sample entropy complexity. Shock. Vib., Article ID 959380, 1–12 (2015). http://dx.doi.org/10.1155/2015/959380
27. Isik, F.: An entropy-based approach for measuring complexity in supply chains. Int. J. Prod. Res. **48**(12), 3681–3696 (2010)

Proactive and Predictive Maintenance of Cyber-Physical Systems

Maxim V. Shcherbakov, Artem V. Glotov and Sergey V. Cheremisinov

Abstract The following chapter describes a concept model for proactive decision support system based on (real-time) predictive analytics and designed for maintenance of cyber-physical systems (CPSs) in order to optimize its downtime. This concept later is referred to as proactive and predictive maintenance decision support systems or P^2M for short. The concept is based on (i) the axioms of predictive decisions making, (ii) the proactive computing principles and (iii) models and methods for intelligent data processing. The aforementioned concept extends an idea of data-driven intelligent systems by using two approaches. The first approach implements predictive analytics, i.e. detection of a pre-failure event (called a proactive event) over a certain time period. This approach is based on the sequence of the following operational processes: to detect–to predict–to decide–to act. The second approach helps to automate maintenance decisions, which allows to exclude operational roles and move to supervisory level positions in the operational management structure. The concept includes the following primary components: ontology, a data warehouse (data lake), data factory as a set of data processing methods, flexible pipelines for data handling and processing and business processes with predictive decision logic for cyber-physical systems maintenance. This concept model is considered as the platform for the design of cyber-physical asset performance management systems.

Keywords Industrial cyber-physical systems · Proactive decision support · Predictive maintenance

M. V. Shcherbakov (✉) · A. V. Glotov · S. V. Cheremisinov
Volgograd State Technical University, Volgograd, Russia
e-mail: maxim.shcherbakov@vstu.ru

A. V. Glotov
e-mail: info@mobilegates.ru

Mobile Gas Turbine Energy Stations JSC Company, Moscow, Russia

A. G. Kravets et al. (eds.), *Cyber-Physical Systems: Advances in Design & Modelling*, Studies in Systems, Decision and Control 259,
https://doi.org/10.1007/978-3-030-32579-4_21

1 Introduction

Many enterprises strive to improve their business by optimizing their equipment, one of the best strategies is to decrease the cost of maintenance and prolong its lifetime. Such a reduction to the cost of maintenance is reflected in the product or service prices for consumers. We can distinguish equipment maintenance in the two following categories/types: "reactive maintenance"—is when the failure has already happened and is pending to be fixed, "preventive equipment maintenance"—the pre-failure event has been detected and requires immediate attention and "predictive equipment maintenance"—is when predictive algorithms determine the optimal time for maintenance in order from occurring to prevent pre-failure events. The predictive maintenance is a very promising approach for equipment lifetime optimization, in the current age of digitalization of manufacturing and production.

In essence, the main idea behind such practice is to predict and perform maintenance before a pre-failure event can be registered. The two main goals in predictive maintenance are to identify pre-failure events and to design a maintenance strategy in advance. Predictive maintenance is rapidly gaining traction in the big world of industry 4.0, where an IoT platform serves as a monitoring and decision-making system. Moreover, these systems including hardware and software over IoT platform and are considered as cyber-physical systems (here and after CPS). Typically, the pipeline of predictive maintenance contains the following components: component for acquiring data, data preprocessing component, component for input features definition, fitting predictive models and implementing models in production for everyday usage. In general, the predictive maintenance system uses a specific set of algorithms. This set is designed based on dynamic data of a system and its environment. If it is not possible to obtain results from dynamic data, we can use historical events of systems failure to construct statistical models such as "survival model". If there are data sets that contain both dynamic data and historical data of failures, we can use machine learning techniques to approximate the remaining lifetime. Lastly, we can consider "reliability analysis" based on degradation models [1].

As mentioned above, there are two critical challenges in preventive maintenance of cyber-physical systems. The first is a subject of interest in predictive analytics, it is how to detect a pre-failure event or in other words how to derive life expectancy of equipment from dynamic/historical data. The second critical challenge is how to cut the cost of ownership by defining the optimal maintenance strategy. Both critical points are examined in the new intelligent decision support systems also known as proactive decision support systems (PDSS). The concept of PDSS consists of two components. The first component is a system that implements the predictive analytics, i.e. detection of a pre-failure event (called a proactive event) over a certain time period or forecasting horizon. The component is based on the following sequence of operations: to detect–to predict–to decide–to act. The second component helps to automate maintenance decisions, which allows to exclude operational roles and leave only supervisory level positions in operational management structure [2].

Recent research and reports show that data-driven techniques help to improve manufacturing, but sometimes it is possible to enrich current approaches by accurate mathematical models [3, 4]. A model with a high level of adequacy is called digital twins (white box) can be combined with data-driven techniques (black box) to obtain a promising approach for predictive maintenance domain. The main contribution of this chapter is a new conceptual model for proactive and predictive maintenance of cyber-physical systems (P^2M model). In summary, a concept of proactive decision support system in enterprise machinery gives way for data-driven intelligent systems by utilizing two primary approaches. First is predictive analytics that can be used in the detection of the pre-failure event and is based on digital twins concept; and second, a decision strategy generation and evaluation that is based on proactive and/or automated maintenance.

2 Background

Detecting and predicting equipment failures is not that new for an enterprise. The issue of detecting and predicting equipment failures was always considered as an essential part of an asset performance management that is based on reliability (also known as risk condition management). The theoretical basis for the reliability theory, do offer a clear mathematical apparatus by using an assumption about reliability function and a priori parameters. Therefore, the challenge is to create new approaches that are applicable for cyber-physical systems.

In the framework of the concept of Industry 4.0, the technologies of data collecting, and data processing brought a new perspective to asset performance management [3–5]. Coupled with new technologies, open data sets were published that enable the creation of new prediction models for equipment state and health detecting. One of such models was published by NASA and is about open data failures in the engine turbines [6]. Typically, the task was reduced to the exploratory time series analysis and forecasting [7–10]. It was also considered as a regression or classification task where linear and non-linear models might be applied. However, the development of the neural network approaches [11] and the boosting approaches has made it possible to obtain more accurate results that surpass those obtained by other methods [12].

The main issue of published approaches is that they were tested on the 'clear' data samples, in other words, the data with a sufficient number of labeled observations and events indicating failure or pre-failure state. In practice, however, there are a set of essential limitations that are imposed on a given data. The first limitation of the data quality is due to the physical and data transferring aspects. The second limitation relates to the frequency of data intervals during the process itself. E.g. the observation might be made only during a certain time interval. Often the changes in the load on the equipment or the behavior of the system in an operating environment will lead to a distortion in the collected data at the time of the measurement. The third limitation is associated with an imbalance in data sampling. The majority class that represents a normal function of the system, has much more observations than that of the minority class which contains observations with failures. Under such conditions, the application of classical approaches, based on data analysis or the combined theoretical principles of the theory of reliability with statistical data processing, might be ineffective.

The books Goodfellow and Bengio [13], Deng and Yu [14], and the review chapters Schmidhuber [14] and LeCun [15] discuss the basic principles of deep neural networks. The main area of application for deep neural networks in the aspect of the synthesis is an image synthesis and sequence analysis (text or speech). The development of neural network processing technologies (seq 2seq) has recently undergone a significant change. For example, the latest architecture from Google Brain called the Universal Transformers has already established itself as an effective tool for sequence analysis [16]. This technology is based on pervasive attention approach that has expanded traditional encoder-decoder architectures [17, 18]. Also, the semantic sequence analysis may be considered as a solution for predictive failure modelling.

Traditionally, the decision-making models and methods are based on rational (multi-criteria) choice and/or optimal solution that is derived from relevant alternatives and is used as an algorithmic base for a decision support system [19]. Above approaches are associated with the standard decision-making scheme: identifying and describing the problem, setting a task with possible alternatives, finding and executing solutions. However, decision-making approaches are rarely used in enterprise information systems. This creates a divide between rational decision-making procedure implementation and existing business processes [20, 21]. Among scientific discussions in the field of decision support systems, there is an open issue on combining the formal models of business processes (in the BPMN notation), and the decision-making process in the DMN notation and its relation to data relevant sources. Such setup creates the need for automated decision-making (with-out human intervention) but bound by a strict and fully defined scheme.

Based on the analysis of the literature and technical reports, the following trends can be defined as current research directions in the field of intellectual decision-making support [2, 22]:

- The trend of increasing efficiency of ubiquitous data-collecting technologies, leading to an increase in the amount of heterogeneous data.
  - Growth in the mobile traffic data indicates an increase in the number of portable data collection and transmission devices.
  - The transition from the Internet of Things (IoT) to the Internet of Everything (IoE).
- The trend in the reduction of the development prices for artificial intelligence technologies (commodity), and consequential integration of AI solutions into business enterprises.
- The trend to replace human decision-making functions by Intelligent software (transfer of human functions to the supervisory level) based on the implementation of Smart Advisor technologies and data mining technologies.

- The tendency to integrate the physical world (offline) and the virtual world (digital technologies) hence the transition to cyber-physical systems.
- The transition from the current level into the level of preventive decisions (predictive analytics) and predictive models control.

As a result, a new generation of decision support systems, using proactive computing, should be considered as a promising technological step forward.

Proactive computing is the title of the chapter by David Tennenhouse, in which he described the idea of "transferring" a human from an operational control level (or human-centered) to a super-system (human as a supervisor), thereby having a positive effect on the efficiency and speed of management [2]. Three principles of proactive computations can be affirmed.

- The first principle is called "get physical", it is an integration of components between real (physical) and virtual worlds.
- The second principle is about the connection between the large number of computers embedded in various environments and is referred to as "get connected".
- The third "get real" principle, is the principle of overcoming the limits in the speed of decision-making and is inherently dependent on the abilities of a person.

In order to achieve an effective design for the new type of decision-making systems, it is necessary to answer the following fundamental questions:

1. How to describe a domain or build a semantic model of a domain? The model should include three things that are coherent with principles of system engineering: the description for the system-of-interest, usage system, and the enabling system.
2. How to collect, store and analyze unstructured data for proactive and predictive decision making?
3. How to build CPS identification models (digital twins) with minimal human involvement?
4. How to generate, how to evaluate and how to choose the optimal maintenance action or proactive and predictive decision making?

3 A Conceptual Model

The proposed conceptual model inherently combines two main notions of computation, predictive and proactive ideas. The first idea utilizes forecasting techniques in order to perform predictive maintenance support decisions. The second one focuses on proactive decision making, e.g. minimizing the intervention of human in regular (typical) decision-making processes. A model is a basic concept for a proactive and predictive decision system utilized for asset performance management. Here and after we will use a new definition 'cyber-physical asset performance management' as an extended asset performance management technology, in order to enable greater utilization and longer lifetime of cyber-physical systems. The following essential components are included in the conceptual model:

$$P^2M = < O, D, DF, FP, BDP, VF, TS > \qquad (1)$$

Here O is an ontology of managed cyber-physical system and its environment also, a matrix of requirements and constraints for the cyber-physical system (this matrix is used in the design phase). The D is a data warehouse such as data-lake, which stores both raw data and structured data with metadata. This also includes the tools for data gathering, the pre-processing and quality evaluation of data. The DF stands for data factory and is a set of programming implementation of algorithms for data processing, it includes the unified API for flexible usage. The FP stands for a flexible pipeline of data processing, according to proactive and predictive maintenance. The BDP is a set of business decision processes represented by BMPN diagrams and extended with decision-making procedures that are interlinked with components from the D component. The VF is a set of verification methods with asset performance management estimators over the observed life cycle. The TS is a representation of the related technology stack.

3.1 Ontology Design

The ontology is a formal representation of a domain as a set of entities and relationships in the domain, the domain its structure and semantic. Typically, the ontology might be represented according to the following model:

$$O = <C, \mathrm{A}, R, T, F, D> \qquad (2)$$

Here C is a set of classes (entities), A is a set of attributes of entities and relationships, R—relations between entities, T is a set of data types, F is a set of constraints for entities, attributes, and relationships, D is a set of relationships instances. Among the basic requirement for an ontology design process, we include the following additional principles.

1. The ontology should include instances of objects (e.g. CPS), stakeholders (which are interested in CPS operation), as well as processes over the whole life cycle.
2. It is reasonable to define the following logic levels of system representation: lower level, a CPS component level; the intermediate level, a cyber-physical system itself; the upper level: the system of systems.
3. The hierarchy of concepts and the relationships between the concepts should include:

    a. Concepts of goals and key performance indicators (indicators of the technical condition of CPS, indicators of technical and economic efficiency of CPS);
    b. Concepts of stakeholders;

   c. Object (CPS) and context concepts;
   d. concepts of predictive maintenance decision making processes over the life cycle of CPS;
   e. Concepts describing the components of a proactive and predictive maintenance system;
   f. Concepts of guidelines and requirements for further integration and usage.

4. The concepts in group "Objects" must be considered according to the system engineering approaches [23]. Models are built for a system including the system-of-interests, the enabling system, the usage system, and the systems in the operational environment, in each level of representation.
5. It is necessary to define the links between concepts with data sources, which determine requirements for data sources, predictive data analysis models and the technological stack of proactive and predictive maintenance [24].

The ontology is developed in accordance with the ontology engineering methodology [25, 26]. This said methodology includes the steps proposed by Sure et al.: (1) Feasibility Study, (2) Kickoff, (3) Refinement, (4) Evaluation, (5) Application and Evolution [27, 28]. Table 1 contains the main ontological concepts (classes) in the proposed conceptual model for predictive and proactive maintenance.

As the ontology can be composed in Protégé, the output XML file can be converted to JSON file. The JSON file can be used as a blueprint for data storing. Figure 1 shows an example of such blueprint.

Table 1 The main concepts of the ontology

Classes	Description
Objectives	A class designed to describe the performance indicators of cyber-physical systems. It includes subclasses: a class of technical condition indicators (health index); class of indicators of technical and economic efficiency
Stakeholders	The concept includes the classes of main stakeholders related to CPS operation. It is important to note the list of stakeholders in its entirety
Objects	A class describing objects: the system of interest (CPS), the supporting system, the usage system, the systems in the operational environment
Processes	A class describing the main processes in the framework of risk-oriented management. The class has subclasses: the process of monitoring and evaluating the health index; the process of forecasting health index of CPS; the risk assessment process; the ownership cost estimation process; scenarios development process for types of technical impact selection
Infrastructure	A class designed to represent components of a platform (system) for risk-oriented management. Subclasses: components; data sources; software implementation of models and data processing methods; methods of predictive analytics; pipelines of data processing; typical technical solutions (data processing pipelines)
Tutorials	A class that characterizes the tutorial of implementing proactive and predictive maintenance for cyber-physical systems

```
[{
  "uri": "/support/HYDRAULIC_SYSTEM",
  "id": "9888-XXX-LL01",
  "classification": "/classifications/support/Component",
  "EQUIPMENT_LIST": {
    "EQUIPMENT": {
      "ID": {
        "type": "string",
        "value": ["YYY_XXX_000"]
      },
      "NAME": {
        "type": "string",
        "value": ["VALVE"]
      }
    }
    ...
  }
  "INSTRUMENT_LIST": {
   "INSTRUMENT": {
     "TAG": {
       "type": "string",
        "value": ["WQ_XX_301"]
     },
     "DESCRIPTION": {
       "type": "string",
       "value": ["PRESSURE SENSOR"]
     }
   }
    ...
  }
}]
```


Fig. 1 An example of a JSON file describing CPS for further management

Since the proactive and predictive maintenance of CPS is based on data processing technologies, the 'infrastructure' component of ontology is the core of the P^2M conceptual model. Table 2 contains the main classes of the concept 'Infrastructure'.

3.2 *Data Collecting and Storing: A Data Lake Concept*

Data volume and variety are core characteristics of CPS. These parameters are considered to be the basic requirements for proactive and predictive maintenance and

Table 2 Main classes of the concept 'Infrastructure'

Classes	Description
Data source	The parent class for representation of data sources
Method	Parental class, that characterizes the data processing methods for predictive and proactive maintenance
Architecture	Parent class that characterizes the components of the proactive and predictive management system. It includes subclasses corresponding to the levels of architecture: the level of data collection, the level of real-time data-processing streams, the level of data and metadata storage, the level of data batch processing, the level of predictive and proactive analytics, the level of decision support and representation level
Interface	Parent class that defines interfaces and protocols for communication among data transfer subsystems (JSON, OWL, XML, OPC)

are in the framework of *data-lake* technology. This approach for data storage architectural design combines raw data warehouse and structured data storage. Also, it contains data processing methods and techniques as ETL components [29].

The essential methods and techniques can be viewed in the list below.

1. Methods for heterogeneous data collection and data quality assessment according to different types of observed systems, e.g. CPS, the enabling systems, the usage systems, and others.
2. The distributed heterogeneous data storage. This allows storing both raw heterogeneous and structured data according to a predetermined data scheme.
3. The method for heterogeneous data stream collection and preprocessing. The data stream handling allows to covert data stream into a proper data scheme for further efficient analysis.
4. The data quality criteria and associated methods for assessing the data. The following groups of criteria can be defined: semantic, syntactic, temporal and pragmatic. These criteria allow to evaluate how the data can be used to make decisions in predictive maintenance. A method for assessing data quality is applied that differs in using metadata interpretation and allows to formally describe the quality of data in the form of *data quality certificates*, which are later used to make decisions on further data usage.

3.3 Processes

Given the general idea of asset management performance, the development of the predictive maintenance for CPS should be based on the following sequence:

1. Determine the context of where an asset or CPS is operating at. The required asset functions and its standards of performance (context and functions) are clearly defined.

2. Determine what is a functional failure for each asset.
3. Determine the reasons that can cause functional failures (types of failures).
4. Determine what will happen as a result of the failure (consequences of failure).
5. Determine the critical impact of a failure.
6. Determine what needs to be done to predict or prevent the failure (scheduled, preventive or predictive actions and their frequencies).
7. Determine which methods for preventive or predictive actions are effective and need to be applied.

When the new cyber-physical system or technological equipment (as a functional unit) is deployed, the formalization process of CPS is performed [30]. The following steps are the part of the process:

- Create a list of primary characteristics of a CPS or its functional unit.
- Identify relevant sensors in the data acquisition system.
- Determine the conditions for controlled and emergency shutdown of CPS. The process of CPS monitoring and diagnostics contains the following operations:
- Get the data on main CPS characteristic values and related values of the systems in the operational environment.
- Obtain a data chunk containing the characteristics of the system in use, and values of characteristics of systems in the operational environment.
- Identify potential failures.
  - Determine the time for potential failure occurrence.
  - Determine what kind of functional failure is responsible.
  - Determine potential failure causes.
  - Notify the user about the results of monitoring.

The process of CPS assessing and predicting technical condition should include the steps listed below:

- Obtain historical data of CPS operation from inception till the present time.
- Calculate reliability indicators. Calculate the remainder of the equipment's lifetime. Estimate the time of actual failure occurrence. Determine what functional failures may occur. Determine the potential causes of failure.
- Calculate the technical condition index (the health index) based on indicators of reliability values.
- Notify the user about the results of the performed calculation of the technical condition index (TCI).

The process of assessing risks and consequences because of functional failure is:

- Set the time bound.
- Determine the consequences of a functional failure.
- Determine the severity of said consequences.
- Perform risk computations while considering planned and predictive maintenance.

The decision-making process for the implementation of technical repercussions to ensure the reliability of CPS:

- Create preventive actions to eliminate the causes of a potential nonconformity or other undesirable events that may lead to failure.
- Create predictive action scenarios for monitoring asset status and predicting the need for preventive action or corrective action.
- Create scenarios of predictive actions for monitoring an asset status and forecasting the need for preventive or corrective activity.
- Determine the possible technical effects and their frequency (preventive action or corrective action).
- Perform risk calculation by taking into account various scenarios of application of preventive action or corrective action.

The final process performs monitoring of maintenance implementation and analysis of the results of performed maintenance.

3.4 *Infrastructure*

Infrastructure is a set of joined components for application of proactive and predictive maintenance of CPS that are based on data processing. As proactive and predictive maintenance of cyber-physical systems is reliant on different methods, the Infrastructure can include various software implementing algorithms.

1. Evaluating the reliability algorithms based on empirical laws of failure distribution and manufacturer's information about the system's reliability. The empirical laws of "time-to-failure" distribution are documented in machinery passport and allow for failure-risk assessment.
2. The algorithms of CPS ranked according to the probability of failure.
3. The algorithms for prediction of the CPS (or its components) dynamic characteristics under specified operating conditions are based on diagnostic results and continuous monitoring. Prediction methods may use sequence-to-sequence approaches, recurrent neural networks, LSTM or time series analysis methods [31].
4. The anomaly detection algorithms based on [32]:
   a. classification using supervised learning methods;
   b. clustering using unsupervised learning methods for similarity check of reference data profile and real data profile.

Building the model to detect proactive nonconformities is based on the predictive model control approach. It also combines the use of predictive dynamic characteristics of the target system and identification of relative anomalies. In case the CPS has an acceptable mathematical model (or digital twin) it is possible to apply an algorithm to detect the difference between real and modeled data.

3.5 Data Processing Pipelines

Before implementing the system based on the P^2M concept, models and methods must be implemented within the framework of predictive analysis techniques. We define the following groups of approach:

1. methods to evaluate reliability indicators;
2. potential failures identification;
3. outstanding effective lifetime estimate;
4. generating rundowns containing maintenance actions.

For every method, a pipeline needs to be created as a sequence of data processing operations. To ensure proper data transferring, interfaces are developed according to the low coupling principle. Methods for reliability index evaluation should meet the following requirements:

- The method of computing the reliability index should be based on, analyzing the deviation trend between actual profile values and mathematical model (or digital twin) profile that is obtained over the given time period.
- The calculation of the reliability index should be carried out autonomously. (without human intervention).
- The method should be implemented as a software package with API, for further deployment into the predictive maintenance decision making process.

The method should be based on, analyzing the deviation trend between actual profile values and mathematical model (or digital twin) profile that is obtained over the given time period.

The method should define:

- the potential failure occurrence timestamp;
- estimated probability of potential failure;
- the type/class of potential failure;
- a list and interpretation of nonconformities (symptoms) which characterize a potential functional failure;
- P–F interval parameters for potential failure;
- a list of reasons for potential failure ranked by probability.

Besides, the methods stated above have possible modifications that can be used.

1. A method for the identification of potential failures based on rules (performance standard).
2. A method for identification of potential failures based on analyzing the changes in the basic characteristics and predefined rules.
3. A method for identification of potential failures based on trend analysis of deviation between actual values and values calculated using the mathematical model (digital twin).
4. A method for identification of potential failures based on failures classification algorithms.

5. A method for identification of potential failures based on anomaly detection approach.

A method for generating maintenance actions should include the following operations:

- Create a list of control parameters.
- Create a list of planned work and their frequencies.
- Create a list of preventive actions to eliminate the cause of a potential nonconformity or other undesirable potential events;
- Create a list of predictive actions for monitoring the state of the CPS and predicting the need for preventive or corrective action;
- Identify possible technical impacts and their frequency (planned preventive, preventive action or corrective action);
- Calculate the cost of possible technical actions and their impact on the system;
- Calculate the risk by considering various scenarios based on technical outcomes.

4 Implementation

The presented concept serves as a theoretical footing for a proactive performance management CPS design. According to the concept, the proactive system should consist of the following subsystems or levels:

1. A subsystem for collecting data from a cyber-physical system and its environment. The subsystem connects directly to CPS to collect data via protocols, e.g. OPC Open Platform Communications protocol or APIs. The subsystem may have different data collectors implemented that are responsible for collecting data from specific data sources.
2. A subsystem for real-time data stream processing. The subsystem that is designed for real-time data streams processing that is being received from data collection subsystems. Also, this subsystem is responsible for generating metadata, modifying the data according to setup and writing to data storage (data lake).
3. Data and metadata storage subsystem. The subsystem is designed for distributed storage of raw data, structured data, and metadata. It also provides access to data backup and restore.
4. Subsystem data factory. The subsystem consists of packages, implementing models and methods of statistical and intelligent data processing (package library), as well as machine learning packages.
5. Data processing pipelines subsystem. The subsystem is designed according to requirements of proactive and predictive analytics subsystems. Data processing pipelines subsystem includes components assembled in a pipeline with a clear interface for data transfer.

6. Proactive and predictive maintenance decision-making subsystem. The subsystem is designed to identify potential failures in CPS operation based on predictive analysis techniques and to generate action in order to mitigate the risk. The subsystem uses data factory and data processing pipelines for calculating and forecasting technical conditions.
7. GUI subsystem. The subsystem is designed to visualize the results using the graphic user interface.
8. Administration subsystem. The subsystem is intended for administration of described above subsystems.

The architecture should be built in independent modules format to allow for seamless integration and deployment. Moreover, the architecture should be built according to the maturity of enterprises to deploy proactive and preventive maintenance systems [33, 34].

5 Conclusions

The presented P^2M concept model helps to design the proactive and predictive maintenance system in order to reduce the cost of ownership for the cyber-physical system in an enterprise. A proactive approach can be considered as a preliminary step for self-adjusting or self-maintenance of a cyber-physical system where human intervention still exists but is on the supervisory level of management.

The P^2M model is considered as an invariant to technological implementation with common interfaces for data transferring between components. This concept is used to creating a prototype of the predictive maintenance decision-making system in gas turbine performance management for better reliability.

Acknowledgements This research was supported by the Russian Fund of Basic Research (grant No. 19-47-340010). Special thanks go to George Sergeev for fruitful discussion and essential remarks.

References

1. Gorjian, N., Ma, L., Mittinty, M., Yarlagadda, P., Sun, Y.: A review on degradation models in reliability analysis. In: Kiritsis, D., Emmanouilidis, C., Koronios, A., Mathew, J. (eds.) Engineering Asset Lifecycle Management, pp. 369–384. Springer, London (2010)
2. Tennenhouse, D.: Proactive computing. Commun. ACM **43**(5), 43–50 (2000). https://doi.org/10.1145/332833.332837
3. Industry 4.0 challenges and solutions for digital transformation and use of exponential technologies. Deloitte. https://www2.deloitte.com/content/dam/Deloitte/ch/Documents/manufacturing/ch-en-manufacturing-industry-4-0-24102014.pdf
4. Industry 4.0—opportunities and challenges of Industrial Internet. PricewaterhouseCoopers. https://www.pwc.nl/en/assets/documents/pwc-industrie-4-0.pdf

5. Tupa, J., Jan, S., Steiner, F.: Aspects of risk management implementation for industry 4.0, Procedia Manufacturing, vol. 11, pp. 1223–1230 (2017). ISSN 2351-9789. https://doi.org/10.1016/j.promfg.2017.07.248
6. NASA Ames Prognostic Center. https://ti.arc.nasa.gov/tech/dash/groups/pcoe/prognostic-data-repository/
7. Canizo, M., Onieva, E., Conde, A., Charramendieta, S., Trujillo, S.: Real-Time Predictive Maintenance for Wind Turbines Using Big Data Frameworks, pp. 1–8 (n.d.)
8. Ciomek, K., Ferretti, V., Ferretti, V.: Predictive analytics and disused railways requalification: insights from a post factum analysis perspective. Decis. Support Syst. (2017). https://doi.org/10.1016/j.dss.2017.10.010
9. Cleary, D., Tax, R.I.: Predictive analytics in the public sector: using data mining to assist better target selection for audit. Electron. J. E-Gov.Ment **9**(2), 132–140 (2011)
10. Yu, L.: Wind turbine data analytics for drive-train failure early detection and diagnostics. Volume 1: Aircraft Engine; Ceramics; Coal, Biomass and Alternative Fuels; Wind Turbine Technology, pp. 721–728 (2011). https://doi.org/10.1115/GT2011-45101
11. Schmidhuber, J.: Deep Learning in Neural Networks: An Overview. https://arxiv.org/abs/1404.7828
12. Chen, T., Guestrin, C.: XGBoost: A Scalable Tree Boosting System. https://arxiv.org/abs/1603.02754
13. Goodfellow, I., Bengio, Y., Courville, A.: Deep Learning. MIT Press (2016). http://www.deeplearningbook.org
14. Deng, L., Yu D.: Deep Learning: Methods and Applications. Microsoft Research. One Microsoft Way. Redmond, WA 98052
15. LeCun, Y., Bengio, Y., Hinton, G.: Deep learning. Nature **521**, 436–444 (2015). https://doi.org/10.1038/nature14539
16. Dehghani, M., et al.: Universal Transformers. https://arxiv.org/pdf/1807.03819.pdf
17. Bai, S., Kolter, J.Z., Koltun, V.: An Empirical Evaluation of Generic Convolutional and Recurrent Networks for Sequence Modeling. https://arxiv.org/pdf/1803.01271
18. Elbayad, M., Besacier, L., Verbeek, J.: Pervasive Attention: 2D Convolutional Neural Networks for Sequence-to-Sequence Prediction. https://arxiv.org/pdf/1808.03867
19. Pons, M.B.: The expected value of perfect information in unrepeatable decision-making. Decis. Support Syst. (2018). https://doi.org/10.1016/j.dss.2018.03.003
20. Arnott, D., Pervan, G.: Eight key issues for the decision support systems discipline. Decis. Support Syst. **44**(3), 657–672 (2008). https://doi.org/10.1016/j.dss.2007.09.003
21. Arnott, D., Pervan, G.: A critical analysis of decision support systems research revisited: the rise of design science. J. Inf. Technol. **29**(4), 269–293 (2014). https://doi.org/10.1057/jit.2014.16
22. VanSyckel, S., Becker, C.: A survey of pervasive computing. UBICOMP Comput. Sci. **3**(6), 1–7 (2014). https://doi.org/10.1145/2638728.2641672
23. ISO/IEC/IEEE 15288: Systems and Software Engineering—System Life Cycle Processes (2015). https://www.iso.org/ru/standard/63711.html
24. Nural, M.V., Cotterell, M.E., Peng, H., Xie, R., Ma, P., Miller, J.A.: Automated predictive big data analytics using ontology based semantics. Int. J. Big Data **2**(2), 43–56 (2015). https://www.ncbi.nlm.nih.gov/pmc/chapters/PMC5898823/
25. Jarvenpaa, E., Siltala, N., Hylli, O., Minna, L.: The development of an ontology for describing the capabilities of manufacturing resources. J. Intell. Manuf. **30**, 959 (2019). https://doi.org/10.1007/s10845-018-1427-6
26. Anufriev, D., Petrova, I., Kravets, A., Vasiliev, S.: Big data-driven control technology for the heterarchic system (building cluster case-study). Stud. Syst. Decis. Control. **181**, 205–222 (2019)
27. Sure, Y., Staab, S., Studer, R.: Ontology engineering methodology. In: Staab, S., Studer, R. (eds.) Handbook on Ontologies, 2nd edn., pp. 135–152. Springer, New York (2009). ISBN 978-3-540-70999-2

28. Kizim A., Matokhina, A., Nesterov, B.: Development of ontological knowledge representation model of industrial equipment. Creativity in Intelligent Technologies and Data Science, CIT&DS 2015, Volgograd, Russia, 15–17 Sep 2015. Proceedings, vol. 535, pp. 354. Springer (2015)
29. Tran, V.P., Shcherbakov, M., Nguyen, T.A.: Yet another method for heterogeneous data fusion and preprocessing in proactive decision support systems: distributed architecture approach. In: Vishnevskiy V., Samouylov K., Kozyrev D. (eds.) Distributed Computer and Communication Networks. DCCN 2017. Communications in Computer and Information Science, vol. 700. Springer, Cham (2017)
30. Kizim, A.V., et al.: Predictive modeling as a basis for monitoring, diagnosis, forecasting and upgrading of a technical system. In: 2017 IEEE 11th International Conference on Application of Information and Communication Technologies (AICT). IEEE, pp. 1–5 (2017)
31. Golubev, A., Shcherbakov, M., Shcherbakova, N., Kamaev, V.: Automatic multi-steps forecasting method for multi seasonal time series based on symbolic aggregate approximation and grid search approaches. J. Fundam. Appl. Sci. **8**(3S), 2529–2541 (2016)
32. Shcherbakov, M.V., Brebels A., Shcherbakova, N.L., Kamaev, V.A., Gerget, O.M., Devyatuch, D.V.: Outlier detection and classification in sensor data streams for proactive decision support systems. In: International Conference on Information Technologies in Business and Industry 2016, Tomsk, Rus. Federation, 21–23 Sep 2016. Chapter № 012143, J. Phys Conf. Ser. **803**(1), 8 (2017)
33. Shcherbakov, M.V., Groumpos, P.P., Kravets A.G.: A method and IR4I index indicating the readiness of business processes for data science solutions. Creativity in Intelligent Technologies and Data Science. Second Conference, CIT&DS 2017 (Volgograd, Russia, September 12–14, 2017). In: Kravets, A., Shcherbakov, M., Kultsova, M., Peter Groumpos; Volgograd State Technical University et al. (eds.) Proceedings. Springer International Publishing AG, Germany, 2017, vol. 754, pp. 21–34. (Ser. Communications in Computer and Information Science)
34. Kravets, A., Kozunova, S.: The risk management model of design department's PDM information system. Commun. Comput. Inf. Sci. **754**, 490–500 (2017)

Conceptual Approach to Building a Digital Twin of the Production System

S. I. Suyatinov

Abstract The digital twin is an important component of the cyber-physical system. This new structure of the production system was the result of the development of information technology. The article shows that, despite the long history and success in the development of production information systems, the concept of building digital twins of production systems is at an early stage. One of the problems in creating digital twins is the need for integration and joint processing of a large amount of heterogeneous information. It is shown that the problem of reflecting the current state of the production system is in many ways similar to the problem of the internal representation of the surrounding world in living systems. It is proposed to choose the theory of the levels of the physiologist N. A. Bernstein as the basis of the conceptual approach to the development of digital twins. The mechanisms of forming models of the external world at every level are outlined. A description of the hierarchical system for processing different types of information and obtaining an invariant representation of the external world are presented. The principles of constructing a virtual image in the organization of motor activity are formulated. The implementation of these principles when building a digital twin of the production process will improve the efficiency of integration methods and joint processing of information.

Keywords Cyber-physical system · Production element · Central nervous system · Virtual image · Big data · Digital twin

1 Introduction

The continuous process of computerization of production, begun in the middle of the last century, has now acquired new opportunities in the field of scientific research. This is the so-called NBIC technology. The technological paradigm of the NBIC is based on the fundamental laws inherent in living systems, and is a synthesis of nano-, bio-, informational, and cognitive technologies [1, 2]. Microminiaturization

S. I. Suyatinov (✉)
Bauman Moscow State Technical University, 2-ya Baumanskaya, 5, Moscow 105005, Russia
e-mail: ssi@bmstu.ru

A. G. Kravets et al. (eds.), *Cyber-Physical Systems: Advances in Design & Modelling*, Studies in Systems, Decision and Control 259,
https://doi.org/10.1007/978-3-030-32579-4_22

of sensors and calculators, as well as information and communication technologies allow recording, transmitting and processing large amounts of information about the smallest details of the production process, accumulating and storing information about products throughout the entire life cycle. Increased computing power, as well as intelligent methods for processing large volumes of information, allowed for real-time control of all hierarchical levels of the production system.

These technological advances formed the basis of the fourth industrial revolution (Industry 4.0). In accordance with the concept of Industry 4.0, enterprises are built on the principle of cyber-physical systems, and the cornerstone of the new industrial concept is the inclusion in the production system of its virtual image in the form of mathematical models [3–6]. The principal feature of the virtual image of the industrial system is its dynamics, i.e. change its state in the rhythm of the production process. That is why the system of numerical simulation of the production process using input measurement information has been called the "digital twin". The digital twin allows not only to monitor the current state of production, but also to predict its future state.

2 Digital Twin in the Structure of Cyber-Physical System

Currently, there is a methodology for constructing a digital twin of a product. The concept of designing digital twins of production systems is at an initial stage. Therefore, the development of approaches to the synthesis of the structure and algorithmic filling of the digital twin is an important problem.

The idea of building a digital twin has a long history. The starting point can be considered the emergence of automated process control systems (APCS) [7]. The development of these systems proceeded along the path of improving the services presented, expanding the range of their application, selecting and standardizing the best solutions. So there were the methodologies realized in the form of the standardized business processes providing information support of administrative decisions at different hierarchical levels of the organization of production [8–11].

In general, production information systems can be divided into three categories. The first is computer-aided design (CAD) systems that allow you to create digital models of designed objects. A feature of these systems is the ability to explore the properties of the designed product, using various options (visual, mathematical, schematic) of its virtual image. This also includes (often used in conjunction with CAD-systems) engineering calculation automation systems (CAE systems) and automation systems for preparing programs for machine tools with numerical control (CAM systems).

Information systems of the second category are designed to control the equipment. These include CNC direct digital control systems, which allow rapid re-adjustment of machine equipment to increase production flexibility, and PHM systems designed to predict the status and control equipment operability. PHM-systems use information from sensors installed directly on the equipment, as well as models of the equipment under test.

The third category includes systems that allow to automate the management of information flows of organizational and technical services and strategic management. The main information systems for managing the production activities of an enterprise are MES (corporate production management systems), ERP (automation of enterprise resource planning) and CRM (customer relationship management system). The strategic management services the OLAP system—a set of technologies for the rapid processing of information, including the dynamic construction of reports in various sections, data analysis, monitoring and forecasting of key business indicators used to analyze and make management decisions.

Figure 1 shows the hierarchical structure of information support for production management. Information support is carried out at all stages of the production process, starting from the design stage. The special integration role of the PLM-system should be noted. It provides the interaction of various automated systems (CAD/CAE, CNC, PHM, ERP, CRM and others) in a single information space.

A distinctive feature of the considered information structure is its ability to generate a large flow of heterogeneous information, which, however, is used fragmentarily within a certain stage of the life cycle and within its hierarchical level. Figuratively speaking, the waste information remains in dead storage.

Production is most efficiently managed using digital twin technology. In essence, digital twin is a structured information that changes in pace and in accordance with

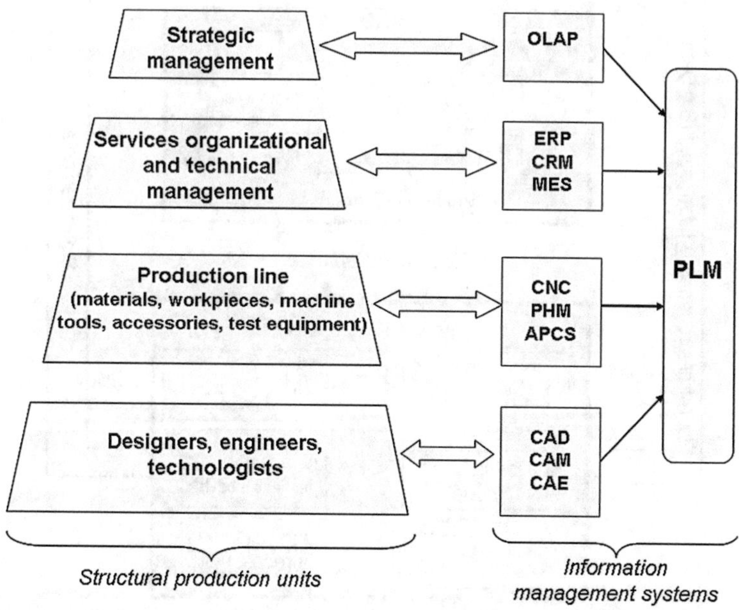

Fig. 1 Information support of production management

changes in the production system itself. The basis of the digital twin technology is a half-a-bit simulation using information (signals) about the current state of the system.

Figure 2 shows the structure of the "production—digital twin" system. The management personnel of top-level use information management systems in their activities. At the lower production level, with the help of sensors D_1, …, D_n, measuring information is collected about the operation of equipment—production elements PE_1, …, PE_n. The digital twin is represented by a set of models. The database (*DB*) of the digital twin stores information in the form of planned targets, technical specifications, technological maps, standards and other similar documentation.

In the analysis and forecasting block, the entire array of available data is processed with specific goals. We will single out the main ones: diagnostics of the state of the production process, identification of "narrow" places; assessment of the state of the

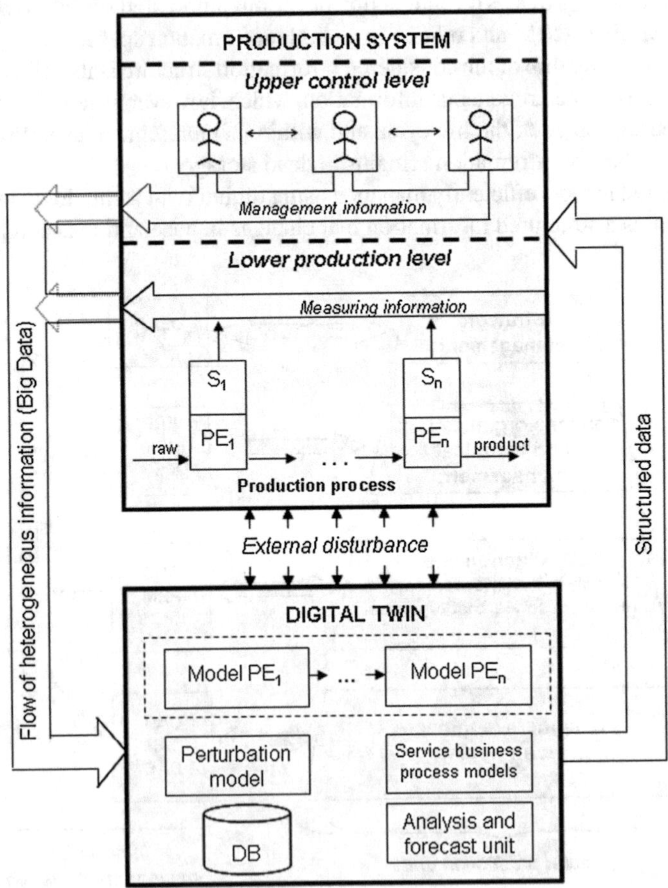

Fig. 2 Production system and its digital twin

equipment in order to organize repair according to the actual state; forecast of the state of the production process under the influence of disturbing factors and re-profiling.

It should be noted that, despite the transparency of the ideas of building a digital twin, their practical implementation is a complex scientific and technical problem. In the first place, this is due to the need to structure large volumes of heterogeneous information, identify the patterns hidden in them and make management decisions. In this regard, it is of interest to have a successful experience in solving similar problems in other areas.

3 Virtual Image in Biological Systems

It is rightly believed that biology is the source of fruitful systemic ideas. In this case, we take into account the fact that all life exists due to the mechanism of internal reflection of the real world. Based on this, we consider the mechanisms of creating the virtual images in biological systems. The existence of an internal model (image) of the outside world is a necessary condition for the survival of a living organism.

In a living organism, the nervous system is a tool for the formation, preservation and use of a model of the external world. In the process of evolution of living matter improved mechanisms for the formation and use of internal reflection of the external world. In the most perfect form, these mechanisms are implemented in the central nervous system (CNS) of a person. In a simplified form, the process of forming and using the model of the outside world can be represented as follows. The process is implemented using the main functional ability of the brain to memorize and predict [12–14]. Using all the available information, the central nervous system in the learning process builds and remembers many virtual models corresponding to different situations, different states of the organism itself and the outside world. Only the best models are remembered.

In the future, faced with the current effects of the environment, the CNS selects the model that best suits the specific case. This model determines human behavior. Thus, for a given purpose of behavior in the CNS formed a program of action and create a model of "how it should be". The model, in particular, predicts the reaction of the senses in the implementation of the program. Then the predicted values are compared with signals from receptors and corrective actions are formed. Thus, we see an analogy with the functions of the digital twin, for example, in the problems of monitoring the state of the equipment. Having studied the available details of the described generalized mechanism, it is possible to use the best solutions in the construction of digital twins of technical systems.

For a number of reasons, these mechanisms are revealed in the most accessible form when studying the process of controlling the movement of a person by the central nervous system. The mechanisms of formation and use of virtual images in the CNS of a person in the organization of movement were studied by Russian neurophysiologist N. A. Bernstein [15, 16]. According to Bernstein, for the implementation of any form of movement (from the simplest form to complex forms), an action program is created in the human CNS, which is a virtual image of a specific behavior (movement).

A virtual image or behavior model is formed in the CNS in the learning process, which is carried out in several stages. Initially, the external and internal motion pattern is formed from the signals of various sensors. At the same time, a person learns to re-encode, according to Bernstein, afferent signals into effector commands. The accumulation of a "re-encryption dictionary" is one of the most important events of this period. In essence, the "re-encryption dictionary" is the set of models from which a virtual image (action program) of the required movement is subsequently formed. A large number of repetitions of the elements of the movement allows you to find models ("re-encryption") that provide the desired response in response to any deviations in any type of movement.

In the case of the digital twin, the "action program" is a production cycle developed on the basis of previously accumulated engineering knowledge. "Re-encryption dictionary" is a model of the stages of the production process, and "corrective actions" is a model of regulation in case of deviation of organizational and technological indicators.

In accordance with Bernstein's theory of sensory corrections, the brain not only sends a specific command to the muscles to perform any movement, but also receives signals from the peripheral sense organs about the results achieved and, on their basis, gives new corrective commands. Thus, there is a process of building movements in which there is not only direct but also continuous feedback between the brain and the executive organs. The entire system forms a closed loop of interactions, known as the Bernstein Ring. The general scheme of the ring is shown in Fig. 3.

The Ring includes motor "exits" (effector), sensory "inputs" (receptor), object of subject action, block of re-encryption, generator of a target program, setting driver and device for comparison. The successive stages of complex movement are recorded in program. At any given moment, some of its particular stage, or element, is being worked out, and the corresponding private program descends into the setting device. From the setting device, the signal Sw is fed to the comparison device. A feedback signal Iw arrives at the same block from the receptor informing about the state of the object. In the comparison device, these signals are compared, and at the output of it, signal Δw is obtained, i.e. mismatch signals between the required Sw and the actual Iw. They get to the re-encryption unit, where correction signals come from, and through intermediate instances (regulator) they get to the effector.

The result of any complex movement depends not only on the actual control signals, but also on a number of additional factors. Disturbing effects make deviations in the planned course of movement, and do not give into preliminary accounting. Therefore, the mechanism of movement is a continuous process of comparing the position and state of the organism with its virtual image.

At the same time, motion control has a multi-level nature, organized on the principle of multi-loop feedback. The predicted values of the receptor signals for the implementation of specified behavior models and the actual outputs of the receptor

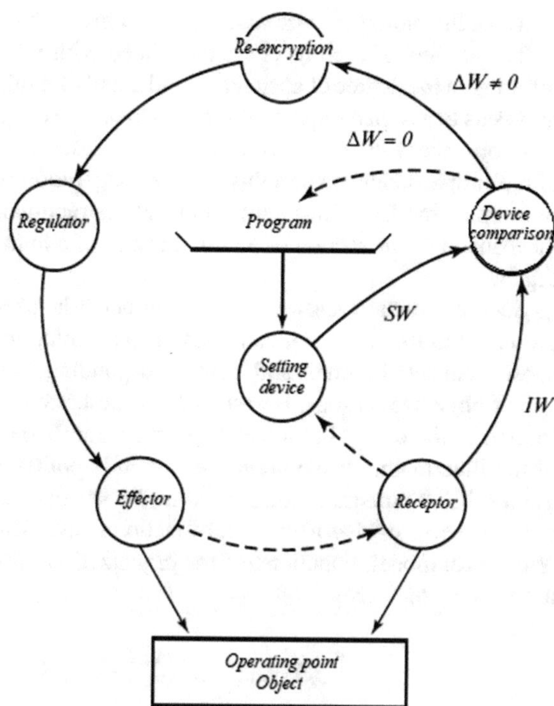

Fig. 3 Bernstein Ring

field at all levels are compared. Thus, multi-model, multi-scale control is implemented. Here it should be noted the special importance Bernstein's theory of movements construction levels for understanding the structure of the virtual image and the mechanisms of processing a large number of heterogeneous information [17–19].

The essence of the theory is that, depending on the source of feedback signals and the content of information (whether they report the degree of muscle tension or relative position of body parts), afferent signals come to different sensory centers of the brain and accordingly switch to motor paths at different levels.

Each level has specific, peculiar only to it, motor manifestations. Each level has its own class of movements. The scientist singled out five levels, denoting them with letters: *A*, *B*, *C*, *D*, and *E*. Levels differ in the degree of detail of the motion representation. Intra-level and inter-level interaction is based on the Bernstein Rings principle.

Level *A* receives signals from muscle proprioceptors, which report the degree of muscle tension, as well as from organs of equilibrium. At level *B*, signals that report the mutual position and movement of body parts are mainly processed. This is the level of analysis of the state "in the space of the body." Level *C* receives all information about the external space; it builds movements adapted to space-time properties (to form, position, length, weight, time, etc.). Level *D* is a cortical, higher level of abstraction. It does not set specific movements, but sets a specific result of movement. Level *E* sets the meaning or purpose of the movement.

Thus, the biological virtual image has a hierarchical structure.

The models of each level are associated with the corresponding sensory fields, differing in the degree of abstraction and detail of motion fragments. Such a structure allows us to assign a sequential set of models and corresponding corrections to them in accordance with a complex movement, represented as a sequential implementation of the simplest elements. In this case, the separation of models by levels allows you to combine models of different detail and abstraction in one image. The hierarchical mechanism of interaction of a virtual image of a human motor reaction is shown in Fig. 4.

Such a structure allows us to put in accordance with the complex movement, presented in the form of a consistent implementation of the simplest elements, the subsequent set of models and their corresponding corrections.

For objective reasons, Bernstein N. A. described the functions of levels D and E located in the cortex in the most general form. It should be noted that the analysis of the functioning of the brain are certain hypotheses. Their accuracy is based on the results of numerous studies. From the standpoint of modern data, the cerebral cortex (neocortex) has a structural-functional organization, similar to the considered multi-level model. Functionally, the organization of the neocortex is in the form of a "sectoral" hierarchy.

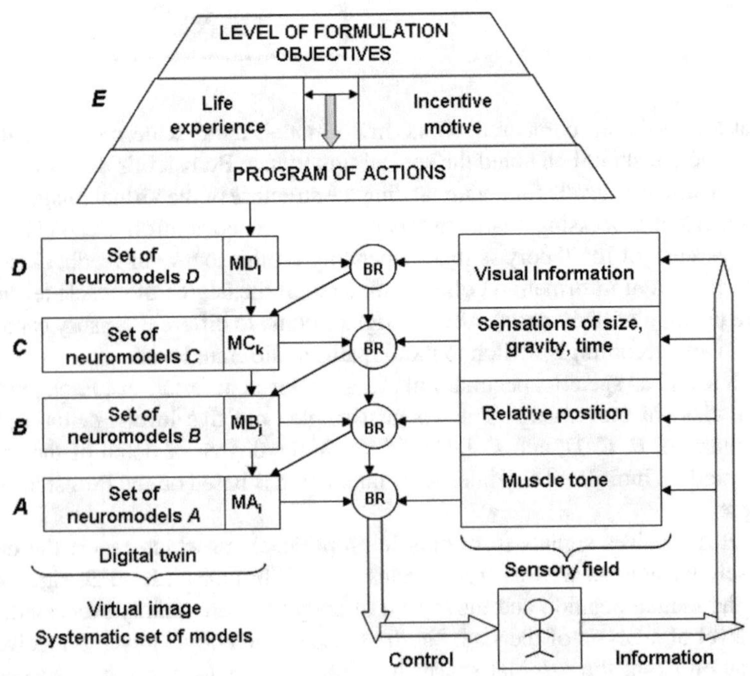

Fig. 4 The interaction of man and his virtual image in the process of organizing movement

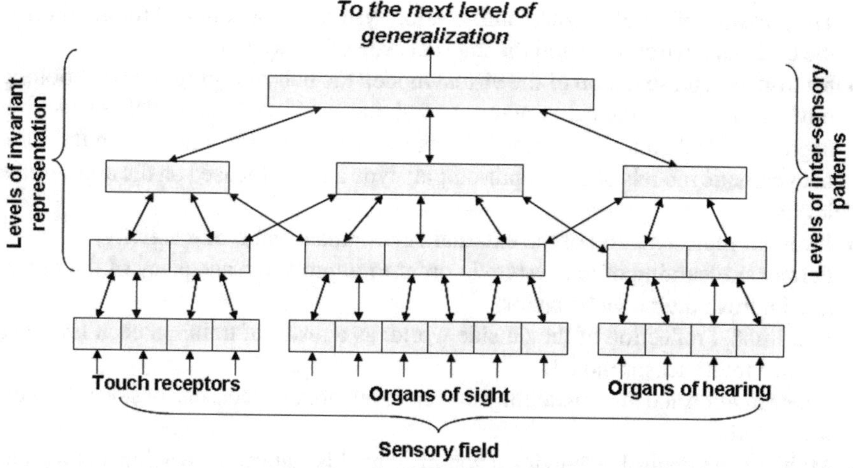

Fig. 5 Multilevel architecture of presentation and processing of information in the neocortex

There are zones of visual, auditory, motor and other perceptions and information processing. Each of these zones has a hierarchical structure. The neurons of the lower level fix individual, particular elements of the image. At the next level, a more generalized image is constructed from particular elements. In this case, a common characteristic of generalized images is their relative invariance to changes in the particular elements of the underlying images. This is how the parts roll up and the formation of a generalized image with the basic properties for this level. Figure 5 shows the simplified architecture of representation and inter-sensory information processing in the neocortex.

The figure shows the next important detail. Numerous studies have established that the levels of "sectoral" zones are interconnected by direct and feedback links. This allows you to form polysensory cybernetic models at different levels, integrating heterogeneous information. In addition, interzone communications allow producing intersensory forecasts.

Currently, the structure and principles of motion control in biological systems are reflected in technical systems with intellectual properties [20–23].

4 Principles of Building a Digital Twin

The analysis of the mechanisms for forming the internal reflection of the external world allows us to formulate the following basic principles of the conceptual approach to building a digital twin of the production system:

- Hierarchical structure of digital twin production system;

- The presence of a multichannel information system (sensory fields) for monitoring the external environment and the internal state of the system;
- Multiscale representation of the virtual model: the behavior program (functioning model) has a hierarchical structure containing consistent submodels of varying degrees of detail and abstract representation, which are implemented in the form of cybernetic models of the "input-output" type and are focused on the appropriate levels;
- Invariant representation of the external environment and motor activity;
- Permanent learning of the system in order to improve the adequacy of the model and improve their own behavior;
- Multimodel reflection of the outside world: as a result of training, each level has its own multi-sensor model;
- Subordination and level matching: the top-level model selects one of several lower-level models;
- Multi-circuit control: achieving a complete goal is equivalent to achieving a set of sub-goals at each hierarchical level;
- Availability of mechanisms for predicting changes in the external world, own behavior and also advanced control.

The stated principles can serve as a conceptual basis for the construction of applied systems using the elements of intellectual control inherent in living organisms.

5 Conclusion

In the process of evolution, nature created and perfected the mechanisms of reflection of the external world in the form of an internal image. The formation and preservation of the internal image is carried out using neural network structures. The hierarchical organization and inter-level relationships of these structures provide the formation of multisensor models and an invariant representation of the image at various levels. The analysis of the mechanisms of formation of the internal image in the organization of the movement made it possible to formulate nine principles of organization of the virtual image. The stated principles can serve as a basis for creating an effective structure of the digital twin.

References

1. Kamensky, E.G.: Context of NBIC-technologies development: institutions, ideology and social myths. Mediterr. J. Soc. Sci. **6**(6), 181–185 (2015)
2. Roco, M.C., Bainbridge, W.S. (eds.): Converging Technologies for Improving Human Performance: Nanotechnology, Biotechnology, Information Technology and Cognitive Science. Kluwer Academic Publishers, The Netherlands (2003)
3. Lee, J., Bagheri, B., Kao, H.A.: A cyber-physical systems architecture for industry 4.0-based manufacturing systems. Manufact. Lett. **3**, 18–23 (2015)

4. Tarassov, V.B.: Enterprise total agentification as a way to industry 4.0: forming artificial societies via goal-resource networks. In: Abraham, A., Kovalev, S., Tarassov, V., Snasel, V., Sukhanov, A. (eds) Proceedings of the Third International Scientific Conference "Intelligent Information Technologies for Industry" (IITI'18). IITI'18 2018. Advances in Intelligent Systems and Computing, vol. 874. Springer, Cham (2019). http://dx.doi.org/10.1007/978-3-030-01818-4_3
5. Koval', V.A., Osenin, V.N., Suyatinov, S.I., Torgashova O.Y.: J. Comput. Syst. Sci. Int. **50**(4), 638–653 (2011). http://dx.doi.org/10.1134/S1064230711040125
6. Protalinsky, O.M., Shcherbatov, I.A., Stepanov P.V.: Identification of the actual state and entity availability forecasting in power engineering using neural-network technologies. J. Phys. Conf. Series. **891**(1) Article 012289 (2017). https://doi.org/10.1088/1742-6596/891/1/012289
7. Antipov, K.V., Maslakov, M.P., Yurenko, K.I.: Improvement of the automated control systems for the development of the metallurgy. Procedia Eng. **129**, 1010–1014 (2015). https://doi.org/10.1016/j.proeng.2015.12.164
8. Bozhko, A.: Math modeling of sequential coherent and linear assembly plans in CAD systems. Global Smart Indus. Conf. (GloSIC) **2018**, 1–5 (2018). https://doi.org/10.1109/GloSIC.2018.8570090
9. Skvortsov, V., Proletarsky, A., Arzybaev, A.: Feature recognition module of the CAPP system. In: Proceedings of the 2019 IEEE Conference of Russian Young Researchers in Electrical and Electronic Engineering, ElConRus (2019). http://dx.doi.org/10.1109/EIConRus.2019.8656655
10. Xu, P., Wang, Z., Li, V.: Prognostics and health management (PHM) system requirements and validation. In: 2010 Prognostics and System Health Management Conference. Macao, pp. 1–4 (2010). https://doi.org/10.1109/phm.2010.5413560
11. Loh, B.K., Koo, K.L., Ho, K.F., Idrus, R.: A review of customer relationship management system benefits and implementation in small and medium enterprises. In: Proceedings of the 12th WSEAS International Conference on Mathematics and Computers in Biology, Business and Acoustics, pp. 247–253 (2011)
12. Tai, L., Liu, M.: Deep-learning in mobile robotics—from perception to control systems: a survey on why and why not. CoRR, abs/1612.07139 (2016)
13. Poldrack, R.A., Farah, M.J.: Progress and challenges in probing the human brain. Nature **526**, 371–379 (2015)
14. LeCun, Y., Bengio, Y., Hinton, G.: Deep learning. Nature **521**(7553), 436 (2015)
15. Bernstein, N.A.: The current problems of modern neurophysiology. In: Sporns, O., Edelman, G.M. (eds.) Bernstein's Dynamic View of the Brain: The Current Problems of Modem Neurophysiology. Motor Control. **2**(4), 285–299 (1998). (Original work published 1945)
16. Bongaardt, R., Meijer, O.G.: Bernstein's theory of movement behavior: historical development and contemporary relevance. J. Mot. Behav. **32**(1), 57–71 (2000)
17. Vitor, L.S., Turvey, M.T.: Bernstein's levels of movement construction: a contemporary perspective. Hum. Movement Sci. **57**, 111–133 (2018). https://doi.org/10.1016/j.humov.2017.11.013
18. Suyatinov, S.: Bernstein's theory of levels and its application for assessing the human operator state. In: Dolinina O. et al. (eds.) Recent Research in Control Engineering and Decision Making. ICIT-2019. Studies in Systems, Decision and Control, vol. 199, pp. 298–312. Springer, Cham (2019). https://doi.org/10.1007/978-3-030-12072-6_25
19. Alexandrov, A.V., Frolov, A.A., Mergner, T., Hettich, G., Frolov, A.M.: Movement control in anthropomorphic robot using a human inspired eigenmovement concept. Russian J. Biomech. **22**(1), 48–61 (2018). https://doi.org/10.15593/RJBiomech/2018.1.05
20. Shen, K., Selezneva, M.S., Neusypin, K.A., Proletarsky, A.V.: Novel variable structure measurement system with intelligent components for flight vehicles. Metrol. Meas. Syst. **24**(2), 347–356 (2017)
21. Buldakova, T.I, Suyatinov, S.I.: The significance of interdisciplinary projects in becoming a research engineer. In: Smirnova, E.V., Clark, R.P. (eds.) Handbook of Research on Engineering Education in a Global Context, pp. 243–253. IGI Global, Hershey, PA (2019). https://doi.org/10.4018/978-1-5225-3395-5.ch022

22. Buldakova, T.I., Dzhalolov, A.S.: Analysis of data processes and choices of data-processing and security technologies in situation centers. Sci. Technical Info. Process. **39**(2), 127–132 (2012). https://doi.org/10.3103/S0147688212020116
23. Zubov, N.E., Li, M.V., Mikrin, E.A., Ryabchenko, V.N.: Terminal synthesis of orbital orientation for a spacecraft. J. Comput. Syst. Sci. Int. **56**(4), 721–737 (2017). https://doi.org/10.1134/S1064230717040190

Deep Neural Networks Application in Models with Complex Technological Objects

Valeriy Meshalkin, Andrey Puchkov, Maksim Dli and Yekaterina Lobaneva

Abstract A method for creation of computer models in complex multiply connected technological objects based on the application of machine learning methods is described. For technological information processing hierarchical neural network structure integrated into cyber-physical systems of control is developed. It allows to monitor an object condition and forecast its development trends. A description for the algorithm and program, which performs the proposed method of model building, is given.

Keywords Cyber-physical systems · Machine learning · Program models · Deep neural networks · Computer vision

1 Problem Statement

The number of information channels in automated process control system (APCS) increases due to the raise of the complexity in the technological process under control, which requires the application of a new paradigm when creating complex control systems, such as a cyber-physical system (CPS) [1]. This system is characterized by the use of multidisciplinary approaches in its operation as well as Big Data methods caused by the increase in APCS complexity [2–4].

V. Meshalkin
D. Mendeleev, University of Chemical Technology, Miusskaya square 9, 125047 Moscow, Russia
e-mail: clogist@muctr.ru

A. Puchkov (✉) · M. Dli · Y. Lobaneva
Moscow Power Engineering Institute (Branch) in Smolensk, National Research University, Energetichesky proyezd 1, Smolensk 2014013, Russia
e-mail: putchkov63@mail.ru

M. Dli
e-mail: MiDli@mail.ru

Y. Lobaneva
e-mail: lobaneva94@mail.ru

A. G. Kravets et al. (eds.), *Cyber-Physical Systems: Advances in Design & Modelling*, Studies in Systems, Decision and Control 259,
https://doi.org/10.1007/978-3-030-32579-4_23

Therefore, a distinctive feature for a number of manufactures is their long service life which leads to the use of outdated technological solutions. Their change is impossible without a general modernization of production. Modernization in its turn involves significant financial expenditure because of downtime as well.

An alternative direction of modernization is reengineering of the information support for technological process by improving the systems for collecting and processing technological information, their duration and the volume distribution in the entire physical process, which is typical for CPS, the use of modern state diagnostics and control algorithms. This procedure has less financial and time expenditures compared with equipment modernization. Thus, the direction to improve control and measuring infrastructure of APCS, based on modern achievements of information technologies, presents and actual research problem which solution can bring tangible advantages for enterprises in a short period.

2 Background and Methods

The proposed methods are based on machine learning [5]. Deep learning using convolutional neural networks (CNN) is among its methods which fined wide application in solving real problems. High results, shown by CNN when recognizing images, lead to a great spectrum of their applications in the subject fields where there is an opportunity to reformulate the initial problem for the task of images recognition. These fields include medicine [6, 7], social engineering [8], text processing [9], gesture recognition [10], automatic identification of vehicles in coating production line based on computer vision [11], vehicles identification in a tunnel surveillance control system [12], cracks recognition in concrete [13] etc. The above mentioned list shows, that CNN can be a universal tool for problem solution of data deep analysis too.

The proposed approach for developing a model of complex technological processes is based on the implementation of CNN ensemble connected to the analysis of data in different points of technological process with subsequent processing of the obtained results of the neural classification in the analytical block. In addition to spatial partitioning of industrial zones the processing is discretized according time, it helps to monitor the processes dynamics.

There is a great diversity of buildings and CNN ensembles work interpretation [14], they are widely used in medical applications, in systems of biometric data control, people activity [15–18], but there is practically no works on generalized modeling and diagnostics of technological processes.

Control technical areas, where CNN is supposed to be used, can contain data of various format data sources. In addition to places with evident presence of video streams, for example[19, in Russian], other forms of signals can be also processed, including in-plant noise control, using various methods of sound waves conversion [20].

The enlarged structure of the proposed model of a complex technological process based on the application of deep neural networks is shown in Fig. 1.

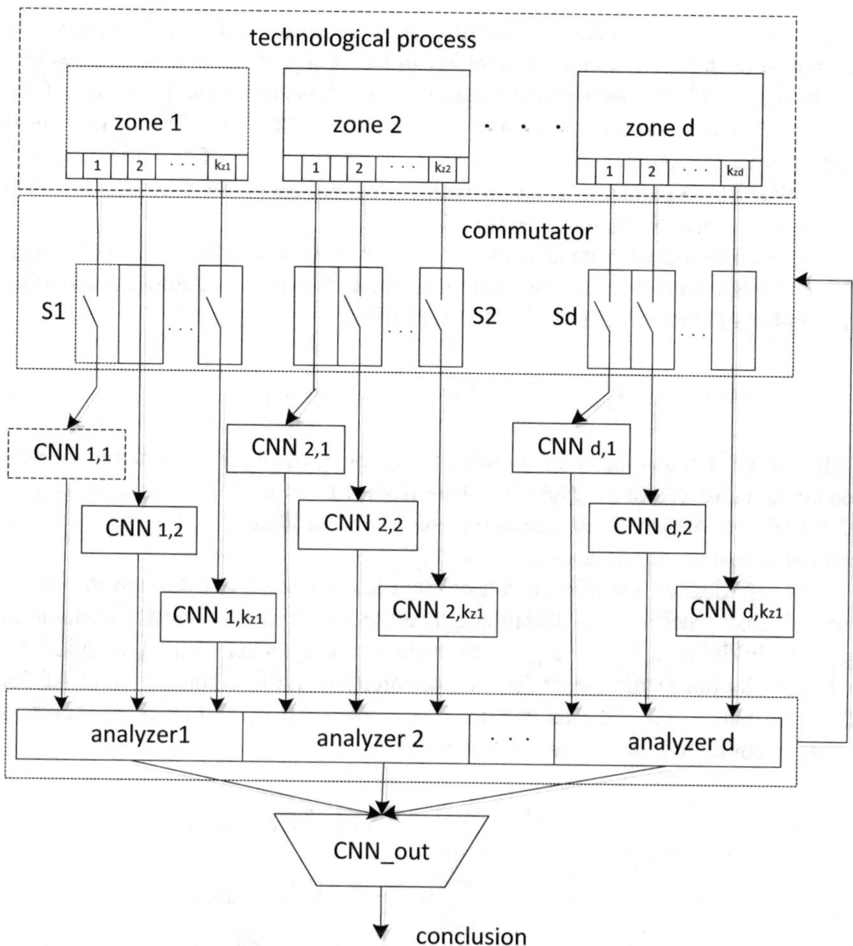

Fig. 1 The structure for the information processing model in CPS

The general algorithmic structure of the proposed model for technological information processing in CPS is as follows. It is supposed, that control and measuring information from technological zones (zone 1, …, d) is presented by multichannel sets of data for each zone. Through the commuter this information is fed to the output of local CNN ensemble which, for each zone, is formed taking into account the form of data representation in information channels and requirements of their further analysis which is carried out in the group of zone analyzers (analyzer 1, …, d). From the zone analyzers output information goes to neural network output (CNN_out) which carries out the estimation and forecast for the state of the entire technological process.

The algorithm for processing information is as follows. Denote the interval of discretization for information channel k_{zd} in technological zone d by $\Delta t_i^{zd} = t_i^{zd} - t_{i-1}^{zd}$. At t_i^{zd} and t_{i-1}^{zd} moments channel commutator k_{zd} is closed and the image of the technological parameter under control is supplied to CNN input. The technological parameter under control is denoted by P^{zd}. The interval value of discretization Δt_i^{zd} is calculated by a particular analyzer with regard to Kotelnikov theorem (it is also known as Nyquist–Shannon theorem) [21].

CNN of information channel k_{zd} recognizes an image on the introduced time step. This procedure consists in forming output vector $V_i^{k_{zd}}$ with dimension corresponding the number of classes n_CL_kzd for each channel:

$$V_i^{k_{zd}} = \left(V_{i,1}^{k_{zd}}, V_{i,2}^{k_{zd}}, \ldots, V_{i,n\_CL\_kzd}^{k_{zd}}\right)^T \tag{1}$$

It is supposed, that the architecture of the output CNN layer is formed in the way the elements of vector (1) contain values from 0 to 1. It characterizes the degree of neural network confidence in parameter P^{zd} controlled according to the image belonging to a particular class at moment t_i^{zd}.

CNN application for information channel k_{zd} is carried out through the whole technological process, but all this time can be divided into fragments with Ti. duration given as initial data for the algorithm operation. Fragments with Ti. duration are identified by the requirements for the periodicity of information flow into APCS. Thus, the matrix of results classification is processed in particular counting analyzers i_T and becomes available to the moment Ti:

$$MV_{i_T}^{k_{zd}} = \begin{pmatrix} V_{1,1}^{k_{zd}}, & V_{1,2}^{k_{zd}}, & \ldots, & V_{1,n\_CL\_kzd}^{k_{zd}} \\ & & \ldots & \\ V_{i_T,1}^{k_{zd}}, & V_{i_T,2}^{k_{zd}}, & \ldots, & V_{i_T,n\_CL\_kzd}^{k_{zd}} \end{pmatrix}^T \tag{2}$$

The obtained set (2) contains the data about CNN confidence dynamics in classification results and can be used when making control decisions in APCS. For this purpose the relation matrix of elements increments (2) to the given discretization Δt_i^{zd} interval is:

$$DV_{i_T}^{k_{zd}} = \begin{pmatrix} \frac{V_{2,1}^{k_{zd}} - V_{1,1}^{k_{zd}}}{\Delta t_i^{zd}} & \ldots & \frac{V_{i_T,1}^{k_{zd}} - V_{Ti-1,1}^{k_{zd}}}{\Delta t_i^{zd}} \\ \frac{V_{2,2}^{k_{zd}} - V_{1,2}^{k_{zd}}}{\Delta t_i^{zd}} & \ldots & \frac{V_{i_T,2}^{k_{zd}} - V_{Ti-1,2}^{k_{zd}}}{\Delta t_i^{zd}} \\ \ldots & \ldots & \ldots \\ \frac{V_{2,n\_CL\_kzd}^{k_{zd}} - V_{1,n\_CL\_kzd}^{k_{zd}}}{\Delta t_i^{zd}} & \ldots & \frac{V_{i_T,n\_CL\_kzd}^{k_{zd}} - V_{Ti-1,n\_CL\_kzd}^{k_{zd}}}{\Delta t_i^{zd}} \end{pmatrix} \tag{3}$$

For the purpose to simplify the notation denote matrix (3) elements as $dv_{i,j}$. Their sense load can be interpreted as analogue of derivatives for continuous functions because they reflect the change of neural confidence in image belonging to a particular class.

The rate of change, in this case it is element $dv_{i,j}$ value, can be used for the forecast of the technological process development. For this purpose matrixes of type (3) are calculated for all measuring channels and all technological zones.

The formation of an output tensor for CNN_out, based on the fragment in i_T timing, is performed by the combination of matrix (3) for all channels and zones

$$TR_{i_T} = \left(DV_{i_T}^{k_{z1}} \quad DV_{i_T}^{k_{zd}} \quad \ldots \quad DV_{i_T}^{k_{z2}} \right) \tag{4}$$

It should be noted, that the number of classes for various information channels and various technological zones can be different (see Fig. 1), therefore, matrix (3) dimension in a general case is different. The number of lines in (3) for all zones is equal as it is determined by the number of counting for discrete time $i = 1, 2, \ldots, i_T$. The number of columns for matrix TR_{i_T} is determined by the number of information channels in each zone, the number of classes identified for each channel and is equal to $(kz1 \cdot n\_CL\_kz1) + (kz2 \cdot n\_CL\_kz2) + \cdots + (kzd \cdot n\_CL\_kzd)$.

When input tensor TR_{i_T} is formed, the multi-stage preprocessing of data ends to ensure the work of output neural network (preprocessing for local CNN was not described as it is not of special interest in this work). Then, it remains to determine the hypotheses space for CNN_out as the application of deep learning deletes the need for construction features to replace complex, contradictory and heavy conveyors with simple learning models which area usually built with the use of several tensor operations [4].

Tensor TR_{i_T} is formed on the basis of data on the transformation dynamics of class membership for information channels parameters reflected in matrix (3). Thus, when forming the space of hypotheses good results to forecast the state of a technological process, values of any parameters for a finished product can be expected with the help CNN_out.

3 Application and Results

To test the proposed model for processing information in CPS of a complex technological process control, a simulation experiment to recognize aluminum alloy ingots images was carried out to determine their aggregate state. The aggregate state is estimated according to the image of a surface observing through a viewing window fitted on the furnace door [19, in Russian]. The received image is shown in Fig. 2 in its left part. For the application under study the general scheme is presented in Fig. 1.

The image for a working surface of aluminum ingots forming by a video camera, has a resolution of 640 × 480 (Fig. 3a). If this resolution is left and used as an example at input of CNN, then the process of learning requires more time considering that the numbers of such examples will be several thousand. Therefore, the dimension of

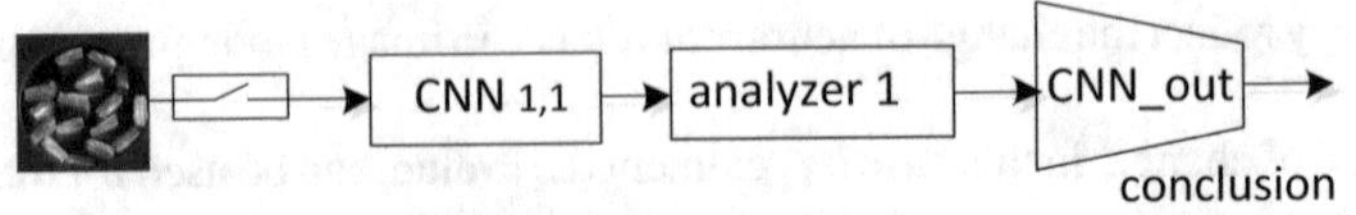

Fig. 2 The structure for processing of melting zone images

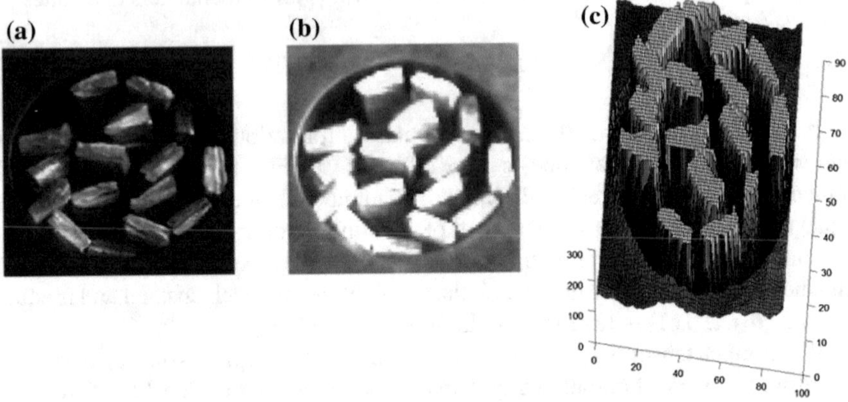

Fig. 3 Melting surface images

an initial image is programmatically reduced to 90 × 90. In addition, it is taken into account, that the melting zone is lighted by an electric arc, so the image brightness can fluctuate when the supply voltage changes in a circuit. To reduce the influence of this factor brightness was normalized (Fig. 3b).

The existing methods of visual identification for aggregate state of a substance (method of triangulation, burst mode, area method) analyze the changes of the surface image area for a melting metal in a three-dimensional space, but not the area of the surface for remelting ingots.

The proposed method for control of an aggregate state based on the deep neural networks also analyzes the three-dimensional surface of ingots. It is ensured by the fact, that the analysis of the melting process in images matrix is characterized by the changes of rises heights in histogram (Fig. 3c) which is taken into account by the neural network when forming the answer. One example from the learning set is a tensor of the second order, the totality of such examples forms the tensor of the third order which is fed to the output of $CNN_{1,1}$. Temporal discretization interval of a video sequence coming from the video camera focused on the bars surfaces is taken equal to one second. Free Video to JPG Converter was used to take shots from the video. To enlarge the number of learning examples the augmentation procedure was applied. During this procedure shifts, zooming, rotations, mirror reflection were implemented for the initial image.

Table 1 Melting time distribution into classes

Class number	Aggregate state	Time span, s
1	Solid	0–269
2	Initial transition	270–279
3	Final transition	280–289
4	Liquid	290–300

The time for melting of an aluminum bars lot is approximately 300 s, but it was divided into intervals corresponding different classes identified by $CNN_{1,1}$ (see Table 1).

The software model for the technological process (aluminum bars melting) under study implementing the recognition of aggregate state transition was performed in Python 3.6 language. IDE Spyder from Anaconda (version for Linux) was chosen as the development environment. Convolutional neural networks were developed using specialized Keras library, which is a superstructure above the tensor computation framework TensorFlow [22, 23].

$CNN_{1,1}$ containing seven alternating convolutional layers and subsampling and one output fully connected layer with four outputs (according to the numbers of recognizable classes), was implemented in the software. The learning sample has a size of 2000 examples (400 examples are from the testing sample).

The analysis of the results for the software work, presented in the upper graph of Fig. 4, shows that within one run at $CNN_{1,1}$ outputs the dynamics of classification during melting process is visible.

The final results of classification can be obtained if the choice of $CNN_{1,1}$ output according to the majority principle is made at every second counting. At multiple run the number of classes, as it is shown in the graph at the bottom of Fig. 4, are correctly recognized by $CNN_{1,1}$ neural network, unstable recognition is occurred only in the zone of transition from one class to the other one. This circumstance can be explained by the complexity of detection for difference in the surfaces on the junction of classes, in addition the relative length of unstable time intervals is not long.

To forecast the dynamics of the process development matrix (3) is calculated, the number of which forms the initial data for CNN_out working. The results of the obtaining images for two melting's are presented in Fig. 5. Figure 5 shows that for different processes of aluminum bars melting the image is different, this is the evidence of possibility to use the methods of texture recognition in CNN_out block to forecast the process development. In this study the experiment to calculate the dynamics was not carried out, but it is planned for further work.

CNN learning was performed on GeForce GTX 1060 video card installed on Asus FX502VM notebook with CPU IntelCore i7-7700HQ. The process control was exercised with the help of TensorBoard which allows to visualize the current accuracy and learning error.

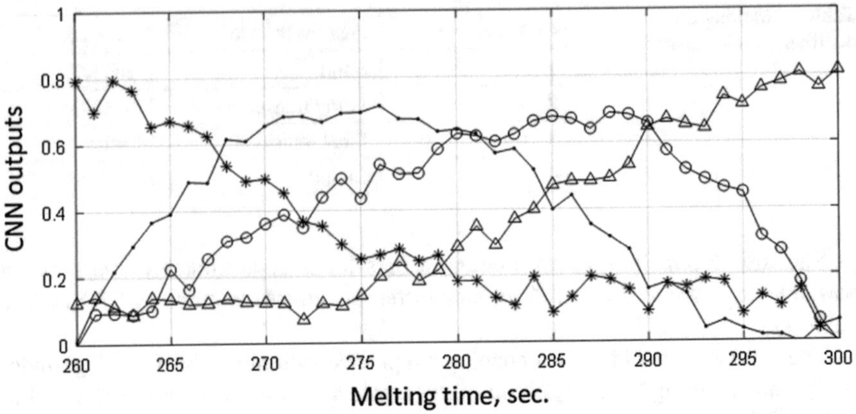

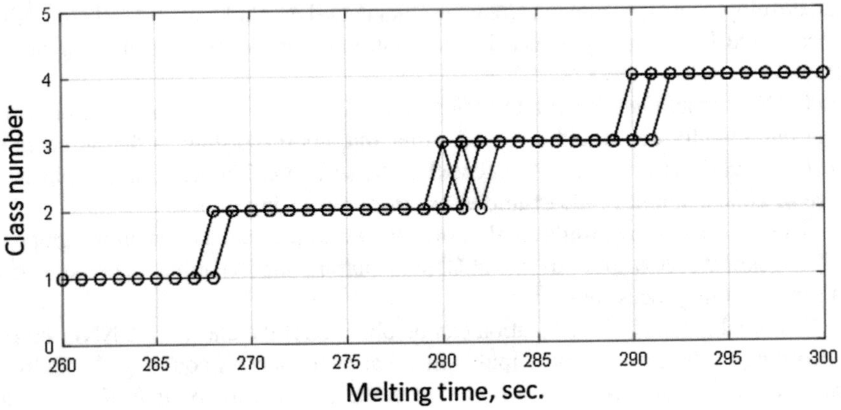

CNN outputs
1
0.8
0.6
0.4
0.2
0
260 265 270 275 280 285 290 295 300
Melting time, sec.

Class number
5
4
3
2
1
0
260 265 270 275 280 285 290 295 300
Melting time, sec.

Fig. 4 Image recognition results

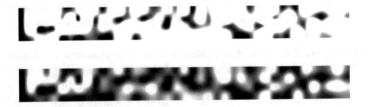

Fig. 5 Matrix dynamics process visualization

4 Conclusion

As a result of the conducted study for the possibility of using deep neural networks as a part of a cyber-physical system of complex technological objects control the following results were obtained:

1. The algorithmic structure for the model of processing information coming from different technological zones of the integrated multi-stage technological process based on the convolutional neural networks ensemble implementation is developed.
2. The results of model experiment for recognition of aluminum aggregate state based on the proposed algorithm are given. The results show that the algorithm, based on the convolutional neural networks, solves the task for classification of metal aggregate state well, excepting some zones in the transition regions from one class to the other one.
3. The method to follow up the dynamics in technological process based on the formation of matrix values changes at the output of the convolutional network, performing the classification for technological process conditions and its further visualization for output convolution neural network recognition is proposed.

Acknowledgements The reported study was funded by RFBR according to the research project № 19-01-00425

References

1. Wolf, W.: Cyber-physical systems. Computer. **9**(3), 88–89 (2009)
2. Lee, J., et al.: Recent advances and trends in predictive manufacturing systems in big data environment. Manuf. Lett. **1**(1), 38–41 (2013)
3. Namiot, D.: On big data stream processing. Int. Journal of Open Info. Technol. **3**(8), 48–51
4. Sleep, S., Gooner, R., Hulland, J.: The big data hierarchy: a multi-stage perspective on implementing big data. In: Obal, M., Krey, N., Bushardt, C. (eds.) Let's Get Engaged! Crossing the Threshold of Marketing's Engagement Era. Developments in Marketing Science: Proceedings of the Academy of Marketing Science. Springer, Cham (2016)
5. Scholle, F.: Deep learning in Python. SPb, Peter, 400 p (2018)
6. Rohit, S., Chakravarthy, S.: BMC Neurosci **12**(Suppl 1), 35 (2011). https://doi.org/10.1186/1471-2202-12-S1-P35
7. Khryashchev, V., Lebedev, A., Stepanova, O., Srednyakova, A.: Using convolutional neural networks in the problem of cell nuclei segmentation on histological images. In: Dolinina, O., Brovko, A., Pechenkin, V., Lvov, A., Zhmud, V., Kreinovich, V. (eds.) Recent Research in Control Engineering and Decision Making. ICIT 2019. Studies in Systems, Decision and Control, vol. 199. Springer, Cham (2019)
8. Severyn, A., Moschitti, A.: Twitter sentiment analysis with deep convolutional neural networks. In: ACM SIGIR Conference on Research and Development in Information Retrieval, pp. 959–962. ACM Press, Santiago (2015)
9. Wang, P., Xu, J., Xu, B., Liu, C., Zhang, H., Wang, F.: Semantic clustering and convolutional neural network for short text categorization. In: 53rd Annual Meeting of the Association for

Computational Linguistics and 7th International Joint Conference on Natural Language Processing, vol. 2, pp. 352–357. ACL Press, Beijing (2015)
10. Ahlawat, S., Batra, V., Banerjee, S., Saha, J., Garg, A.K.: Hand gesture recognition using convolutional neural network. In: Bhattacharyya, S., Hassanien, A., Gupta, D., Khanna, A., Pan, I. (eds.) International Conference on Innovative Computing and Communications. Lecture Notes in Networks and Systems, vol. 56. Springer, Singapore (2019)
11. Xiang, L., et al.: Automatic vehicle identification in coating production line based on computer vision. In: International Conference on Computer Science and Engineering Technology, pp. 260–267. World Scientific Publication Co. Pvt. Ltd (2016)
12. Chen, H.T., et al.: Multi-camera vehicle identification in tunnel surveillance system. In: IEEE International Conference on Multimedia & Expo Workshops, pp. 1–6. IEEE (2015)
13. Cha, Y.J., Choi, W.: Vision-based concrete crack detection using a convolutional neural network. In: Caicedo, J., Pakzad, S. (eds.) Dynamics of Civil Structures, vol. 2. Conference Proceedings of the Society for Experimental Mechanics Series. Springer, Cham (2017)
14. Frazão, X., Alexandre, L.A.: Weighted Convolutional Neural Network Ensemble. In: Bayro-Corrochano, E., Hancock, E. (eds.) Progress in Pattern Recognition, Image Analysis, Computer Vision, and Applications. CIARP 2014. Lecture Notes in Computer Science, vol. 8827. Springer, Cham (2014)
15. Fan, Y., Lam, J.C.K., Li, V.O.K.: Multi-region Ensemble Convolutional Neural Network for Facial Expression Recognition. In: Kůrková V., Manolopoulos Y., Hammer B., Iliadis L., Maglogiannis I. (eds.) Artificial Neural Networks and Machine Learning—ICANN 2018. ICANN 2018. Lecture Notes in Computer Science, vol. 11139. Springer, Cham (2018)
16. Kori, A., Soni, M., Pranjal, B., Khened, M., Alex, V., Krishnamurthi, G.: Ensemble of fully convolutional neural network for brain tumor segmentation from magnetic resonance images. In: Crimi, A., Bakas, S., Kuijf, H., Keyvan, F., Reyes, M., van Walsum, T. (eds.) Brainlesion: Glioma, Multiple Sclerosis, Stroke and Traumatic Brain Injuries. BrainLes 2018. Lecture Notes in Computer Science, vol. 11384. Springer, Cham (2019)
17. Koitka, S., Friedrich, C.M.: Optimized convolutional neural network ensembles for medical subfigure classification. In: Jones G. et al. (eds.) Experimental IR Meets Multilinguality, Multimodality, and Interaction. CLEF 2017. Lecture Notes in Computer Science, vol. 10456. Springer, Cham (2017)
18. Kasnesis, P., Patrikakis, C.Z., Venieris, I.S.: PerceptionNet: a deep convolutional neural network for late sensor fusion. In: Arai K., Kapoor, S., Bhatia, R. (eds.) Intelligent Systems and Applications. IntelliSys 2018. Advances in Intelligent Systems and Computing, vol. 868. Springer, Cham (2019)
19. Shkundin, S. Z., Kolistratov, M.V., Belobokova, Y.A.: Algorithms performance testing to determe changes of a metal aggregate state. Syst. Administrator. **10**(191), 90–93 (2018)
20. Fu, G.: A novel isolated speech recognition method based on neural network. In: Zhong, Z. (ed.) Proceedings of the International Conference on Information Engineering and Applications (IEA) 2012. Lecture Notes in Electrical Engineering, vol. 220. Springer, London (2013)
21. Ahlswede, R., Ahlswede, A., Althöfer, I., Deppe, C., Tamm, U.: Shannon's model for continuous transmission. In: Ahlswede, A., Althöfer, I., Deppe, C., Tamm, U. (eds.) Transmitting and Gaining Data. Foundations in Signal Processing, Communications and Networking, vol. 11. Springer, Cham (2015)
22. Ramasubramanian, K., Singh, A.: Deep Learning using Keras and TensorFlow. In: Machine Learning Using R. Apress, Berkeley, CA (2019)
23. Srinivasa, K.G., Siddesh, G.M., Srinidhi, H.: Advanced Analytics with TensorFlow. In: Network Data Analytics. Computer Communications and Networks. Springer, Cham (2018)

Intelligent Technologies in the Diagnostics Using Object's Visual Images

Sergey Orlov and Roman Girin

Abstract The problem of complex industrial equipment diagnostics using images in different spectral ranges is considered. An intelligent method for technical states classification according to images of a control object is proposed. Considered a neural network analyzer designed as a two-branch neural network. Convolutional neural network processes simultaneously three object's images obtained in the visual, ultraviolet and infrared bands. The properties of the dataset for learning the neural network are investigated using the dimensionality reduction methods. Examples of the developed method and the neural network analyzer application for monitoring various industrial facilities are given.

Keywords Technical diagnostics · Artificial neural network · Deep learning · Infrared thermography · Ultraviolet light inspection

1 Background

The complexity of industrial equipment and increased requirements for reliability pose the on-line diagnostic task based on the analysis of a large number of monitored parameters. However, traditional methods aimed at measuring a limited set of signals and their subsequent analysis often do not allow to detect the rapid development of failures and emergency situations. Nowadays, measurement methods and devices allow you to record significant amounts of information about an object, for example, controlled object images in different spectral ranges: visible, infrared (IR) [1, 2] and ultraviolet light (UV) [3].

In the electronic component testing, the LOC-in Thermography is used [4]. To do this, the object under investigation is affected by low frequency thermal waves and the response is measured on the surface. LOC-in Thermography method increases

S. Orlov (✉) · R. Girin
Samara State Technical University, 244 Molodogvardeyskaya str., Samara 443100, Russia
e-mail: orlovsp1946@gmail.com

R. Girin
e-mail: romangirin@gmail.com

A. G. Kravets et al. (eds.), *Cyber-Physical Systems: Advances in Design & Modelling*, Studies in Systems, Decision and Control 259,
https://doi.org/10.1007/978-3-030-32579-4_24

on the order of the thermal imager sensitivity, making it easier to identify small areas with a slight overheating.

Another new method for electronic device inspection use Raman IR-Thermography [5, 6]. But in this case there is a problem of operational analysis and decision making, as the operator, or analyst does not have time to evaluate such a large volume of video data.

The authors develop an approach associated with intellectualization the process of analyzing information about the object technical states [7, 8]. This approach is based on the use the artificial neural networks (ANN) for processing the received information in real time. A fundamental property of ANN in this case is the ability of the neural network deep learning [9, 10]. For this, high-performance computing resources are used. Then, a trained neural network may be implemented as a program running on relatively simple computer.

The article is devoted to the study of the intellectual method of diagnosis and to the use the image neural network analyzer in various spectral ranges.

2 Problem of the Technical Object's Diagnostics by Their Images in Various Spectral Ranges

The most widely used methods are found in infrared thermography, and they are often used in conjunction with the object visible images analysis. In industrial diagnostics also used UV control [3], which is focused primarily on the discovery the phenomena associated with the electrostatic field, creepage, corona and arc discharges in high voltage equipment. Joint processing of all three types of images was restrained by the lack of effective methods and means of automatic analysis.

The authors propose to perform the processing of a complex image as a whole, which includes three components: visible, infrared and ultraviolet.

Another problem, in most cases, is due to the absence of a representative dataset of actually measured images in different ranges. This is due to the fact that in the practice of diagnosing many objects the measurement databases were not used and the results of previous tests were not generalized. To solve this problem, we propose in the preliminary stage of neural networks learning to use mathematical models of controlled objects. On the basis of such models, simplified images set is formed that adequately represent the object.

Denote the sets of model images: $R_1^M(x, y)$—visible images, $R_2^M(x, y)$—ultra-violet images, $R_3^M(x, y)$—infrared images (thermograms), x, y—coordinates of the observed surface of the monitored object. Joint analysis of images by a group of experts makes it possible to form classes of the object's technical states, corresponding to certain defects, failures or emergency conditions. For controlled object we denote the set $D_k,\ k = \overline{0, K}$ of possible states. This set includes K inoperable states, corresponding to classified failures, and one D0 state for normal operation of the equipment.

In addition, each class will correspond to a subset of model images:

$$V_k^M(x, y) = \{\{R_{1j}^M(x, y)\}, \{R_{2j}^M(x, y)\}, \{R_{3j}^M(x, y)\}\}, \ j \in J_k,$$

where J_k—an index set of model images corresponding to the inoperable state k.

It should be noted that in the general case the classification problem in this formulation is an incorrect inverse problem. In order to achieve one-to-one correspondence between model types of images and the status of the controlled object it is proposed to analyze the surface video images simultaneously with the set S (t) of additionally measured parameters. As a rule, it is possible to measure some subset of object parameters using built-in measurement channels.

The class D_k is determined by the set of images in different spectra and additionally calculated object parameters:

$$D_k : \{(R_{1j}^k, R_{2j}^k, R_{3j}^k, S_i^k) \in \Omega \wedge (E_k = 1)\}, \ j \in J_k, i \in I_k,$$

where

- I_k an index set of parameters corresponding to the inoperable state k,
- Ω domain of parameters in case of failure,
- E_k expert opinion on the compliance of specified images and additional parameters to the k-th failure type

It is required to find the neural network operator N_k, which establishes the relationship between the vector D_k and measured vector DM.

Then the cooperative processing of complex model images and the S vector in the neural network regularizes the inverse classification problem based on information on additional parameters.

Thus, the set of classes of complex model images is formed in the database of the diagnostics system.

3 Intelligent Diagnostic Method Using the Artificial Neural Networks

The proposed intelligent method of improving the classification accuracy in the diagnosis is implemented through three basic procedures:

1. The complex model images are constructed, supplemented by a set of the object measured parameters to obtain an exact concordance with faults, as well as the formation a diagnostics knowledge base.
2. Using the neural network analyzer in the diagnostic system in the form of a two-branch neural network (2BNN) consisting of a deep convolutional network for image processing and a fully connected neural network for processing additional object parameters.

3. Training 2BNN with a complex model image dataset and further performing the classification of the technical states in the diagnostic process for decision making about maintenance service.

More detailed actions to be taken in the implementation of the proposed method are disclosed below:

1. For a certain class of controlled objects on the basis of mathematical models, the construction of two-dimensional computational images corresponding to various technical states of the object is performed. Together with the selected limited set of measured values of the object parameters, they form a set of complex model images.
2. On the resulting set, subsets of variable design images and measured parameters are built, covering possible deviations of the monitored object's states from the nominal values and characterizing faults.
3. The diagnostic knowledge base is formed containing complex model image set.
4. A neural network analyzer is introduced into the diagnostic system structure. It is a two-branch multilayer neural network for processing images and a vector of additionally measured parameter values.
5. Network 2BNN is being trained on a complex model image dataset from the knowledge base. To verify the learning quality and the efficiency of the chosen structure of the diagnostic system, a multidimensional analysis of the object's state classification is performed.
6. The diagnostic system performs the procedure of measuring real images of the surface and real additional parameters of the object under test.
7. The measured real multispectral image of the object is fed to the convolutional network input (the main branch in the 2BNN), and the additionally measured parameters are fed to the inputs of a fully connected network (auxiliary branch in the 2BNN).
8. As a result, the neural network analyzer performs the classification of object technical states and the type of failure is determined. Then the DSS makes a conclusion about the possibility of its further operation.
9. According to the results of images measurements, the complex model images correction can be performed, or real images are added to the dataset for retraining.

The intelligent diagnostic system structure using a neural network analyzer is shown in Fig. 1.

The measuring system MS registers a three-part video stream: images in the visible, ultraviolet and infrared ranges. In addition, the MS registers additional object parameters $S(t)$. The neural network analyzer processes the time series of additionally measured parameters. Thus, the result of the object technical state classification appears at the analyzer output. The decision support system DSS generates object modes control signals and, under certain conditions, initiates the correction of mathematical models to retrain the neural network.

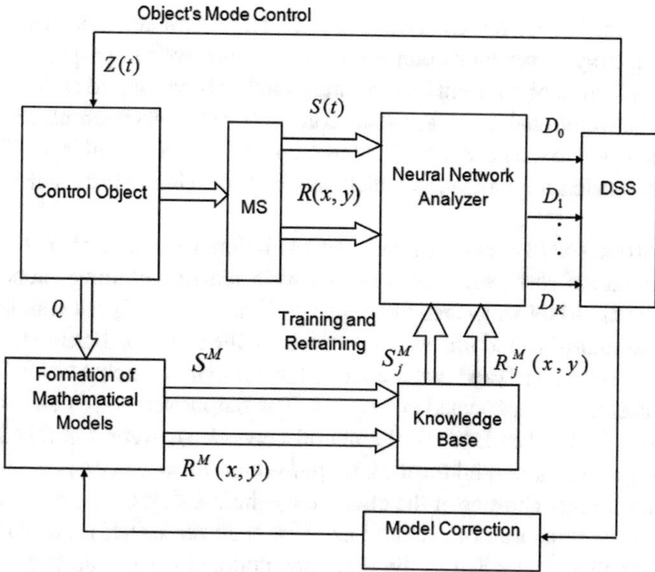

Fig. 1 Intelligent diagnostic system using images analysis

4 Neural Network Analyzer

The diagnostic system uses a neural network analyzer, which is a two-branch neural network 2BNN consisting of a convolutional branch and an auxiliary branch made as a fully connected network (Fig. 2).

The main branch is the deep convolutional neural network ANN 1 which built according to some architectural principles proposed by LeCun [11] also combines

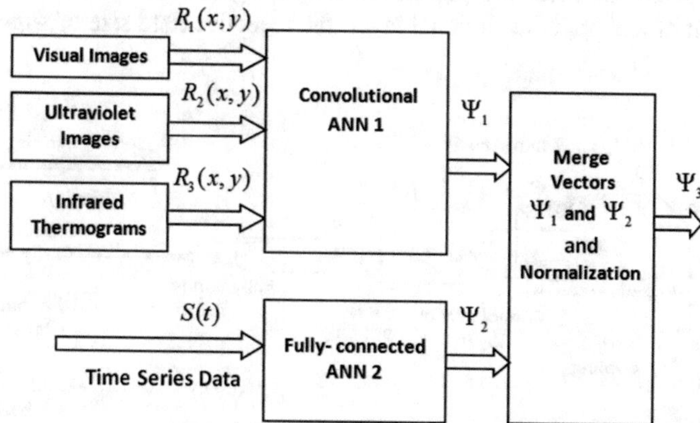

Fig. 2 General scheme of neural network analyzer 2BNN

some feature introduced by Krizhevsky in [12]. The neural network analyzer implements the sparsity of synaptic connections in the same way as proposed in [10]. In this way, a selection of different features map subsets is performed at the entrance to the second convolutional layer. As for the convolutional layers neurons, each of them is fully connected to the perceptual field input of the convolutional layer. The neural network uses batch normalization which was first introduced Joffe and Szegedy in [13].

The network ANN 1 consists of several convolutional layers combined into a feed-forward neural network. Such networks are well treated multidimensional signals. The input of the ANN 1 receives three images, which are analyzed simultaneously, as a complex multi-layered image. As proposed in the developed method, additional parameters $S\ (t)$ are analyzed that are fed to the auxiliary branch ANN 2. The ANN 2 is formed as a fully connected perceptron. The output vectors Ψ_1 and Ψ_2 are the input signals for the last layer of the neural network analyzer. On this layer, this vectors merge, normalize and form the output vector $\Psi_3 = (D_0, D_1, \ldots, D_K)$, whose components are probabilities of the classified technical states.

The detailed structure of the neural network analyzer 2BNN is shown in Fig. 3. The complex image supplied to the first convolutional layer is an array of 225 × 225 × 5, where 5 is the depth of the entire array. This depth is determined from the following considerations. Three images are received at the input, while the visible image is a color RGB image with a depth of 3. The infrared image has a depth of 1, since the thermal imager transmits an array of measured surface temperature points in the format "Radiometric JPEG" with 14-bit encoding. The UV image is a monochrome image in the CCIR standard; the image depth is also equal to 1.

The receptive field size for the first convolutional layer is 11 × 11. The first layer neurons extract certain "features" from the complex image. The number of features maps after the first layer is taken equal to 9. The output signal of the first convolutional layer is transmitted to the layer that performs the selection of the maximum value from the receptive field (Max pooling layer). The receptive field size of this layer is 4 × 4 pixels and the stride is 3 pixels. The output signal of this layer is transmitted to the input of another convolutional layer, the receptive field size of which is 5 ×

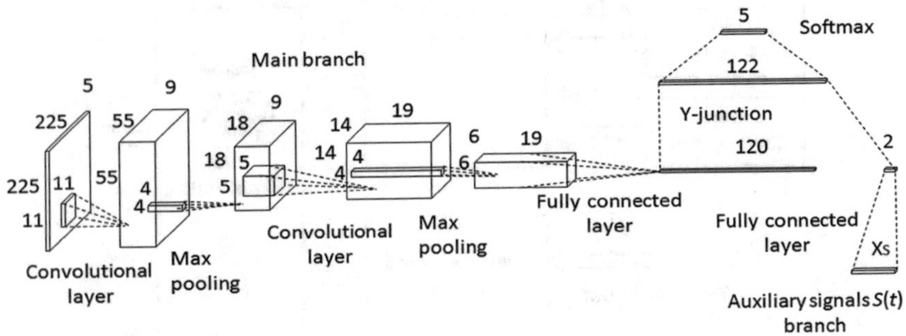

Fig. 3 The detailed structure of the 2BNN

5 pixels and the stride is 1 pixel. The number of feature map generated by this layer is 19. The second Max pooling layer with receptive field 4 × 4 prepares data for the third convolutional layer with receptive field 6 × 6. Since the input signal size of the third layer coincides in size its receptive field, the third layer is essentially fully connected and the length of its output vector is 120.

The ANN 2 branch processes data on the additionally measured parameters $S\ (t)$ in parallel. To do this, the fully connected layer is used. The combination of the output signals of both branches occurs on the Y-junction layer. The output normalization is performed on the interval [0, 1] in the neurons final layer. This neural network analyzer classifies five states: operable D_0 and four inoperable D_1, D_2, D_3, D_4.

5 The Training of Neural Network Analyzer

The training dataset is prepared as follows:

- a complex model image set is being formed, possibly supplemented with real images obtained earlier;
- experts classify the resulting of image set in accordance with recognizable technical states;
- the process of neural network learning is carried out using the method of error back propagation.

The quality of the learning is crucial for efficient operation of the neural network analyzer and all diagnostic system in general [14, 15]. The use of model images in ANN learning instead of real images leads to a method error δ_M in determining the object technical states. The main factors affecting its value are:

- the computational method error for solving the mathematical model equations, as well as the discretization and rounding errors,
- variability of the object design parameters,
- influence of the neural network architecture,
- influence of the data processed by the neural network,
- the method and neural network training parameters.

The relative method error δ_M is defined as:

$$\delta_M = \sqrt{\delta_{Str}^2 + \delta_D^2 + \delta_{Tr}^2},$$

where

δ_{Str} method error determined by the ANN structure,
δ_D the error determined by the noise in the input data of the neural network analyzer,
δ_{Tr} the error determined by the neural network learning method.

Let us estimate the error of data noise. The type of data and the algorithm used to generalize them are important for machine learning. By noise we shall mean the presence of any region in the n-dimensional sign space of the dataset, where there are complex images of several classes simultaneously, inseparable by the hyperplane. The location of the dataset element in this space is determined by the attribute values to which the additionally recorded parameters correspond, as well as the features extracted from the images by the convolutional network. The feature space dimension is determined by the output vector length and is equal to 122.

Analyzing the data in the dataset, we can determine the proportion of the images that are in noisy areas. This makes it possible to estimate the accuracy threshold of an ANN trained in such a dataset, since images related to noise are likely to introduce an error into the neural network operation. To estimate the dataset and separation of individual noisy areas the Dimensionality Reduction methods have been used [16, 17]. Three dimensionality reduction techniques were used: Principal Component Analysis (PCA), Multidimensional Scaling (MDS), t-Distributed Stochastic Neighbor Embedding (t-SNE) [18]. Figure 4 shows the results of the dataset dimension reducing in the diagrams for the five technical states classes. Charts are built using the tools Skikit-learn [19].

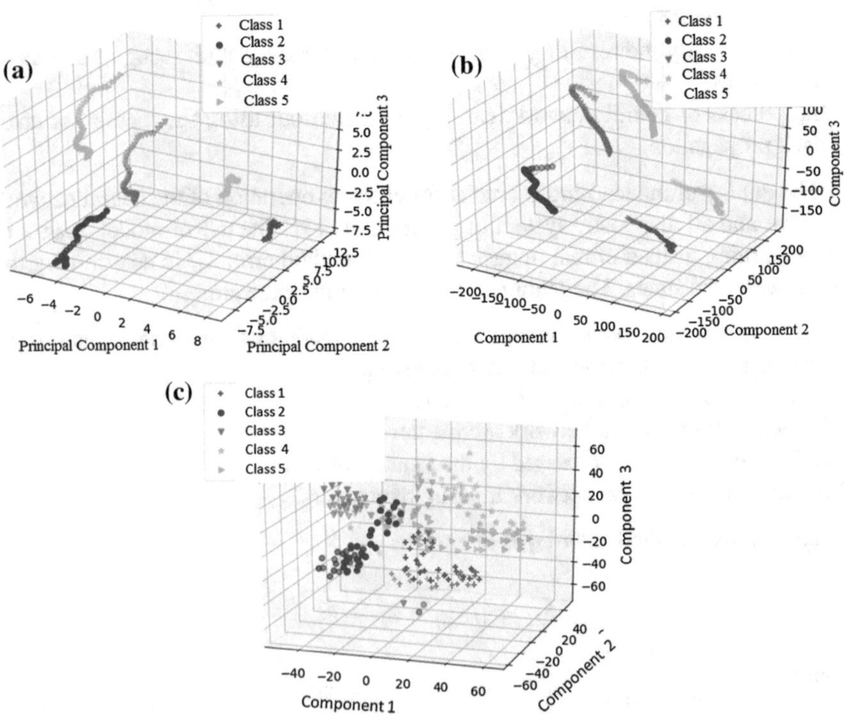

Fig. 4 Diagrams for five classes: **a** PCA chart; **b** MDS chart; **c** t-SNE chart

6 Implementation and Results

The developed intellectualization method and neural network analyzer were used in a number of the technical diagnostic systems.

1. The photosensitive matrix chip diagnosis. Monitoring was carried out by chip's surface thermograms with simultaneous measurement of a number of electrical signals. The use of a neural network analyzer for the classification of failures according to thermograms made it possible to increase accuracy up to 97.5% and reduce data processing time by 20% due to the technical states analysis in real time.
2. Catenary of contact wire monitoring. The control is performed in the movement of the computerized track-test car. The diagnostic system used the thermal imager Micro Epsilon TIM600 and ultraviolet camera CoroCAM 6D. The thermal imager registers the thermal field of the high voltage equipment and the ultraviolet camera registers corona and arc discharges on the catenary equipment. The classification accuracy was obtained at 98% with a reduction in data processing time up to 30%.
3. Rail track monitoring. In the computerized track-test car there is a system for obtaining railroad bed video images using four video cameras. The neural network analyzer was used when rail fastenings have been monitored, an example of which is shown in Fig. 5.

Dataset was composed the rail fastenings images in good condition and the missing rail fastenings. It was analyzed by dimension reduction to identify distinguishing features. Then the input of the learned neural network received images which gradually closed black rectangle considered most characteristic for fastening elements. Similar experiments are often referred to as the "occlusion test" [20]. According to the neural network processing results can be seen, which elements of the image are making the greatest contribution in favor of a classification outcome. An example of model images for the described experiment is shown in Fig. 6.

Figure 7 shows over each image the vector of values obtained at the neural network analyzer outputs. The first element of the vector is the probability of the rails fastening presence on the image, the second element is the probability of the rails fastening absence.

The diagrams presented in Fig. 8 show that the classes considered in this task are

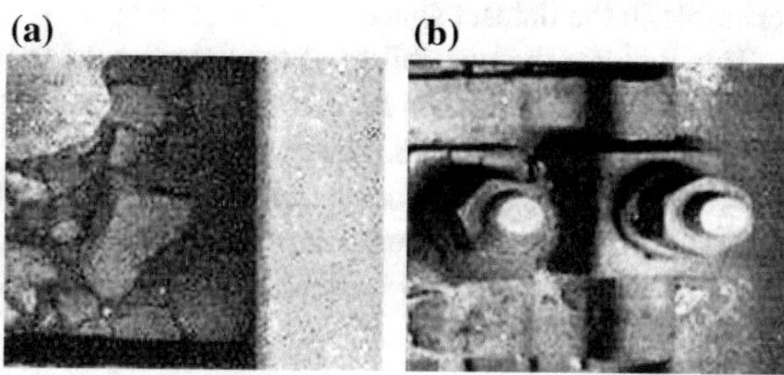

Fig. 5 Rail fastening images: **a** railway bed without fastening, **b** rail fastening is operable

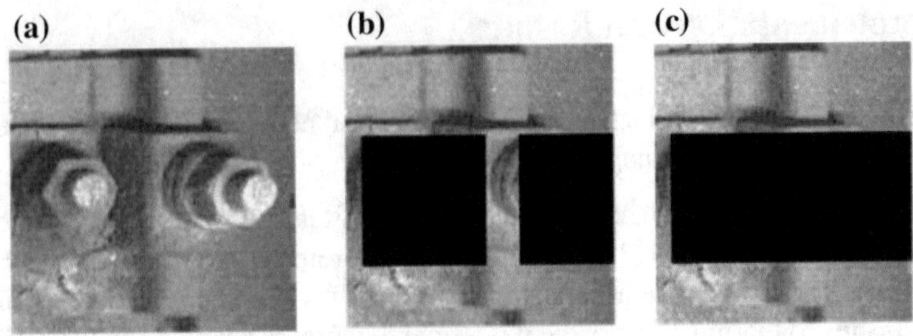

Fig. 6 Images of rails with closed elements on the images: **a** the original image

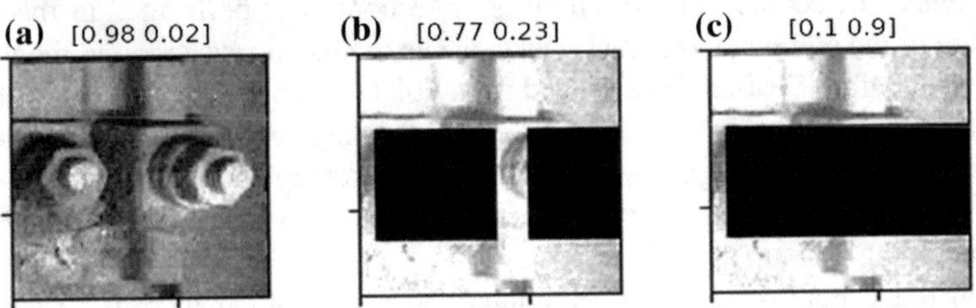

Fig. 7 Classification results of modified images

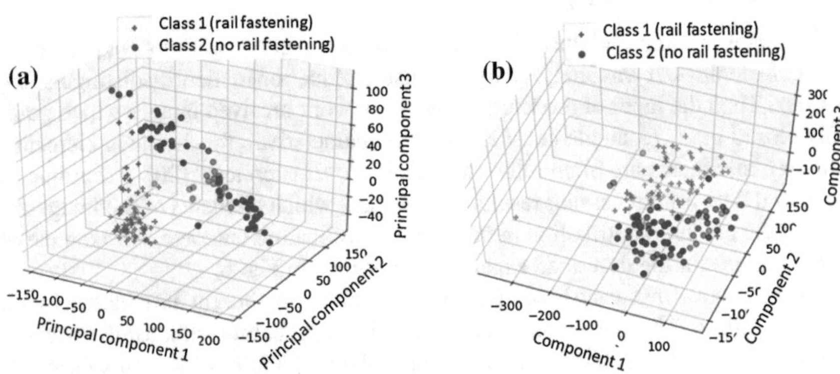

Fig. 8 Dimensionality reduction diagrams: **a** PCA; **b** t-SNE

separable in the dataset space.

This justifies the possibility of the neural network analyzer learning to solve the classification problem. The relative number of images that fall into common areas for classes does not exceed 1%. The neural network analyzer was trained on a dataset of 50,000 images, 25,000 in each of two classes: serviceable rail fastenings and missing rail fastenings. The convoluted neural network with a learning rate of 0.00005 and a number of epochs equal to 100 achieved a classification accuracy of 90%.

7 Conclusion

Intelligent technologies allow improving the accuracy of detecting faulty states and reducing the processing time of control data. The obvious advantage is the possibility to obtain a new quality of diagnosis through the rapid analysis the large amount of information contained in the images of the different spectral ranges. The development of modern neurochips [21] opens up new possibilities for building miniature parallel computing structures for the neural networks implementation. In this case, they can be installed on air drone to monitor the extended objects, such as railways, oil and gas pipelines and high-voltage power transmission lines.

References

1. Maldague Xavier, P.V.: Nondestructive Evaluation of Materials by Infrared Thermography. Springer-Verlag, London (1993)
2. Lanzoni, D.: Infrared Thermography. Electrical and Industrial Applications. CreateSpace Independent Publishing Platform (2015)
3. Yi, H., Kai, L.: Inspection and Monitoring Technologies of Transmission Lines with Remote Sensing. ACADEMIC PRESS (2017)
4. Breitenstein, O., Warta, W., Schubert, M.: Lock-in Thermography. Basics and Use for Evaluating Electronic Devices and Materials. Springer International Publishing (2018)
5. Sarua, A., Ji, H., Kuball, M., Uren, M.J., Martin, T., Hilton, K.P., Balmer, R.S.: Integrated Raman-IR thermography for monitoring of self-heating in AlGaN/GaN transistor structure. IEEE Trans. Electron Dev. **53**(10), 2438–2447 (2006)
6. Kuball, M., Sarua, A., Pomeroy, J.W., Falk, A., Albright, A., Uren, M.J., Martin, T.: Integrated Raman-IR thermography for reliability and performance optimization, and failure analysis of electronic devices. In: Conference Proceedings from the 33rd International Symposium for Testing and Failure Analysis, ISTFA 2007, USA, ASM International (2007)
7. Orlov, S.P., Vasilchenko, A.N.: Intelligent measuring system for testing and failure analysis of electronic devices. In: Proceedings of the XIX IEEE International Conference on Soft Computing and Measurements, SCM'2016, May 25–27, 2016, vol. 1, pp. 401–403. Saint-Petersburg, Russia (2016)
8. Orlov, S.P., Girin, R.V.: The use of neural networks for testing and failure analysis of electronic devices. In: Proceedings of the 2nd International Scientific-Practical Conference "Fuzzy Technologies in the Industry (FTI 2018)" 23–25 October, 2018. CEUR-WS.org/ vol. 2258/paper 21, pp. 160–167. Ulyanovsk, Russia (2018)
9. Hykin, S.: Neural Networks. A Comprehensive Foundation, 2nd edn. Prentice Hall (1999)
10. Norvig, P., Rassell, S.: Artificial Intelligence: A Modern Approach, 3rd edn. Pearson (2010)
11. LeCun, Y., Bottou, L., Bengio, Y., Haffner, P.: Gradient-based Learning Applied to Document Recognition, pp. 306–351. IEEE Press (1998)
12. Krizhevsky, A., Sutskever, I., Hinton, G.: ImageNet classification with deep convolutional neural networks. In: NIPS'12 Proceedings of the 25th International Conference on Neural Information Processing Systems, vol. 1, pp. 1097–1105 (2012)
13. Ioffe, S., Szegedy, C.: Batch Normalization: accelerating Deep Network Training Reducing Internal Covariate Shift. Cornell University. (2015). Library.htpps://arxiv.org/abs/1502.03167v3
14. Goodfellow, I., Bengio, Y., Aaron, C.: Deep Learning. The MIT Press (2016)
15. Tilouche, S., Basseto, S., Nia, V.P.: Classification algorithms for virtual metrology. In: Proceedings of the 2014 IEEE International Conference on Management of Innovation and Technology, Singapore, pp. 495–499 (2014)

16. Vidal, R., Yi, Ma., Sastry, S.S.: Generalized Principal Component Analysis. Springer-Verlag, New York (2016)
17. Borg, I., Groenen, P.: Modern Multidimensional Scaling: theory and Applications, 2nd edn. Springer, New York, NY (2005)
18. Van der Maaten, L., Hinton, G.: Visualizing data using t-SNE. J. Mach. Learn. Res. **9**, 2579–2605 (2008)
19. Machine Learning in Python. Decomposing signals in components. https://scikit-learn.org/stable/modules/decomposition.html#decompositions. Accessed 10 Nov 2018
20. Zeiler, M.D., Fergus, R.: Visualizing and understanding convolutional networks, computer vision—ECCV 2014. Lect. Notes Comp. Sci. **8689**, 818–833 (2014)
21. Intel Nervana Neural Network Processors. https://www.intel.ai/nervana-nnp/#gs.6nbk85. Accessed 19 Feb 2019

Modeling Cyber-Physical System Object in State Space (on the Example of Paver)

Andrey Prokopev, Zhasurbek Nabizhanov, Vladimir Ivanchura and Rurik Emelyanov

Abstract We has considered results of theoretical description of the cyber-physical system object's model—asphalt paver with a compacting working body of increased efficiency based on the state space method are considered. The working body includes a tamper, screed and pressure bar. The mathematical model of the process of interaction of the object with the compacting road-building material takes into account the masses of the main structural elements of the working body and pavement. A rheological model of a viscoelastic Kelvin–Voight body is using to describe the compacted material. Suitability of developed mathematical model experimentally confirmed by simulation modelling of the system using program MATLAB/Simulink.

Keywords Cyber-physical system · Paver · Working body · Process of compaction · Road construction mixture · Rheological model · State space

1 Introduction

Innovative advances in the information technologies field predetermine the extension of cyber-physical system (CPS) to the most directions of human activities, which are formed at the interface of the internet, things and services [1]. Internet of Things technology is the technological base of the cyber-physical system [2]. Examples of modern scientific and technical development are cyber-physical system road construction's control, intelligent construction, intelligent compaction (IC)

A. Prokopev (✉) · Z. Nabizhanov · V. Ivanchura · R. Emelyanov
Siberian Federal University, 79 Svobodny, Krasnoyarsk 660041, Russia
e-mail: prok1@yandex.ru

Z. Nabizhanov
e-mail: jasur150691@yandex.ru

V. Ivanchura
e-mail: ivan43ura@yandex.ru

R. Emelyanov
e-mail: ert-44@yandex.ru

A. G. Kravets et al. (eds.), *Cyber-Physical Systems: Advances in Design & Modelling*, Studies in Systems, Decision and Control 259,
https://doi.org/10.1007/978-3-030-32579-4_25

[3, 4], continuous compaction control (CCC) [5, 6], intelligent neural network system of automated control of vibratory roller [7, 8], unmanned vibration rollers [9, 10], automated road construction kit "paver–vibratory rollers" [11]. Improving the CPS objects—vibratory rollers, asphalt pavers, is an important scientific direction [12–15].

As it is known [16], cybernetic physics investigates physical systems by cybernetic methods. The design task of GPS objects begins with choosing models of control object and control's aim. Input and output of the system are given in cybernetic models, because of the essential role in the formation of feedback coupling [16]. GPS project effectiveness depends on the quality of the simulated model of processes [16]. Thus, simulated models of GPS objects, which based on modern software, have special meaning. For example, software package MATLAB/Simulink makes the possible implementation of technology of model-based design, and the creation of the hybrid model consisted of the computer model and real physical objects.

The article is concerned with the theoretical description of the cybernetic system object—asphalt paver's with a working body of increased efficiency. The result of the working process implementation paver is preliminarily compacted smooth asphalt surface with a specific profile. Theoretical investigations of compaction of different road construction materials (ground coat and asphalt mixes) are considered and provided by many Russian [17] and foreign scientists [18–23]. Limitations of existing mathematical models of an investigated process are computational problems in the design of control systems, in the investigation of the dynamic model, to which the considered model is related.

It is recommended that the state space method should be used in order to fix those limitations and efficiency enhancement of theoretical investigation of control object based on modern software MATLAB, this method allows realize clear formalization and automation of computational procedure [24]. Description of systems in state space allows finding and investigating such features, which could be hidden in the case of using classical methods for frequency response analysis and description of terms "input–output". Matrix notation, which is used in the state space method, has the advantage at the numerical solution, moreover explicitness of mathematical formulation and solutions itself are not getting worse even for MIMO-systems (Multiple Inputs Multiple Outputs), which describe the behavior of complex production package [25].

2 The Mathematical Formulation of the Problem

Several flavors of paver working bodies are known [26]. The working body of improved efficiency, which consists of such compacting devices as tamper—finishing plate—pressure bar, is taken for investigation. The process of mixture compaction carried out with continuous contact of plate and mixture. Mixture compaction is characterized by distortion of a mix by tamper with kinematic gear (primary compaction is achieved in 4–6 actions), by finishing and pressure bar.

Finishing plate ensures the improvement of road surface structure and fixing realized degree of surface compaction. Being static finishing plate make vibratory actions using tight coupling with a tamper. The parametric value of vibrations depends on tamper weight and vibrational frequency. Pressing plate creates periodic impulse loading with a frequency of 50–70 Hz and pressure in hydraulic system loading of 5–15 Pa [26].

Frequency, amplitude, velocity, acceleration are primary dynamic parameters of vibrations of compacting devices and compacted areas particles. Vibration amplitude of any compactor depends on stress-strain properties of compacted material and can be changed in the process of its compaction [27].

Following assumptions were made during the preparation of the mathematical model of the compaction process:

1. structural elements of the machine have infinite stiffness;
2. working body operates in shock-free mode;
3. compacted layer possesses viscoelastic properties;
4. viscoelastic properties of vibration damper are linear;
5. vertical component of vibration is considered only;
6. inertial properties of the compacted medium are taken into account.

The scheme of the dynamic model of the mixture compaction process by paver working body is in Fig. 1.

On the scheme, Fig. 1, the following notations are used: m_1—finishing plate weight, kg; m_2—pressure bar weight, kg; m_3—tamper weight, kg; m_4—mixture weight under tamper, kg; m_5—mixture weight under vibrating plate, kg; k_5—coefficient of elastic resistance of compacted mixture under plate, N/m; c_1—damping coefficient of compacted mixture under plate, Ns/m; k_2—coefficient of elastic resistance of vibrator buffer, N/m; c_2—damping coefficient of vibrator buffer, Ns/m; k_3—coefficient of elastic resistance of compacted mixture under tamping beam,

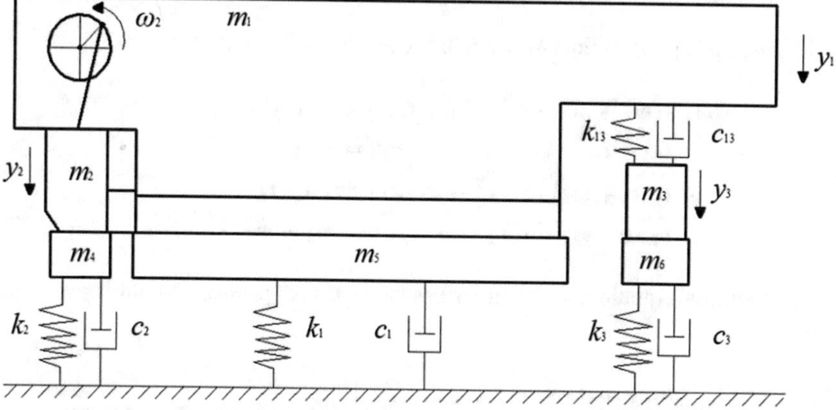

Fig. 1 The dynamic model of mixture compaction process by paver working body (tamper, finishing plate, one pressure bar)

N/m; c_3—damping coefficient of compacted mixture under tamping beam, Ns/m; y_1, y_2, y_3—movement of working body elements, consequently, m.

As follows from the analysis of mixture compaction process by paver working body (Fig. 1), on the ground of Newton's second law, the mathematical model of vibrating system "tamper—finishing plate—pressure bar—mixture" was made, which shows simultaneously both vibration dynamics of constructional element and rheological properties of mixture compacted.

Differential equations of tamper motion

$$(m_2 + m_4) \cdot \ddot{y}_2 + c_2 \cdot \dot{y}_2 + k_2 \cdot y_2 = F_2 + m_4 \cdot g, \tag{1}$$

where, F_2—the power of tamper pusher, N.

Taking into account the relative motion concept, we will receive an additional equation

$$y_2 = y_1 + e \cdot \sin(\omega_2 \cdot t), \tag{2}$$

where e—excentricity radius of tamper, m; ω_2—angular frequency of revolution of tamper gear, rad/s.

Putting up Eq. (2) to Eq. (1) with transformations, we will receive the following equation

$$\begin{aligned} F_2 = {} & (m_2 + m_4) \cdot \ddot{y}_1 + c_2 \cdot \dot{y}_1 + k_2 \cdot y_1 - (m_2 + m_4) \cdot e \cdot \omega_2^2 \cdot \sin(\omega_2 \cdot t) \\ & + k_2 \cdot e \cdot \sin(\omega_2 \cdot t) + c_2 \cdot e \cdot \omega_2 \cdot \sin(\omega_2 \cdot t + \pi/2) - m_4 \cdot g. \end{aligned} \tag{3}$$

Differential equations of motion of finishing plate

$$\begin{aligned} & (m_1 + m_5) \cdot \ddot{y}_1 + (c_1 + c_{13}) \cdot \dot{y}_1 - c_{13} \cdot \dot{y}_3 - k_1 \cdot y_1 - k_{13} \cdot y_3 \\ & = -F_3 - F_2 + (m_1 + m_5) \cdot g. \end{aligned} \tag{4}$$

Putting up Eq. (3) to Eq. (4), we will receive the following equation

$$\begin{aligned} & (m_1 + m_2 + m_4 + m_5) \cdot \ddot{y}_1 + (c_1 + c_2 + c_{13}) \cdot \dot{y}_1 - c_{13} \cdot \dot{y}_3 \\ & + (k_1 + k_2 + k_{13}) \cdot y_1 - k_{13} \cdot y_3 = -F_3 \\ & + ((m_2 + m_4) \cdot e \cdot \omega_2^2 - k_2 \cdot e) \cdot \sin(\omega_2 \cdot t) \\ & - c_2 \cdot e \cdot \omega_2 \cdot \sin(\omega_2 \cdot t + \pi/2) + (m_1 + m_4 + m_5) \cdot g. \end{aligned} \tag{5}$$

A differential equation, which describes the motion of pressure bar during mixture compaction

$$\begin{aligned} & (m_3 + m_6) \cdot \ddot{y}_3 - c_{13} \cdot \dot{y}_1 + (c_3 + c_{13}) \cdot \dot{y}_3 \\ & - k_{13} \cdot y_1 + (k_3 + k_{13}) \cdot y_3 = F_3 + (m_3 + m_6) \cdot g, \end{aligned} \tag{6}$$

where F_3—the power of hydraulic exciter affecting pressure bur, N.

The mathematical model of investigation object was found during transformations

$$\begin{cases} (m_1 + m_2 + m_4 + m_5) \cdot \ddot{y}_1 + (c_1 + c_2 + c_{13}) \cdot \dot{y}_1 - c_{13} \cdot \dot{y}_3 \\ \quad +(k_1 + k_2 + k_{13}) \cdot y_1 - k_{13} \cdot y_3 = -F_3 \\ \quad +((m_2 + m_4) \cdot e \cdot \omega_2^2 - k_2 \cdot e) \cdot \sin(\omega_2 \cdot t) \\ \quad -c_2 \cdot e \cdot \omega_2 \cdot \sin(\omega_2 \cdot t + \pi/2) + (m_1 + m_4 + m_5) \cdot g; \\ (m_3 + m_6) \cdot \ddot{y}_3 - c_{13} \cdot \dot{y}_1 + (c_3 + c_{13}) \cdot \dot{y}_3 - k_{13} \cdot y_1 \\ \quad +(k_3 + k_{13}) \cdot y_3 = F_3 + (m_3 + m_6) \cdot g. \end{cases} \tag{7}$$

3 The Mathematical Model in the State Space

The state-space method allows to present control system (7) as follows [24, 27]:

$$\dot{x}(t) = A(t) \cdot x(t) + B(t) \cdot u(t); \tag{8}$$

$$y(t) = C(t) \cdot x(t) + D(t) \cdot u(t), \tag{9}$$

where, $x(t)$—state vector of dimension $(n \times 1)$, which elements are state variables of systems of nth-order, $x(t) = [x_1(t), x_2(t), \ldots, x_n(t)]^T$; $y(t)$—output vector of dimension $(p \times 1)$, which elements are response variables of the system, $y(t) = \left[y_1(t), y_2(t), \ldots, y_p(t)\right]^T$; $u(t)$—input vector of dimension $(r \times 1)$, which elements are input variables of the system, $u(t) = [u_1(t), u_2(t), \ldots, u_r(t)]^T$; $A(t)$—the coefficient matrix of the system $(n \times n)$; $B(t)$—input matrix $(n \times m)$; $C(t)$—output matrix $(p \times n)$, here p—output value number; $D(t)$—detour matrix $(p \times m)$, which define direct dependence of output on input.

For further transformations, the equation system was led to the following form

$$\ddot{y}_1 = \frac{\begin{bmatrix} -(c_1 + c_2 + c_{13}) \cdot \dot{y}_1 + c_{13} \cdot \dot{y}_3 + (k_1 + k_2 + k_{13}) \cdot y_1 \\ +k_{13} \cdot y_3 - F_3 + ((m_2 + m_4) \cdot e \cdot \omega_2^2 - k_2 \cdot e) \cdot \sin(\omega_2 \cdot t) \\ +c_2 \cdot e \cdot \omega_2 \cdot \sin(\omega_2 \cdot t + \pi/2) + (m_1 + m_4 + m_5) \cdot g \end{bmatrix}}{m_1 + m_2 + m_4 + m_5};$$

$$\ddot{y}_3 = \frac{\begin{bmatrix} c_{13} \cdot \dot{y}_1 - (c_3 + c_{13}) \cdot \dot{y}_3 + k_{13} \cdot y_1 - (k_3 + k_{13}) \cdot y_3 \\ +F_3 + (m_3 + m_6) \cdot g \end{bmatrix}}{m_3 + m_6}. \tag{10}$$

The movement and speed of the compacting devices uniquely determine the state of the object. We introduce the designations of the state variables of the object: x_1—vertical movement of finishing plate, $x_1 = y_1$; x_2—the velocity of vertical movement of finishing plate, $x_2 = \dot{y}_1$; x_3—the vertical movement of the pressure bar, $x_3 = y_3$; x_4—the velocity of vertical movement of the pressure bar, $x_4 = \dot{y}_3$.

To simplify the study of the object model, the system of Eq. (10), taking into account the accepted designations, is reduced to a system of differential equations in the normal form of Cauchy

$$
\begin{aligned}
x_2 &= \dot{y}_1;\\
\dot{x}_2 &= \frac{\begin{bmatrix} -(c_1 + c_2 + c_{13}) \cdot x_2 + c_{13} \cdot x_4 + (k_1 + k_2 + k_{13}) \cdot x_1 \\ +k_{13} \cdot x_3 - F_3 \\ +((m_2 + m_4) \cdot e \cdot \omega_2^2 - k_2 \cdot e) \cdot \sin(\omega_2 \cdot t) \\ +c_2 \cdot e \cdot \omega_2 \cdot \sin(\omega_2 \cdot t + \pi/2) + (m_1 + m_4 + m_5) \cdot g \end{bmatrix}}{m_1 + m_2 + m_4 + m_5};\\
x_4 &= \dot{y}_3;\\
\dot{x}_4 &= \frac{\begin{bmatrix} c_{13} \cdot x_2 - (c_3 + c_{13}) \cdot x_4 + k_{13} \cdot x_1 - (k_3 + k_{13}) \cdot x_3 \\ +F_3 + (m_3 + m_6) \cdot g \end{bmatrix}}{m_3 + m_6}.
\end{aligned}
\tag{11}
$$

Moving state variables to the appropriate vector and matrix produces the following results. Model of the process under study in the state space, in vector-matrix form

$$\dot{x}(t) = A(t) \cdot x(t) + B(t) \cdot u(t);$$

the coefficient matrix of the system $A(t)$

$$A(t) = \begin{bmatrix} 0 & 1 & 0 & 0 \\ -\frac{k_1+k_2+k_{13}}{K} & -\frac{c_1+c_2+c_{13}}{K} & \frac{k_{13}}{K} & \frac{c_{13}}{K} \\ 0 & 0 & 0 & 1 \\ \frac{k_{13}}{m_3+m_6} & \frac{c_{13}}{m_3+m_6} & -\frac{k_3+k_{13}}{m_3+m_6} & -\frac{c_3+c_{13}}{m_3+m_6} \end{bmatrix};$$

here $K = m_1 + m_2 + m_4 + m_5$; input matrix $B(t)$

$$B(t) = \begin{bmatrix} 0 & 0 \\ \frac{1}{m_1+m_2+m_4+m_5} & 0 \\ 0 & 0 \\ 0 & \frac{1}{m_3+m_6} \end{bmatrix};$$

$$y(t) = C(t) \cdot x(t) + D(t) \cdot u(t);$$

output matrix $C(t)$

$$C(t) = \begin{bmatrix} 1\,0\,0\,0 \\ 0\,1\,0\,0 \\ 0\,0\,1\,0 \\ 0\,0\,0\,1 \end{bmatrix};$$

input vector $u(t)$

$$u(t) = \begin{bmatrix} -F_3 + ((m_2 + m_4) \cdot e \cdot \omega_2^2 - k_2 \cdot e) \cdot \sin(\omega_2 \cdot t) \\ -c_2 \cdot e \cdot \omega_2 \cdot \sin(\omega_2 \cdot t + \pi/2) + (m_1 + m_4 + m_5) \cdot g \\ F_3 + (m_3 + m_6) \cdot g \end{bmatrix}.$$

Matrix $D(t)$ usually equal to zero, as in physical systems, all channels between inputs and outputs typically have dynamic links [28].

4 Investigation of the Mathematical Model

The simulated model was developed on the language MATLAB/Simulink, Fig. 2, to estimate mathematical model adequacy of mixture compaction process by paver working body.

For the computer experiment of the investigated process, the initial data from the scientific publication are used [27]

$k_1 = 4 \times 10^6\,\text{N/m}$; $k_2 = 8.5 \times 10^5\,\text{N/m}$; $k_3 = 8.5 \times 10^6\,\text{N/m}$;
$k_{13} = 1.1 \times 10^7\,\text{N/m}$; $c_1 = 3200\,\text{N} \cdot \text{s/m}$; $c_2 = 1200\,\text{N} \cdot \text{s/m}$; $c_3 = 1200\,\text{N} \cdot \text{s/m}$;
$c_{13} = 1200\,\text{N} \cdot \text{s/m}$; $m = 21.6\,\text{kg}$; $m_1 = 682\,\text{kg}$; $m_2 = 71.3\,\text{kg}$; $m_3 = 250\,\text{kg}$;

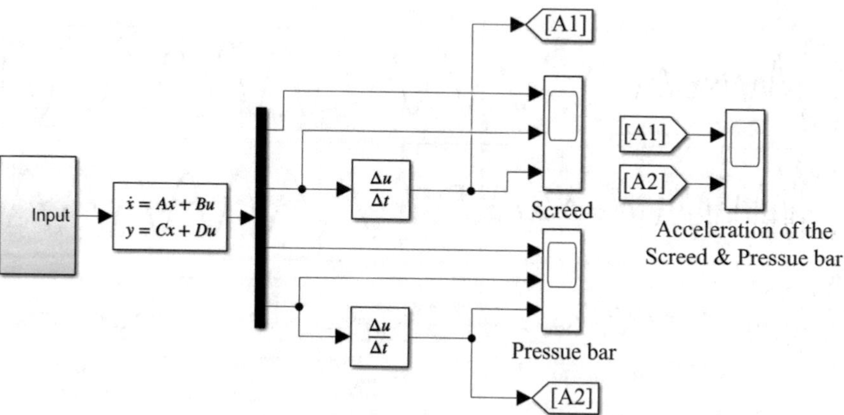

Fig. 2 Simulated model on the language MATLAB/Simulink

$m_4 = 0.1 \cdot m_2;\ m_5 = 0.2 \cdot m_1;\ m_6 = 0.2 \cdot m_3;\ r = 0.03\,\text{m};\ e = 0.006\,\text{m};$
$f_2 = 15\,\text{Hz};\ f_3 = 25\,\text{Hz};\ F_3 = 9000\,\text{N}.$

As a result of computer modeling, the parameters of the working process of the finishing plate and the pressure bar are obtained when compacting the mixture: displacement; velocity; acceleration (Fig. 3, Fig. 4).

The obtained dependences of the transition process correspond to the nature of the vibrating process. The results have similar values to those published in the article [27]. The acceleration amplitude of the finishing plate corresponds to the real data, taking into account the given conditions of the working process. These acceleration amplitudes of the finishing plate, after applying the spectral analysis method, can be used for continuous compaction control.

Transfer-function identification of control object.

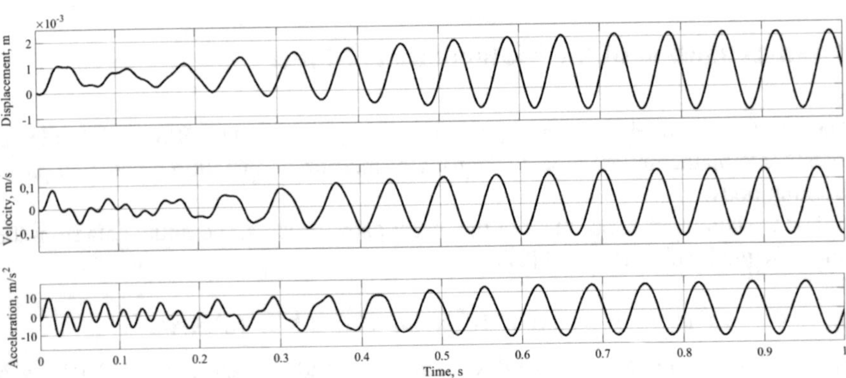

Fig. 3 The characteristic curve of vibrating process parameters of paver finishing plate

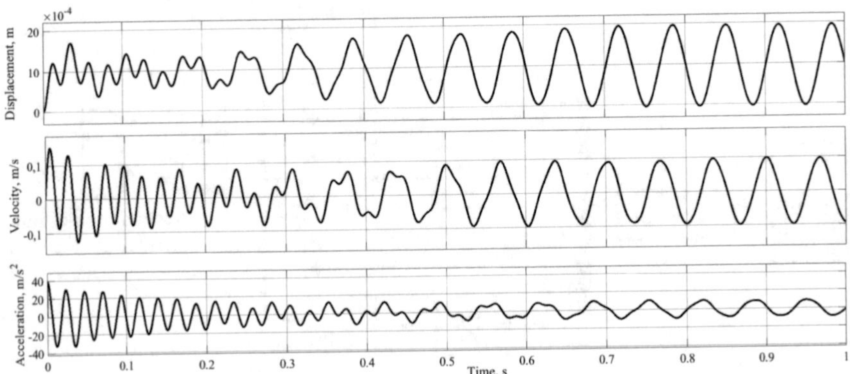

Fig. 4 The characteristic curve of vibrating process parameters of the pressing plate

The task of determining the transfer function (TF) is performed for the research of the mathematical model with the use of block modeling software. As a result of MATLAB commands implementation, we received the transfer function of the input with the controlled variable is the angular frequency of vibration.

The transfer function of the controlled variable—a movement of finishing plate

$$W(s) = \frac{0.001115 \cdot s^2 + 0.00892 \cdot s + 72.48}{s^4 + 14.24 \cdot s^3 + 8.272 \cdot 10^4 \cdot s^2 + 4.491 \cdot 10^5 \cdot s + 6.99 \cdot 10^8};$$

the transfer function of the controlled variable—velocity of the finishing plate

$$W(s) = \frac{0.001115 \cdot s^3 + 0.00892 \cdot s^2 + 72.48 \cdot s}{s^4 + 14.24 \cdot s^3 + 8.272 \times 10^4 \cdot s^2 + 4.491 \times 10^5 \cdot s + 6.99 \times 10^8}.$$

The obtained TF correspond strictly to the correct form since the degree of the numerator is less than the degree of the denominator. In further investigations when designing the controller, it is necessary to take into account the high order of the mathematical model of the control object.

5 Conclusion

A mathematical model of the object of the cyber-physical system in the state space – asphalt paver with a working body of increased efficiency is obtained. An experimental check of its operability is executed. The models of the transfer functions at the input with the controlled variable are determined. Transfer functions have a fourth-order, which complicates the task of designing the control system regulator.

The results of the work are the stage of research work in the field of design of the cyber-physical system of road construction works.

References

1. Anderl, R., Eigner, M., Sendler, U., Stark, R.: Smart Engineering—Interdisziplinäre Produktentstehung. acatech Diskussion. Springer, 58 p. (2012). https://doi.org/10.1007/978-3-642-29372-6
2. Serpanos, D., Wolf, M.: Internet-of-Things (IoT) Systems. Architectures, Algorithms, Methodologies. Springer, Cham, 95 p. (2018). https://doi.org/10.1007/978-3-319-69715-4_1
3. Xu, Q., Chang, G.K.: Adaptive quality control and acceptance of pavement material density for intelligent road construction. Autom. Constr. **62**, 78–88 (2015). https://doi.org/10.1016/j.autcon.2015.11.004
4. Chang, G.K., Mohanraj, K., Stone, W.A., Oesch, D.J., Gallivan, V.: Leveraging intelligent compaction and thermal profiling technologies to improve asphalt pavement construction quality: A case study. Trans. Res. Rec. J. Trans. Res. Board **2672**(26), 48–56 (2018). https://doi.org/10.1177/0361198118758285

5. Pistrol, J., Villwock, S., Völkel, W., Kopf, F., Adam, D.: Continuous compaction control (CCC) with oscillating rollers. Procedia Eng. **143**, 514–521 (2016). https://doi.org/10.1016/j.proeng.2016.06.065
6. Hu, W., Shu, X., Huang, B., Woods, M.: Field investigation of intelligent compaction for hot mix asphalt resurfacing. Front. Struct. Civ. Eng. **11**(1), 47–55 (2017). https://doi.org/10.1007/s11709-016-0362-x
7. Barman, M., Nazari, M., Imran, S.A., Commuri, S., Zaman, M., Beainy, F., Singh, D.: Quality control of subgrade soil using intelligent compaction. Innovative Infrastruct. Solutions **1**(1), 23 (2016). https://doi.org/10.1007/s41062-016-0020-0
8. Barman, M., Imran, S.A., Nazari, M., Commuri, S., Zaman, M.: Use of intelligent compaction in detecting and remediating under-compacted spots during compaction of asphalt layers. In: Hossain Z., Zhang J., Chen C. (eds.) Solving Pavement and Construction Materials Problems with Innovative and Cutting-edge Technologies. GeoChina 2018. Sustainable Civil Infrastructures, pp. 131–141. Springer, Cham (2019). https://doi.org/10.1007/978-3-319-95792-0_11
9. Fang, X., Bian, Y., Yang, M., Liu, G.: Development of a path following control model for an unmanned vibratory roller in vibration compaction. Adv. Mech. Eng. **10**(5), 1–16 (2018). https://doi.org/10.1177/1687814018773660
10. Bian, Y., Fang, X., Yang, M., Zhao, Z.: Automatic rolling control for unmanned vibratory roller based on fuzzy algorithm. J. Tongji Univ. (Nat. Sci.) **45**(12), 1831–1838 (2017). https://doi.org/10.11908/j.issn.0253-374x.2017.12.013
11. Zhu, S., Li, X., Wang, H., Yu, D.: Development of an automated remote asphalt paving quality control system. Transp. Res. Rec. **2672**(26), 28–39 (2018). https://doi.org/10.1177/0361198118758690
12. Liu, D.H., Li, Z.L., Lian, Z.H.: Compaction quality assessment of earth-rock dam materials using roller integrated compaction monitoring technology. Automat. Constr. **44**, 234–246 (2014). https://doi.org/10.1016/j.autcon.2014.04.016
13. Kenneally, B., Musimbi, O.M., Wang, J.: Finite element analysis of vibratory roller response on layered soil systems. Comput. Geotech. **67**, 73–82 (2015). https://doi.org/10.1016/j.compgeo.2015.02.015
14. Li, J., Zhang, Z., Xu, H.: Dynamic characteristics of the vibratory roller test-bed vibration isolation system: simulation and experiment. J. Terramech. **56**, 139–156 (2014). https://doi.org/10.1016/j.jterra.2014.10.002
15. Xu, Q., Chang, G.K.: Adaptive quality control and acceptance of pavement material density for intelligent road construction. Automat. Constr. **62**, 78–88 (2016). https://doi.org/10.1016/j.autcon.2015.11.004
16. Kyung-Joon, P., Zheng, R., Liu, X.: Cyber-physical systems: milestones and research challenges. Comput. Commun. **36**, 1–7 (2012). https://doi.org/10.1016/j.comcom.2012.09.006
17. Mikheyev, V.V., Saveliev, S.V.: Modelling of deformation process for the layer of elastoviscoplastic media under surface action of periodic force of arbitrary type. J. Phys. Conf. Ser. **944**(1), 012079 (2018). https://doi.org/10.1088/1742-6596/944/1/012079
18. Rinehart, R.V.: Instrumentation of a roller compactor to monitor vibration behavior during earthwork compaction. J. Autom. Constr. **17**(2), 144–150 (2008)
19. White, D., Thompson, M.: Relationships between in situ and roller-integrated compaction measurements for granular soils. J. Geotech. Geoenviron. Eng. **134**(12), 1763–1770 (2008). https://doi.org/10.1061/(ASCE)1090-0241(2008)134:12(1763)
20. Bejan, S.: The roller-ground dynamic interaction in the compaction process through vibrations for road construction. Rom. J. Trans. Infrastruct. **5**(2), 1–9 (2016). https://doi.org/10.1515/rjti-2016-0044
21. Beainy, F., Commuri, S., Zaman, M.: Dynamical response of vibratory rollers during the compaction of asphalt pavements. J. Eng. Mech. **140**(7), 04014039 (2014). https://doi.org/10.1061/(asce)em.1943-7889.0000730
22. Imran, S.A., Commuri, S., Barman, M., Zaman, M., Beainy, F.: Modeling the dynamics of asphalt-roller interaction during compaction. J. Constr. Eng. Manag. **143**(7), 1763–1770 (2017). https://doi.org/10.1061/(ASCE)CO.1943-7862.0001293

23. Li, S., Hu, C.: Study on dynamic model of vibratory roller-soil system. IOP Conf. Ser. Earth Environ. Sci. **113**, 012187 (2018). https://doi.org/10.1088/1755-1315/113/1/012187
24. Derusso, P.M., Roy, R.J., Close, Ch.M.: State Variables for Engineers, 608 p. John Wiley & Sons, New York (1965)
25. Strejc, V.: State Space Theory of Discrete Linear Control, 426 p. John Wiley & Sons (1981)
26. WIRTGEN GROUP: Concentrating on the essentials: high quality paving. RoadNews **7**, 28–47 (2019). https://media.voegele.info/media/03_voegele/aktuelles_und_presse/roadnews_magazin/roadnews_07/RadNews_07__en.pdf
27. Sun, J., Xu, G.: Dynamics modeling and analysis of paver screed based on computer simulation. J. Appl. Sci. **13**(7), 1059–1065 (2013). https://doi.org/10.3923/jas.2013.1059.1065
28. Phillips, C.L., Harbor R.D.: Feedback Control Systems, 784 p. Pearson (2010)

Accelerometer Data Based Cyber-Physical System for Training Intensity Estimation

Igor D. Kazakov, Nataliya L. Shcherbakova, Adriaan Brebels and Maxim V. Shcherbakov

Abstract The correctness of athlete's behavior can be controlled by health care cyber-physical system containing distributed (mobile) sensors and intelligent data processing. Such cyber-physical systems determine a concrete set of events, such as jumps or falls and identify events parameters, e.g. height and duration. The proposed accelerometer data based cyber-physical system differs from existed ones by an original method for detection of various types of athlete's behavior. A proposed cyber-physical system contains on three modules: the data acquisition module, the data processing module and the processed data visualization module. A method for jump recognition is based on high frequency accelerometer data. The system is developed using Android Studio, R Studio development environments. The results provided by accelerometer data based cyber-physical system might be used for coaches and doctor in sports medicine for decisions regarding the optimal load in future training sessions. Use cases including different experimental setup shows the efficiency of the proposed system.

Keywords Cyber-physical systems · Accelerometer data · Health-care system · Jump detection

I. D. Kazakov · N. L. Shcherbakova · M. V. Shcherbakov (✉)
Volgograd State Technical University, Volgograd 400005, Russia
e-mail: maxim.shcherbakov@gmail.com

I. D. Kazakov
e-mail: igorkazakov1997@gmail.com

N. L. Shcherbakova
e-mail: natalia.shchrbakova@gmail.com

A. Brebels
Katholieke Universiteit Leuven, Geel 2440, Belgium
e-mail: adriaan.brebels@portacapena.com

A. G. Kravets et al. (eds.), *Cyber-Physical Systems: Advances in Design & Modelling*, Studies in Systems, Decision and Control 259,
https://doi.org/10.1007/978-3-030-32579-4_26

1 Introduction

Basically, a cyber-physical system (CPS) created for improvement production in different domains via combination of physical and informational elements. The CPS contains on computing elements for interacting sensors which monitor cyber and physical components and actuators in operational environment. CPSs use sensors to collect data about environment to gain a deeper knowledge of the environment, which enables a more accurate actuation [1].

Recently, the usage of CPS is dramatically increasing in different domains. For example, more health care systems are created as CPS to get more benefits for an end-user. These systems can improve remote monitoring of physical parameters of patients in real time to reduce the need for hospitalization or to improve care for the disabled and the elderly. Also CPS systems are used to study the functions of the human body, improve the exchange of real-time information between equipment, control systems and improve the efficiency of these processes due to the automatic monitoring and control of the entire process [2–6].

Sports achievements and health of people involved in sports directly depend on the improvement of training methods and the development of knowledge about the physical abilities of a person. An analysis of the biomechanical structure of the athlete's motor actions will enable further progress in its technical and physical training.

The contribution of the chapter is an accelerometer data based cyber-physical system which differs from existed ones by an original method for detection of various types of athlete's behavior. The system is developed using Android Studio, R Studio development environments for high frequency data analysis.

2 Background

Modern trends in the development and use of cyber-physical health monitoring systems have been investigated in many chapters [7–11].

The biomechanical structure of human movements when playing sports in three-dimensional space and the asymmetry of movements is considered in [12].

On this basis the method of assessing the degree of tension and non-relaxation of skeletal muscles, which allows you to classify the basic and special exercises of athletes on the speed-strength parameters, was proposed.

To determine the characteristics of the device in space, in most cases the accelerometer, compass and gyroscope are used, often installed on modern devices. Compass and gyroscope do not allow to detect movement, but provide an opportunity to determine the orientation and tilt of the device in space. Changing the position of the device is carried out using an accelerometer [13, 14].

The works also confirm the possibility of assessing and further analyzing human physical activity using data obtained from an accelerometer [15–17].

The available devices that collect data on the human physical activity in space are primarily smartphones on the Android OS [18]. Android OS applications, such as GPS-based pilots for GPS pilots [19], barometer-based SyPressure Pro (Barometer) [20], Altimeter–Altimeter [21], and altimeters of varying degrees of accuracy [22], allow you to determine the height of the device seas, but do it with insufficient accuracy (>10 cm). Further, additional information processing is required to obtain the height of the jump.

Devices such as a WOO-tracker [23] or Vert [24] vertical jump gauge allow you to determine the jump height, but do not count the number of jumps and have a high cost, being separate devices.

3 A Cyber-Physical System

The cyber-physical system processes data obtained from different sources and based on existing approaches [25, 26] it is advisable to distinguish three modules: the data acquisition module, the data processing module and the processed data visualization module.

Human interaction with the system takes place with the modules of data acquisition and visualization of the processed data.

The data acquisition module saves the structured data from the accelerometer of the phone to a file. When receiving data from the accelerometer, a string is formed that contains the following data, separated by the symbol '!':

- acceleration along the axis OX
- acceleration along the OY axis
- acceleration along the OZ axis
- Timestamp

Example: "9.811235! 0.0! 0.0! 12"

The data processing module is developed in the R programming language. The module is divided into two parts:

- jump definition
- jump height determination

The output of this module is a file containing the height of each jump and the timestamp of the landing.

The data visualization module receives a file from the data processing module as input and builds a histogram of heights from the data from the file.

Data acquisition from the accelerometer and visualization of the processed data is used to receive data for further processing, with which the user has direct contact (installed on the athlete's mobile device running the Android OS). Also, this program receives the processed data from the cloud storage and presents them in the form of a histogram. Data processing is performed by the program on the cloud storage. The processing program is implemented in the programming language R. Figure 1 shows a cyber-physical system scenario for assessing the intensity of training based on accelerometer data.

Pseudocode jump recognition:

1. Download the file "FILE" with the original data;
2. Convert "FILE" to table "df" with four fields: "X", "Y", "Z", "Timestamp";
3. Define the vector "LIST":
   - 3.1. For each vector "X", "Y", "Z":
     - 3.1.1. Find the mean value of the vector;
     - 3.1.2. If the mean value of the vector is 9.81 ± 2:
       - 3.1.2.1. Assign this vector to the "LIST" vector;
4. Define the vector "A":
   - 4.1. For each value in the "LIST" vector:
     - 4.1.1. If the value is 9.81 ± 0.5:
       - 4.1.1.1. Add a value to the vector "A";
5. Determine the vector "MAX" containing the indices of all peaks of the vector "LIST":
   - 5.1. Create a vector "sign";
   - 5.2. For each "i" value of the vector "LIST":
     - 5.2.1. Find the difference sign "i + 1" and "i" values of the vector "LIST";
     - 5.2.2. Add the result to the sign vector;
     - 5.2.3. Increase the value of "i" by 2;
   - 5.3. For each "i" value of the vector "sign":
     - 5.3.1. If "i" value is greater than "i + 1"
       - 5.3.1.1. Add the value of the vector "LIST" with the index "i" to the vector "MAX";
     - 5.3.2. Increase "i" by 2;
6. Create a table "JUMPS", consisting of two fields: "start", "end";
7. Define the "JUMPS" table:
   - 7.1. For each "i" value of the vector "MAX":
     - 7.1.1. The value with the index "i" is added to the "start" field of the table "JUMPS";
     - 7.1.2. The value with the index "i + 1" is added to the "end" field of the table "JUMPS";
     - 7.1.3. Increase the value of "i" by 2;
8. For each row of the table "JUMPS" calculate the height of the jump:
   - 8.1. Assign "START" the index of the first value from the "LIST" vector nearest to the acceleration of free fall to the right of the "start" of the "JUMPS" table;

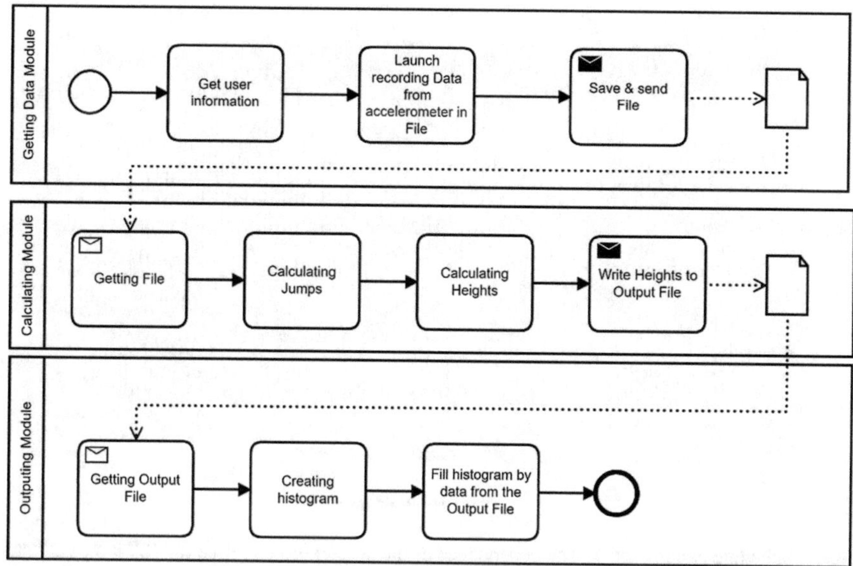

Fig. 1 A cyber-physical system scenario for assessing the intensity of training based on accelerometer data

8.2. Assign the "END" index of the first value from the "LIST" vector nearest to the acceleration of free fall to the right of the "end" of the "JUMPS" table;
8.3. Calculate the time "time" from "START" to "END";
8.4. Calculate the average acceleration "a" from "START" to "END";
8.5. Calculate the height "h" by applying the formula "h" = "a" * "t" ^ 2/2;
8.6. Add jump height to the "HEIGHTS" list;

9. Create a new "OUTPUT" file;
10. Write the vector values "HEIGHTS" to the file.

4 Results and Discussion

To confirm the effectiveness of the system, the following series of experiments were carried out.

Experiment 1. Identification of the jump by counting the acceleration maxima (Fig. 2).

Hypothesis. During the jump, the athlete's body experiences two overloads. The first—at the time of acceleration of the body for separation from the floor. The second—at the time of landing and stabilization. Thus, the number of jumps can be

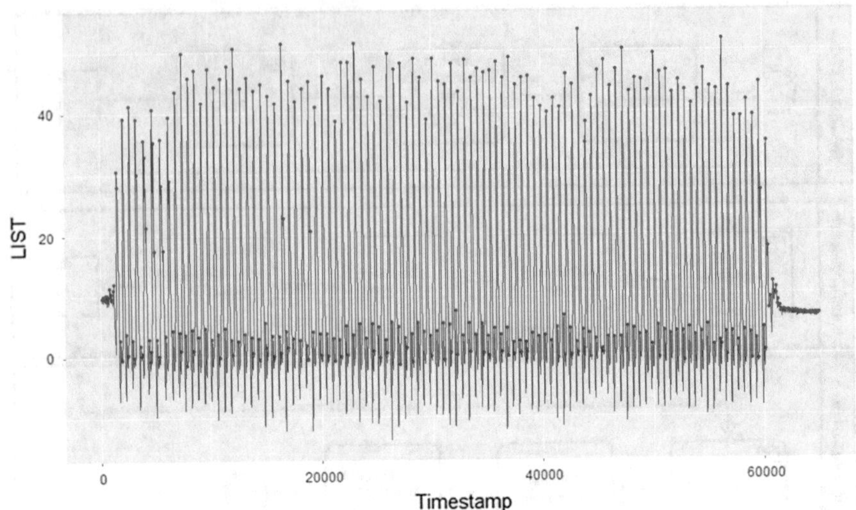

Fig. 2 Schedule experiment 1. The dependence of the acceleration on time for 100 jumps athlete, where x-axe represents timestamps and y-axe represents acceleration

calculated by the following formula N/2, where N is the number of maxima along the entire acceleration graph.

Input data. File with 100 jumps athlete.

Result. 384 local maxima were found. Using the hypothesis of the first experiment we get 192 jumps.

Conclusion. In the calculation of the maxima fall noise maxima that need to be cut off.

Experiment 2. Removing noise to determine maxima (Fig. 3).

Hypothesis. The average acceleration of the processed data vector is equal to the free fall acceleration rounded to hundredths. The maxima that fall within the segment (9.81 $\pm$ 3) m/s^2 will be considered noise and not taken into account in the calculations.

Input data. Acceleration data file 100 athlete jumps.

Result. 384 local maxima were found. Applying the hypotheses of the first and second experiments, we obtain 292 local maxima and 146 jumps.

Conclusion. From the acceleration graph, it is clear that some maxima were found incorrectly. It is necessary to expand the search range of the maximum.

Experiment 3. Increasing the time range of the search maxima (Figs. 4, 5).

Hypothesis. The program code of the maximum search involves editing the maximum search range to determine them along the graph. If you increase the range, you can achieve a more accurate determination of the necessary maxima.

Input: Acceleration data file 100 athlete jumps.

Result. Found 116 local maxima. Applying the hypotheses of the first and second experiments, we obtain 101 local maxima and 50 jumps.

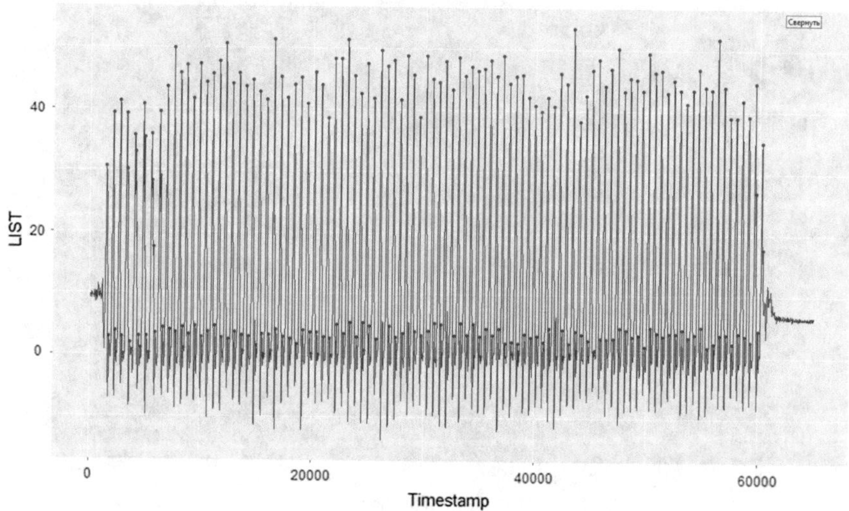

Fig. 3 Schedule of experiment 2. The dependence of the acceleration on time for 100 jumps athlete, where x-axe represents timestamps and y-axe represents acceleration

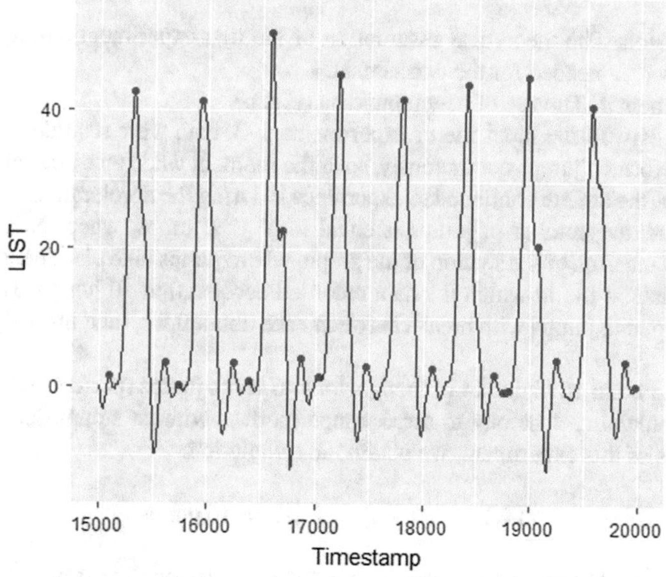

Fig. 4 Experiment Graph 2. Time interval from 15 to 20 s, where x-axe represents timestamps and y-axe represents acceleration

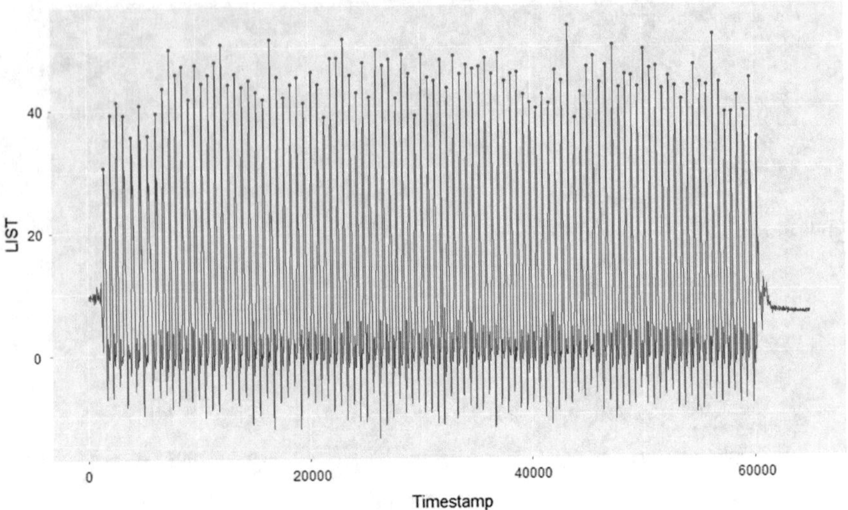

Fig. 5 Schedule of experiment 3. The dependence of the acceleration on time for 100 jumps athlete, where x-axe represents timestamps and y-axe represents acceleration

Conclusion. The theoretical assumption of the first experiment requires adjustment. New data needed for the experiment.

Experiment 4. The use of new input data (Fig. 6).

Hypothesis. In the input file of experiments 1–3, data were recorded where the athlete performed jumps continuously, so at the moment when an overload occurred on landing, the athlete continued to experience it during the acceleration for the next jump. Thus, the number of jumps is equal to N − 1 jumps, where N is the total number of maxima of the section of the graph, where jumps take place continuously.

Input data: a file in which the data recorded acceleration 10 jumps. The athlete made 7 separate jumps with breaks between each about 0.5 s, then made 3 jumps in a row.

Result. Found 18 highs. 14 of them belong to seven jumps (perfect with a break), and the remaining 4 belong to three jumps (perfect without stopping). Using the hypothesis of this experiment, we obtain the calculation:

$$(14/2) + (4 - 1) = 7 + 3 = 10\,\text{jumps}.$$

Conclusion. When counting jumps, it is necessary to separate areas where jumps are performed continuously and areas where an athlete has time to stabilize between two jumps (there is a break between jumps for more than 0.5 s).

Determining the height of jumps (Fig. 7).

To determine the height of the jump, we use the kinematic formula for uniformly accelerated motion, taking into account the average acceleration calculated by the selected time range:

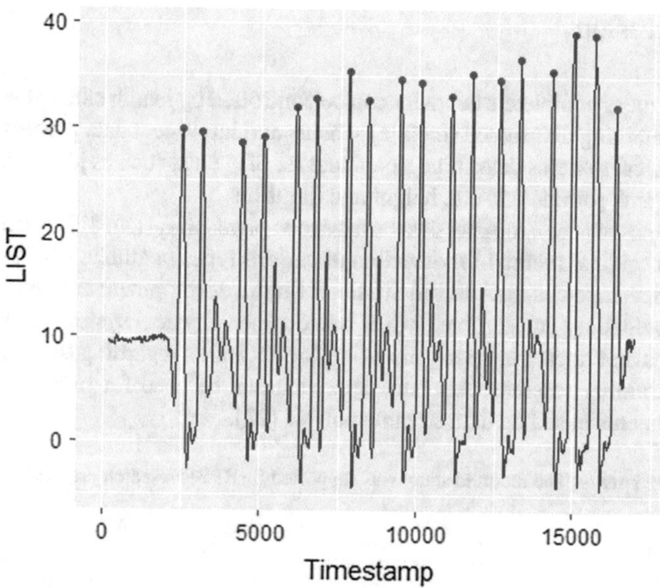

Fig. 6 Schedule experiment 4. The dependence of the acceleration on time for 10 jumps athlete, where x-axe represents timestamps and y-axe represents acceleration

Fig. 7 Graph height of 100 jumps

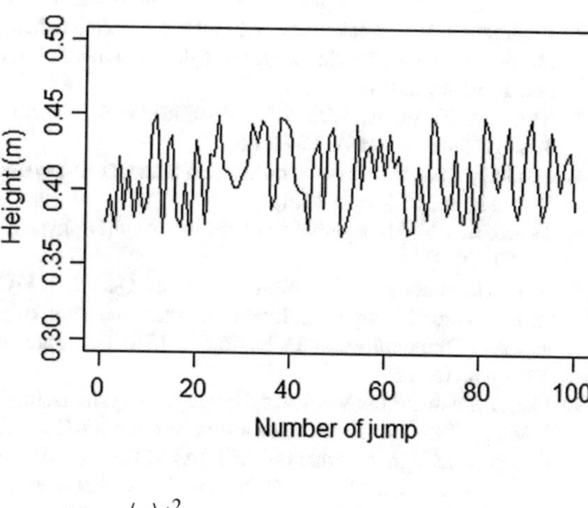

$$h = \frac{\langle a \rangle t^2}{2}$$

Therefore, we need to find the average acceleration (<a>) and the time of the jump (t). For this we need to find the pivot points: the starting point of the jump and the landing point.

5 Conclusion

The correctness of athlete's behavior can be controlled by health care cyber-physical system containing distributed (mobile) sensors and intelligent data processing. Such cyber-physical systems determine a concrete set of events, such as jumps or falls and identify events parameters, e.g. height and duration.

The proposed accelerometer data based cyber-physical system differs from existed ones by an original method for detection of various types of athlete's behaviour. The system is developed using Android Studio, R Studio development environments. The results provided by accelerometer data based cyber-physical system might be used for coaches and doctor in sports medicine for decisions regarding the optimal load in future training sessions. Also this approach can be used for proactive decision support systems based on data stream analysis [27].

Acknowledgements The reported study was supported by RFBR research project 19-47-340010.

References

1. Zanni, A.: Cyber-physical systems and smart cities. IBM Developer. https://developer.ibm.com/chapters/ba-cyber-physical-systems-and-smart-cities-iot/. Last Accessed 05 Apr 2019
2. Khakhanov, V.I., Khakhanov, V.I., Obrizan, V.I., Mishchenko, A.S., Filippenko, I.V.: Cyber-physical systems as technologies of cyber-administration (analytical review). Electr. Comput. Sci. **1**, 39–45 (2014)
3. Kuj, S.A, Tsvetkov, V.Y.: Network-centric management and cyber-physical systems. Educ. Resour. Technol. **2**(19), 86–92 (2017)
4. Schöder, T.: Cyber-physical production management system SEPIA. Electrotechnical Comput. Syst. **13**(89), 197–202 (2014)
5. Tsvetkov, V.Y.: Management with the use of cyber-physical systems. Prospect. Sci. Educ. **3**(27), 55–60 (2017)
6. Abid, H., Phuong, L.T.T., Wang, J., Lee, S., Qaisar, S: V-Cloud: Vehicular cyber-physical systems and cloud computing. In: 4th International Symposium on Applied Sciences in Biomedical and Communication Technologies, ISABEL, Barcelona (2011). https://doi.org/10.1145/2093698.2093863
7. Lee, I., Sokolsky, O.: Medical cyber-physical systems. In: 47th Design Automation Conference, DAC, pp. 743–748. Electronic Design Automation Consortium (EDAC), ACM Special Interest Group on Design Automation (SIGDA), IEEE-CEDA, Anaheim, CA (2010)
8. Hackmann, G., Guo, W., Lu, C., Yan, G., Dyke, S.: Cyber-physical codesign of distributed structural health monitoring with wireless sensor networks. In: 1st ACM/IEEE International Conference on Cyber-Physical Systems, pp. 119–128. ACM Special Interest Group on Embedded Systems (SIGBED), IEEE Technical Committee on Real-Time Systems (TCRTS), Stockholm (2010)
9. Jezewski, J., Horoba, K., Wrobel, J., Pawlak, A., Czabanski, R., Jezewski, M.: Selected design issues of the medical cyber-physical system for telemonitoring pregnancy at home. Microprocess. Microsyst. **46**(Part A), 35–43 (2016)
10. Pawlak, A., Jezewski, J., Horoba, K.: Dependable medical cyber-physical system for home telecare of high-risk pregnancy. Ada User J. **36**(4), 254–258 (2015)

11. Gerget, O., Devyatykh, D., Shcherbakov, M.: Data-driven approach for modeling of control action impact on anemia dynamics based on energy-informational health state criteria. In: Kravets A., Shcherbakov M., Kultsova M., Groumpos P. (eds.) Creativity in Intelligent Technologies and Data Science. CIT&DS 2017. Communications in Computer and Information Science, vol. 754. Springer, Cham (2017)
12. Stepanov, V.S.: Asymmetry of motor actions of athletes in three-dimensional space. SPb (2001)
13. Pestov, E.A.: Mobile device motion recognition. Int. J. Open Inf. Technol. **1**, 10–35 (2013)
14. Butakov, N.A: The applicability of inertial navigation systems in mobile devices. Int. J. Open Inf. Technol. **2**(5), 24–31 (2014)
15. Loginov, S.I.: The possibility of assessing the physical activity of a person using accelerometers motion sensors. (literature review). Bull. N. Med. Technol. **XIV**(1), 149–151 (2007)
16. Kazantsev, A.G., Lavrov, D.N.: Personality identification by gait based on wavelet-parameterization of accelerometer readings. Math. Struct. Model. **23**, 31–37 (2011)
17. Syretsky, G.: A: Mems-sensors of orientation and motion parameters. Interexpo Geo-Siberia **2**(2), 27–32 (2012)
18. del Rosario, Michael B., Redmond, Stephen J., Nigel, H.: Lovell tracking the evolution of smartphone sensing for monitoring human movement. Sensors **15**, 18901–18933 (2015)
19. Gaggle. https://play.google.com/store/apps/details?id=com.geeksville.gaggle&hl=en_US. Last Accessed 05 Apr 2019
20. SyPressure Pro (Barometer). https://play.google.com/store/apps/details?id=sy.android.sypressurepro#?t=W251bGwsMSwxLDIxMiwic3kuYW5kcm9pZC5zeXByZXNzdXJlcHJvIl0. Last Accessed 05 Apr 2019
21. Height indicator—altimeter. https://play.google.com/store/apps/details?id=com.exatools.altimeter, last accessed 2019/04/05
22. Accurate altimeter. https://apkpure.com/ru/accurate-altimeter/com.arlabsmobile.altimeterfree. Last Accessed 05 Apr 2019
23. WOO-tracker. http://shop.woosports.ru/. Last Accessed 05 Apr 2019
24. Vertical jump parameter meter Vert. https://medgadgets.ru/shop/izmeritel-parametrov-vertikal-nogo-pryzhka-vert.html. Last Accessed 05 Apr 2019
25. Tran, V.P., Shcherbakov, M., Nguyen, T.A. Yet another method for heterogeneous data fusion and preprocessing in proactive decision support systems: distributed architecture approach. In: Vishnevskiy V., Samouylov K., Kozyrev D. (eds.) Distributed Computer and Communication Networks. DCCN 2017. Communications in Computer and Information Science, vol. 700. Springer, Cham (2017)
26. Tra, V.P., Shcherbakov, M., Sai, V.C.: On-the-fly multiple sources data analysis in AR-based decision support systems. In: Vishnevskiy V., Kozyrev D. (eds.) Distributed Computer and Communication Networks. DCCN 2018. Communications in Computer and Information Science, vol. 919. Springer, Cham (2018)
27. Shcherbakov, M., Brebels, A., Shcherbakova, N., Kamaev, V., Gerget, O., Devyatykh, D.: Outlier detection and classification in sensor data streams for proactive decision support systems. In: Conference on Information Technologies in Business and Industry 2016, Journal of Physics: Conference Series, vol. 803 (1), Tomsk (2017). http://dx.doi.org/10.1088/1742-6596/803/1/012143

Assembly and Service Robotic Space Module. Mathematical Model of the Reduced System

Pavel P. Belonozhko

Abstract Despite the significant achievements of the last decades in the field of space robotics, the task of automated Assembly and maintenance of large space objects continues to be relevant. At the same time, it is advisable to consider the set of serviced facilities and maintenance facilities of robotics in the future as a single cyber-physical system. Its key element is the assembly and service robotic space module (ASRSM). An important feature of the ASRSM as an element of the cyber-physical system is the potential variety of possible modes of controlled motion. The mentioned feature is of fundamental importance in the development of a complex of Autonomous robotic means interacting with a complex technical object in extreme conditions. The study of the characteristics of dynamic regimes ASRSM is advantageously carried out with the use of model problems involving the study of simplified models with the subsequent generalization of the results. It provides both theoretical and practical interest to mechanical design scheme ASRSM of the "movable base—massless single-stage handling mechanism payload". It is shown that in the absence of external forces, a nonlinear oscillatory system with one degree of freedom can be put in correspondence with this system. This system is described by an independent Routh equation, and, in accordance with the terminology adopted in analytical mechanics, is called reduced. The methodical features of the mathematical description of the reduced system for the model problem are considered. It is shown that the Routh function considered as the Lagrange function of the reduced system can be excluded from the term corresponding to zero gyroscopic force and being a full derivative in time from some function of positional velocity and coordinate. In the absence of a control moment, an integral of energy can be written in the hinge, which has the form of the sum of the kinetic and potential energy of the reduced system, and determines the family of phase trajectories of the system's own motions. The considered problem is of both applied and methodological interest. Qualitative generalization of the obtained results in the case of spatial reduced systems with several degrees of freedom is relevant from the point of view of using their own inertial motions in the construction of control.

P. P. Belonozhko (✉)
Bauman Moscow State Technical University, 5, 2-ya Baumanskaya Str., Moscow 105005, Russia
e-mail: byelonozhko@mail.ru

A. G. Kravets et al. (eds.), *Cyber-Physical Systems: Advances in Design & Modelling*, Studies in Systems, Decision and Control 259,
https://doi.org/10.1007/978-3-030-32579-4_27

Keywords Assembly and service robotic space modules · Proper inertial motion by degree of freedom of manipulator · Reduced system · Routh equation · Reduced system · Routh equation · Lagrange function · Phase portrait

1 Introduction

Robotic assembly and maintenance of large space objects is one of the promising areas of development of both space technology and space robotics [1–19]. For Fig. 1 as an example, the use of an autonomous grouping of assembly and service robotic space modules for the assembly of a large-size space object is presented [3].

Management of such assembly and service group of robots assumes a significant amount of sensor data processed in real-time mode; intensive interaction between physical and computational processes; use of current information about the state of the system to optimize control processes.

Thus, the combination of mounted and operated by advanced space infrastructure and set of supporting tools for space robotics (Fig. 1) naturally regarded as a single cyber-physical system.

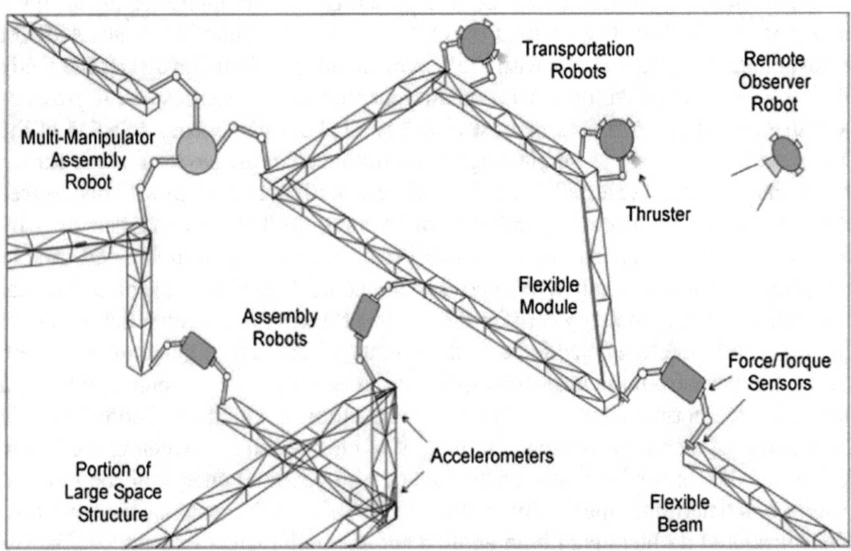

Fig. 1 1—large-size space object, assembled with the help of autonomous grouping of assembly and service robotic space modules [3]

2 Assembly and Service Robotic Space Module as an Element of Cyber-Physical System

As noted in the introduction, the concept of assembly and service autonomic robotic space modules (ASRSM), carrying out the capture of fragments of the assembled structure, delivering them to the Assembly site and installing them in the normal position with the help of the manipulator Fig. 1. Important design features of ASRSM [1]:

- the presence of a movable base;
- the presence of one or more manipulators.

Today there is an experience of experimental orbital testings of such devices (ETS-VII, Orbital Express).

Can be allocated to different modes of functioning of Autonomous ro-botirovna space module [1]: the controlled movement of a module without a load; the module docking to the base station or the mounting structure without the use of a manipulator, docking module to the base station or Monti-financed project design by using the manipulator; a grip manipulator mounted relative to the base station unit; the capture of the manipulator in inertial free space of the block; the controlled movement of a module with a load held by the manipulator; controlled the movement of goods by means of a manipulator; controlled movement of the module with the load fixed on the base; connection of the unit to the mounted structure.

Taking into account the variety of possible modes of movement, an important principle of the organization of the movement of robots—ensuring compliance with the free and forced movements of the manipulator-is relevant in relation to the ASRSM.

3 Model Problem

Consider the flat motion of a system of two solids: the space module 1 (base) and the movable load 2, connected by an ideal single-stage massless manipulator (Fig. 2).

Masses of bodies 1 and 2—m_1 and m_2 respectively, J_1 and J_2—moments of body inertia 1 and 2 in respect to the centers of mass C_1 and C_2, l_1 and l_2—distance between centers of mass and a joint. The motion is viewed in respect to the nonrotating coordinate system XCY originating in the system's center of mass C, which will be inertial should there be no external forces and moments applied to the system. Position relative to XCY is defined by angle φ_1, which describes absolute motion of the platform, and joint angle q, which describes motion of the load in respect to the platform (Fig. 1). Control moment is applied to the joint bounding platform and load M.

Kinetic energy of the system

$$T = T(\dot{\varphi}_1, \dot{q}, q) = \frac{1}{2} a_{\dot{\varphi}_1} \dot{\varphi}_1^2 + a_{\dot{\varphi}_1 \dot{q}} \dot{\varphi}_1 \dot{q} + \frac{1}{2} a_{\dot{q}} \dot{q}^2, \tag{1}$$

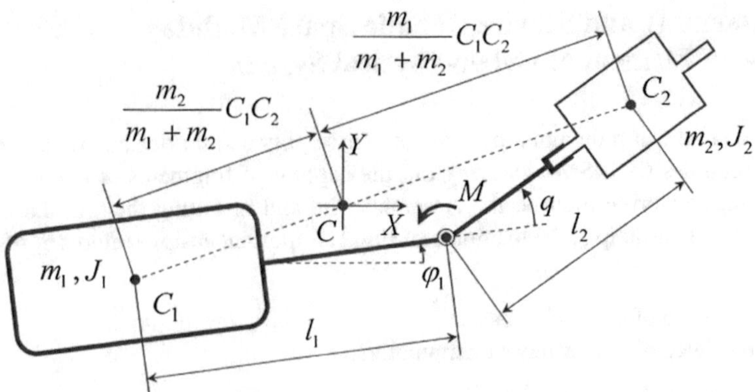

Fig. 2 2—the system: space module—the manipulator moves the load

where

$$
\begin{aligned}
a_{\dot{\varphi}_1} &= a_{\dot{\varphi}_1}(q) = J_1 + J_2 + \tilde{m}l_1^2 + \tilde{m}l_2^2 + 2\tilde{m}l_1l_1\cos q,\\
a_{\dot{\varphi}_1\dot{q}} &= a_{\dot{\varphi}_1\dot{q}}(q) = J_2 + \tilde{m}l_2^2 + \tilde{m}l_1l_1\cos q,\\
a_{\dot{q}} &= a_{\dot{q}}(q) = J_2 + \tilde{m}l_2^2,\\
\tilde{m} &= \frac{m_1m_2}{m_1+m_2}.
\end{aligned}
\tag{2}
$$

The point is the derivative of time t.

Let the forces and moments external to the system be absent. Then the coordinate q is positional, and the coordinate φ_1 is cyclic, and the cyclic integral takes place

$$
\frac{\partial T}{\partial \dot{\varphi}_1} = a_{\dot{\varphi}_1}\dot{\varphi}_1 + a_{\dot{\varphi}_1\dot{q}}\dot{q} = K = const,
\tag{3}
$$

reflecting the fact of constancy of the kinetic moment of the system K in the absence of external moments.

Express $\dot{\varphi}_1$ from (3)

$$
\dot{\varphi}_1 = f_{\dot{\varphi}_1} = f_{\dot{\varphi}_1}(\dot{q}, q, K) = -\frac{a_{\dot{\varphi}_1\dot{q}}}{a_{\dot{\varphi}_1}}\dot{q} + \frac{1}{a_{\dot{\varphi}_1}}K.
\tag{4}
$$

Substitute (4) into (1)

$$
\begin{aligned}
T^* &= T^*(\dot{q}, q, K) = T(\dot{\varphi}_1, \dot{q}, q)|_{\dot{\varphi}_1 = f_{\dot{\varphi}_1}}\\
&= \frac{1}{2}a_{\dot{\varphi}_1}f_{\dot{\varphi}_1}^2 + a_{\dot{\varphi}_1\dot{q}}f_{\dot{\varphi}_1}\dot{q} + \frac{1}{2}a_{\dot{q}}\dot{q}^2.
\end{aligned}
\tag{5}
$$

4 The Routh Function of the Original System and the Lagrange Function of the Reduced System

Let's write down the Routh function

$$R = R(\dot{q}, q, K) = T^*(\dot{q}, q, K) - f_{\dot{\varphi}_1}(\dot{q}, q, K) \cdot K, \tag{6}$$

$$R = \frac{1}{2}\left(a_{\dot{q}} - \frac{a^2_{\dot{\varphi}_1\dot{q}}}{a_{\dot{\varphi}_1}}\right)\dot{q}^2 + \frac{a_{\dot{\varphi}_1\dot{q}}}{a_{\dot{\varphi}_1}}K\dot{q} - \frac{1}{2}\frac{1}{a_{\dot{\varphi}_1}}K^2. \tag{7}$$

We assume that the Routh function (7) is a Lagrange function of some reduced mechanical system with one degree of freedom. The positional coordinate q of the original system is the generalized coordinate of the reduced system. The independent equation of dynamics of the controlled relative motion of the initial system is the Routh equation for the reduced system

$$\frac{\mathrm{d}}{\mathrm{d}t}\left(\frac{\partial R}{\partial \dot{q}}\right) - \frac{\partial R}{\partial q} = M. \tag{8}$$

Thus, in the task of controlling the movement of cargo relative to the base, the above system can be considered as an object of control. For a qualitative analysis of the eigen dynamics of the reduced system, we analyze the structure of the Routh function.

Let's write (7) taking into account designations (2)

$$\begin{aligned} R = R(\dot{q}, q, K) &= \frac{1}{2}\dot{q}^2\frac{(J_1 + \tilde{m}l_1^2)(J_2 + \tilde{m}l_2^2) - \tilde{m}^2 l_1^2 l_2^2 \cos^2 q}{(J_1 + J_2 + \tilde{m}l_1^2 + \tilde{m}l_2^2 + 2\tilde{m}l_1 l_2 \cos q)} \\ &+ \dot{q}K\frac{(J_2 + \tilde{m}l_2^2 + \tilde{m}l_1 l_2 \cos q)}{(J_1 + J_2 + \tilde{m}l_1^2 + \tilde{m}l_2^2 + 2\tilde{m}l_1 l_2 \cos q)} \\ &- \frac{1}{2}K^2\frac{1}{(J_1 + J_2 + \tilde{m}l_1^2 + \tilde{m}l_2^2 + 2\tilde{m}l_1 l_2 \cos q)}. \end{aligned} \tag{9}$$

We introduce notations for the terms of the right part (7)

$$R_2 = \frac{1}{2}\left(a_{\dot{q}} - \frac{a^2_{\dot{\varphi}_1\dot{q}}}{a_{\dot{\varphi}_1}}\right)\dot{q}^2, \qquad R_1 = \frac{a_{\dot{\varphi}_1\dot{q}}}{a_{\dot{\varphi}_1}}K\dot{q}, \qquad R_0 = -\frac{1}{2}\frac{1}{a_{\dot{\varphi}_1}}K^2. \tag{10}$$

We introduce also an auxiliary designation

$$\alpha_1 = J_1 + \tilde{m}l_1^2,$$
$$\alpha_2 = J_2 + \tilde{m}l_2^2$$

$$\beta = \tilde{m} l_1 l_2. \tag{11}$$

Then the term R_1 included in the Routh function can be represented as

$$R_1 = \dot{q} K \frac{\alpha_2 + \beta \cos q}{\alpha_1 + \alpha_2 + 2\beta \cos q}. \tag{12}$$

Integrate the right side (12)

$$\begin{aligned} F_{R_1}(t) &= \int \dot{q} K \frac{\alpha_2 + \beta \cos q}{\alpha_1 + \alpha_2 + 2\beta \cos q} \mathrm{d}t = \frac{1}{2} K \int \frac{\alpha_2 - \alpha_1 + \alpha_1 + \alpha_2 + 2\beta \cos q}{\alpha_1 + \alpha_2 + 2\beta \cos q} \mathrm{d}q \\ &= \frac{1}{2} q K + \frac{1}{2} K \int \frac{\alpha_2 - \alpha_1}{\alpha_1 + \alpha_2 + 2\beta \cos q} \mathrm{d}q. \end{aligned} \tag{13}$$

The integration constant in (13) is omitted. Use the substitution

$$tg \frac{q}{2} = z. \tag{14}$$

Then

$$\cos q = \frac{1 - z^2}{1 + z^2}, \quad \mathrm{d}q = \frac{2}{1 + z^2} \mathrm{d}z. \tag{15}$$

We have for the second term of the right part (13)

$$\begin{aligned} \frac{1}{2} K \int \frac{\alpha_2 - \alpha_1}{\alpha_1 + \alpha_2 + 2\beta \cos q} \mathrm{d}q &= \frac{1}{2} K \int \frac{\alpha_2 - \alpha_1}{\alpha_1 + \alpha_2 + 2\beta \frac{1 - z^2}{1 + z^2}} \frac{2}{1 + z^2} \mathrm{d}z \\ &= K \int \frac{\alpha_2 - \alpha_1}{(\alpha_1 + \alpha_2)\left(1 + z^2\right) + 2\beta\left(1 - z^2\right)} \mathrm{d}z \\ &= K \int \frac{\alpha_2 - \alpha_1}{(\alpha_1 + \alpha_2 + 2\beta) + (\alpha_1 + \alpha_2 - 2\beta) z^2} \mathrm{d}z \\ &= K \frac{\alpha_2 - \alpha_1}{\alpha_1 + \alpha_2 - 2\beta} \int \frac{1}{z^2 + \frac{\alpha_1 + \alpha_2 + 2\beta}{\alpha_1 + \alpha_2 - 2\beta}} \mathrm{d}z. \end{aligned} \tag{16}$$

We introduce the notation for the obviously positive expression in the denominator of the integrand

$$a^2 = \frac{\alpha_1 + \alpha_2 + 2\beta}{\alpha_1 + \alpha_2 - 2\beta} = \frac{J_1 + J_2 + \tilde{m}(l_1 + l_2)^2}{J_1 + J_2 + \tilde{m}(l_1 - l_2)^2}. \tag{17}$$

We also denote

$$b = \frac{\alpha_2 - \alpha_1}{\alpha_1 + \alpha_2 - 2\beta} = \frac{J_2 + \tilde{m} l_2^2 - J_1 - \tilde{m} l_1^2}{J_1 + J_2 + \tilde{m}(l_1 - l_2)^2}. \tag{18}$$

Then the integral (16) is

$$K\frac{\alpha_2-\alpha_1}{\alpha_1+\alpha_2-2\beta}\int\frac{1}{z^2+\frac{\alpha_1+\alpha_2+2\beta}{\alpha_1+\alpha_2-2\beta}}\mathrm{d}z = Kb\int\frac{\mathrm{d}z}{z^2+a^2} = K\frac{b}{a}arctg\frac{z}{a}. \tag{19}$$

Given (19), the function (13) takes the form

$$F_{R_1} = F_{R_1}(q(t)) = F_{R_1}(t) = \frac{1}{2}Kq(t) + K\frac{b}{a}arctg\left(\frac{tg\frac{q(t)}{2}}{a}\right) + C_{R_1}, \tag{20}$$

where C_{R_1} is a constant of integration. Thus, the term R_1 functions Routh is the full time derivative of function (20)

$$R_1 = \frac{\mathrm{d}}{\mathrm{d}t}F_{R_1}, \tag{21}$$

and can be omitted, since the Lagrange function of the mechanical system is determined to the full time derivative of some function [20].

Then the Lagrange function of the reduced system can be written as

$$\begin{aligned} L_R = {} & \frac{1}{2}\dot{q}^2\frac{(J_1+\tilde{m}l_1^2)(J_2+\tilde{m}l_2^2)-\tilde{m}^2l_1^2l_2^2\cos^2 q}{(J_1+J_2+\tilde{m}l_1^2+\tilde{m}l_2^2+2\tilde{m}l_1l_2\cos q)} \\ & -\frac{1}{2}K^2\frac{1}{(J_1+J_2+\tilde{m}l_1^2+\tilde{m}l_2^2+2\tilde{m}l_1l_2\cos q)}. \end{aligned} \tag{22}$$

We modify the Lagrange function (22) by adding to it a constant T_K defined by the expression

$$T_K = \frac{1}{2}K^2\frac{1}{(J_1+J_2+\tilde{m}l_1^2+\tilde{m}l_2^2+2\tilde{m}l_1l_2)}. \tag{23}$$

Get

$$\begin{aligned} L_R^* = L_R + T_K = {} & \frac{1}{2}\dot{q}^2\frac{(J_1+\tilde{m}l_1^2)(J_2+\tilde{m}l_2^2)-\tilde{m}^2l_1^2l_2^2\cos^2 q}{(J_1+J_2+\tilde{m}l_1^2+\tilde{m}l_2^2+2\tilde{m}l_1l_2\cos q)} \\ & -\frac{1}{2}K^2\frac{1}{(J_1+J_2+\tilde{m}l_1^2+\tilde{m}l_2^2+2\tilde{m}l_1l_2\cos q)} \\ & +\frac{1}{2}K^2\frac{1}{(J_1+J_2+\tilde{m}l_1^2+\tilde{m}l_2^2+2\tilde{m}l_1l_2)}. \end{aligned} \tag{24}$$

The modified Lagrange function (24) is the difference between the kinetic and potential energy of the reduced system [16, 18, 19]

$$
\begin{aligned}
L_R^* &= E_k - E_p,\\
E_k &= \frac{1}{2}\dot{q}^2 \frac{(J_1 + \tilde{m}l_1^2)(J_2 + \tilde{m}l_2^2) - \tilde{m}^2 l_1^2 l_2^2 \cos^2 q}{(J_1 + J_2 + \tilde{m}l_1^2 + \tilde{m}l_2^2 + 2\tilde{m}l_1 l_2 \cos q)},\\
E_p &= \frac{1}{2}K^2 \frac{1}{(J_1 + J_2 + \tilde{m}l_1^2 + \tilde{m}l_2^2 + 2\tilde{m}l_1 l_2 \cos q)}\\
&\quad - \frac{1}{2}K^2 \frac{1}{(J_1 + J_2 + \tilde{m}l_1^2 + \tilde{m}l_2^2 + 2\tilde{m}l_1 l_2)}.
\end{aligned} \tag{25}
$$

5 Equation of Dynamics of the Reduced System

The equation of dynamics of the reduced system is written in the form

$$
\frac{\mathrm{d}}{\mathrm{d}t}\left(\frac{\partial L_R^*}{\partial \dot{q}}\right) - \frac{\partial L_R^*}{\partial q} = M. \tag{26}
$$

Substituting (25) in (26), we obtain

$$
\begin{aligned}
&\ddot{q}\frac{(J_1 + \tilde{m}l_1^2)(J_2 + \tilde{m}l_2^2) - \tilde{m}^2 l_1^2 l_2^2 \cos^2 q}{(J_1 + J_2 + \tilde{m}l_1^2 + \tilde{m}l_2^2 + 2\tilde{m}l_1 l_2 \cos q)}\\
&+ \dot{q}^2 \frac{\tilde{m}l_1 l_2 \sin q (J_1 + \tilde{m}l_1^2 + \tilde{m}l_1 l_2 \cos q)(J_2 + \tilde{m}l_2^2 + \tilde{m}l_1 l_2 \cos q)}{(J_1 + J_2 + \tilde{m}l_1^2 + \tilde{m}l_2^2 + 2\tilde{m}l_1 l_2 \cos q)^2}\\
&+ K^2 \frac{\tilde{m}l_1 l_2 \sin q}{(J_1 + J_2 + \tilde{m}l_1^2 + \tilde{m}l_2^2 + 2\tilde{m}l_1 l_2 \cos q)^2} = M.
\end{aligned} \tag{27}
$$

From (27) it is easy to see that the reduced system is a nonlinear oscillatory system. The minimum potential energy E_p of the reduced system occurs when $q = 0$ (the hinge is located on a straight line passing through the centers of mass of bodies), this position of the system is the "lower" position of stable equilibrium in which the potential energy is zero, the total energy is kinetic E_k. The maximum potential energy E_p of the reduced system takes place at $q = \pi$, this position of the system is the "upper" position of the unstable equilibrium in which the kinetic energy is zero, the total energy is equal to the potential energy.

In the absence of control action in the hinge ($M = 0$) there are own ballistic movements of the reduced system. Setting the values of the coordinate $q^{(0)}$ and velocity $\dot{q}^{(0)}$ at the initial time in accordance with (25) is determined by the value of the total energy $E_k + E_p$ of the reduced system $M = 0$, due to the condition remaining constant. In this case, there is an energy integral, which is an equation of a family of phase trajectories.

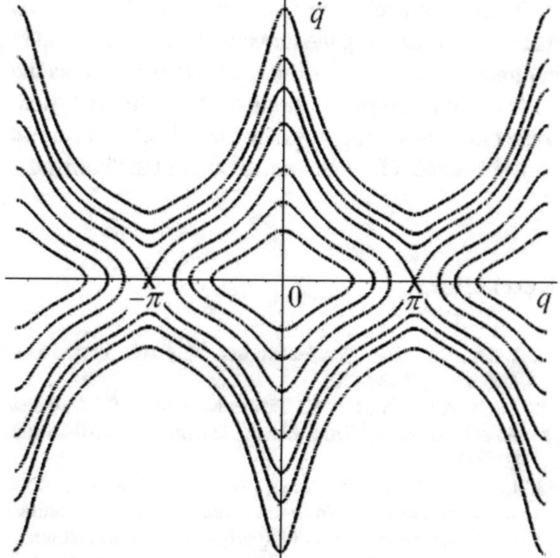

Fig. 3 3—phase portrait of the equation of natural motions of the reduced system

$$E_k + E_p = const. \tag{28}$$

A general view of the phase portrait is shown in Fig. 3.

As shown in Fig. 3, two kinds of own movements can be distinguished: oscillations and circulations. The corresponding groups of phase trajectories are separated by a separatrix. The value of the total energy of the reduced system corresponds to the simulation movement along the separatrix

$$\begin{aligned} E_{p\max} = & \frac{1}{2}K^2 \frac{1}{\left(J_1 + J_2 + \tilde{m}l_1^2 + \tilde{m}l_2^2 + 2\tilde{m}l_1l_2\right)} \\ & - \frac{1}{2}K^2 \frac{1}{\left(J_1 + J_2 + \tilde{m}l_1^2 + \tilde{m}l_2^2 + 2\tilde{m}l_1l_2\right)}, \end{aligned} \tag{29}$$

equal to the maximum possible value of potential energy in the “upper” position of unstable equilibrium.

6 Conclusion

According to the results of the analysis of trends in the development of space robotics, the actual problem of automated assembly of large-size space objects with the help of assembly and service robotic space modules is highlighted. It is shown that the set of serviced space objects and serving ASRSM is a cyber-physical system.

By the example of the plane motion of the system of two hinged bodies it is shown that when controlling the movement of the payload relative to the space module, the nonlinear oscillatory system can be considered as the control object.

The use of proper motions of the reduced system in the synthesis of control in some cases may be appropriate [18, 19, 21]. In particular, the case of impulse control with subsequent free movement along the "ballistic" trajectory is possible.

References

1. Belonozhko, P.P.: Perspective assembly and service robotized space modules. Robot. Tech. Cybern. **2**, 18–23 (2015)
2. Flores-Abad, A., Ma, O., Pham, K., Ulrich, S.: A review of space robotics technologies for on-orbit servicing. Prog. Aerosp. Sci. **68**, 1–26 (2014). https://doi.org/10.1016/j.paerosci.2014.03.002
3. Gradovtsev, A.A., Kondrat'ev A.S., Timofeev A.N.: Robotic support for objects of advanced space infrastructure. In: International Conference "Extreme Robotics", pp. 23–25 November, The Central Research and Experimental Design Institute for Robotics and Engineering Cybernetics, Saint Petersburg (2011). Available at: http://er.rtc.ru/old/docs/2011/ER_PlenarDoclad.pdf. Accessed 06 Mar 2018
4. Vovk A.V., Legostaev V.P., Lopota V.A.: Promising concepts and technologies for creating space technology based on mec-natroniki and microsystem technology. In: Proceedings of the Academy of Sciences. Power industry, no. 3, pp. 3–11 (2011)
5. Management of movable objects. Bibliographic Index, vol. 1. Space projects. Moscow, Institute of Control Sciences n.a. V.A. Trapeznikov of the Russian Academy of Sciences, 2011, 268 p
6. Lysyj, S.R.: Scientific and technical problems and prospects for the development of special-purpose (space) robotics. Extreme robotics. In: Proceedings of the International Scientific and Technical Conference. Saint Petersburg: Polytechnic service, pp. 29–32 (2015)
7. Dalyayev, IYu., Shardyko, I.V., Kuznetsova, E.M.: The prospect of creating robotic service satellites for maintenance and extending the life of a spacecraft. Robot. Tech. Cybern. **3**, 27–31 (2015)
8. Moosavian, S., Ali, A., Papadopoulos, E.: Free-flying robots in space: an overview of dynamics modeling, planning and control. Robotica **25**, 537–547 (2007). https://doi.org/10.1017/S0263574707003438
9. Glumov, V.M., Rutkovskiy, V.Y., Sukhanov, V.M.: Analysis of the flight control features of the space robotized module near the surface of the orbital station. I. Controlling the orientation of the module. In: Proceedings of the Academy of Sciences. Theory and Management Systems, no. 2, pp. 162–169 (2002)
10. Glumov, V.M., Rutkovskiy, V.Y., Sukhanov, V.M.: Analysis of the flight control features of the space robotized module near the surface of the orbital station. II. Manage the trajectories of the module. In: Proceedings of the Academy of Sciences. Theory and Management Systems, no. 3, pp. 140–148 (2002)
11. Rutkovskiy, V.Y., Sukhanov, V.M., Glumov, V.M.: Control of a multimode space robot when performing manipulative operations in the external environment. Autom. Remote Control (11), 96–111 (2010)
12. Rutkovskiy, V.Y., Sukhanov, V.M., Glumov, V.M.: Some problems of control of free-flying space manipulator robots. I. Mechatron. Autom. Manag. **10**, 52–59 (2010)
13. Rutkovskiy, V.Y., Sukhanov, V.M., Glumov, V.M.: Some problems of control of free-flying space manipulator robots. II. Mechatron. Autom. Manag. **12**, 54–65 (2010)
14. Artemenko, Y.N., Karpenko, A.P., Belonozhko, P.P.: Features of manipulator dynamics modeling into account a movable platform. In: Gorodetskiy A.E. (ed.) Smart electromechanical

systems. Studies in Systems, Decision and Control 49, pp. 177–190. Springer International Publishing, Switzerland (2016). https://doi.org/10.1007/978-3-319-27547-5_17

15. Artemenko, Y.N., Karpenko, A.P., Belonozhko, P.P.: synthesis of control of hinged bodies relative motion ensuring move of orientable body to necessary absolute position. In: Gorodetskiy, A.E., Kurbanov, V.G. (eds.) Smart Electromechanical Systems: The Central Nervous System. Studies in Systems, Decision and Control 95, pp 231–239.Springer International Publishing, Switzerland (2017). https://doi.org/10.1007/978-3-319-53327-8_16
16. Belonozhko, P.P.: Methodical features of acquisition of independent dynamic equation of relative movement of one-degree of freedom manipulator on movable foundation as control object. In: Gorodetskiy, A.E., Kurbanov, V.G. (eds.) Smart Electromechanical Systems: The Central Nervous System. Studies in Systems, Decision and Control 95, pp. 261–270. Springer International Publishing, Switzerland (2017). https://doi.org/10.1007/978-3-319-53327-8_19
17. Artemenko, Y.N., Karpenko, A.P., Belonozhko, P.P.: synthesis of the program motion of a robotic space module acting as the element of an assembly and servicing system for emerging orbital facilities. In: Gorodetskiy, A.E., Tarasova I.L. (eds.) Smart Electromechanical Systems: Group Interaction. Studies in Systems, Decision and Control, vol. 174, pp. 217–227. Springer International Publishing Switzerland (2019). ISBN 978-3-319-99759-9_18
18. Belonozhko, P.P.: Robotic assembly and servicing space module peculiarities of dynamic study of given system. In: Gorodetskiy, A.E., Tarasova I.L. (eds.) Smart Electromechanical Systems: Group Interaction. Studies in Systems, Decision and Control, vol. 174, pp. 287–296. Springer International Publishing Switzerland (2019). ISBN 978-3-319-99759-9_23
19. Belonozhko, P.P.: Own inertial movements of a robotic space module. The dynamics of the reduced system. In: Industrial Engineering: Proceedings of the IV International Scientific and Technical Conference, pp. 281–286. Publishing Center SUSU, Chelyabinsk (2018). ISBN 978-5-696-05021-8
20. Landau, L.D., Lifshits, E.M.: Theoretical physics: textbook. benefit. In: Mechanics, The science. Physics Mathematical Literature, vol. 10, 4th edn, 216 p. (1988). ISBN 5-02-013850 (vol. I)
21. Formalsky, A.M.: Motion Control of Unstable Objects. In: Fizmatlit, M. (ed.), 232 p. (2014). ISBN 978-5-92221-1460-8

Printed in the United States
By Bookmasters